13.00

Advanced Mathematics for Engineers and Scientists

HARPERCOLLINS COLLEGE OUTLINE

Advanced Mathematics for Engineers and Scientists

Paul C. DuChateau, Ph.D.
Colorado State University

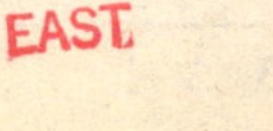

HarperPerennial
A Division of HarperCollins*Publishers*

This book is dedicated to my father and to the memory of my mother.

An American BookWorks Corporation Production

Project Manager: William Hamill

Editor: Robert A. Weinstein

Library of Congress Catalog Card Number: 91-58267

ISBN: 0-06-467151-8

92 93 94 95 96 ABW/RRD 10 9 8 7 6 5 4 3 2 1

Contents

Preface

The old saying that you can't tell a book by its cover is especially true for a title like *Advanced Mathematics for Engineers and Scientists*. A title like that leaves an author considerable freedom with respect to level and content and I feel obligated therefore to use this preface to explain what you will find in this book.

The book is written in such a way that it may be used either as a primary text or as a supplemental reference for a course in applied mathematics at the junior/senior or beginning graduate level. The core chapters are devoted to linear algebra and ordinary and partial differential equations. These are the topics required for the analysis of discrete, semidiscrete, and continuous mathematical models for physical systems. The solved problems in each chapter have been selected to illustrate not only how to construct solutions to various algebraic and differential equations but how to interpret those solutions in order to gain insight into the behavior of the system modeled by the equation.

The book separates naturally into the following sections:

Linear Algebra and Calculus: Chapters 1, 2, and 3

Ordinary Differential Equations: Chapters 4 to 6

Partial Differential Equations: Chapters 7 to 11

Approximation Methods: Chapters 12 and 13

Chapters 1 and 2 are devoted to the two motivating problems of linear algebra: systems of linear equations and algebraic eigenvalue problems. Along with more traditional material, these chapters include topics like; least squares solutions of overdetermined systems with applications to regression analysis, and eigenvalue calculations for large, banded matrices.

Chapter 3 collects a number of results from multivariable calculus and shows how some of them can be used to derive the equations mathematical physics. The models derived include the Navier Stokes equations of fluid dynamics and Maxwell's equations for electromagnetic fields. Other solved problems show how certain simplifying assumptions reduce the full Navier

Stokes equations to the equations for the propagation of acoustic waves while other assumptions lead to the equations for potential flow.

Chapters 4, 5, and 6 are devoted to linear and nonlinear ordinary differential equations. The emphasis in these chapters is on the applications of the results to analysis of physical systems. In particular, chapter 5 provides and introduction to dynamical systems.

Chapters 7 and 8 provide the basic solution techniques used in chapter 9 for solving linear problems in partial differential equations. Chapter 10 treats nonlinear problems in partial differential equations including discussions of shock waves and expansion fans, and a brief introduction to solitary wave solutions for nonlinear differential equations.

Chapter 11 introduces functional optimization with applications to variational formulation of boundary value problems in differential equations. This material provides the theoretical foundation for the finite element method which is briefly described in chapter 13. Chapter 12 discusses finite difference methods, the other commonly used method for approximating solutions to differential equations.

Paul DuChateau

1

Systems of Linear Algebraic Equations

This chapter provides an introduction to the solution of systems of linear algebraic equations. After a brief discussion of matrix notation we present the Gaussian elimination algorithm for solving linear systems. We also show how the algorithm can be extended slightly to provide the so-called LU factorization of the coefficient matrix. This factorization is nearly equivalent to computing the matrix inverse and is an extremely effective solution approach for certain kinds of problems. The solved problems provide simple BASIC computer programs for both the Gaussian elimination algorithm and the LU decomposition. These are applied to example problems to illustrate their use. The solved problems also include examples of physical problems for which the mathematical models lead to systems of linear equations.

The presentation of the solution algorithms is rather formal, particularly with respect to explaining what happens when the algorithms fail in the case of a singular system of equations. To provide a clearer understanding of these and other matters we include a brief development of some abstract ideas from linear algebra. We introduce the four fundamental subspaces associated with a matrix A: the row and column spaces, the null space and the range of A. The relationships that exist between these subspaces and the corresponding subspaces for the transpose matrix A^T *provide the key to understanding the solution of systems of linear equations in the singular as well as the nonsingular case. The solved problems expand on the ideas set forth in the text. For example, Problem 1.15 gives a physical interpretation for the abstract solvability condition that must be imposed on the data vector in a singular system. Problems 1.16 through 1.20 discuss the notion of a "least squares" solution for an overdetermined system and apply this to least squares fits for experimental data.*

We should perhaps conclude this introduction with the disclaimer that this chapter is meant to be only an introduction to the numerical and abstract aspects of systems of linear algebraic equations. While the Gaussian elimination algorithm does form the core of many of the more sophisticated solution algorithms for linear systems, the version provided here contains none of the enhancements that exist to take advantage of special matrix structure, nor does it contain provisions to compensate for numerical instabilities in the system. Such considerations are properly the subject of a more advanced course on numerical linear algebra. This chapter seeks only to provide the foundation on which more advanced treatments can build.(1.1)

SYSTEMS OF SIMULTANEOUS LINEAR EQUATIONS

Terminology

Consider the following system of m equations in the n unknowns $x_1, \ldots, x_n$:

$$\begin{aligned} a_{11}x_1 + a_{12}x_2 + \cdots + a_{1n}x_n &= b_1 \\ &\vdots \\ a_{m1}x_1 + a_{m2}x_2 + \cdots + a_{mn}x_n &= b_m \end{aligned} \tag{1.1}$$

Here $a_{11}, \ldots, a_{mn}$ denote the *coefficients* in the equations and the numbers $b_1, \ldots, b_m$ are referred to as the *data*. In a specific example these quantities would be given numerical values and we would then be concerned with finding values for $x_1, \ldots, x_n$ such that these equations were satisfied. An n-tuple of real values $\{x_1, \ldots, x_n\}$ which satisfies each of the m equations in (1.1) is said to be a *solution* for the system. The collection of all n-tuples which are solutions is called the *solution set* for the system. For any given system, one of the following three possibilities must occur:

(a) The solution set contains a single n-tuple; then the system is said to be *nonsingular.*

(b) The solution set contains infinitely many n-tuples; then the system is said to be *singular*, or, more precisely, *singular dependent* or *underdetermined.*

(c) The solution set is empty; in this case we say the system is *singular inconsistent* or *overdetermined.*

EXAMPLE 1.1

(a) Consider the system

$$x_1 + x_2 = 4$$
$$x_1 - x_2 = 0$$

Each equation in this simple system defines a separate function whose graph is a straight line in the x_1x_2 plane. The lines corresponding to these equations are seen to intersect at the unique point (2, 2). Thus, the solution set for this system consists of the single 2-tuple [2, 2]. The system is nonsingular.

(b) The system
$$x_1 + x_2 = 4$$
$$2x_1 + 2x_2 = 8$$
produces just a single line in the x_1x_2 plane (alternatively, there are two lines that coincide). The solution set for the system contains infinitely many 2-tuples corresponding to all of the points on the line. That is, every 2-tuple that is of the form $[x_1, 2 - x_1]$ for any choice of x_1 is in the solution set. The system is singular. More precisely, the system is singular dependent (underdetermined). The equations of the system are not independent equations.

(c) The system
$$x_1 + x_2 = 4$$
$$2x_1 + 2x_2 = 0$$
produces a pair of parallel lines in the x_1x_2 plane. There are no points that lie on both lines and consequently the solution set for the system is empty. The system is singular. More precisely, the system is singular inconsistent (overdetermined). The equations of the system are contradictory.

Matrix Notation

VECTORS

In order to discuss systems of equations efficiently it will be convenient to have the notion of an n-vector. We shall use the notation X to denote an array of n numbers arranged in a column, and X^{T} will denote the same array arranged in a row,

$$X = \begin{bmatrix} x_1 \\ \vdots \\ x_n \end{bmatrix}, X^{\mathsf{T}} = [x_1, \ldots, x_n]$$

We refer to these respectively as *column vectors* and *row vectors*. The purpose of this distinction will become apparent.

INNER (DOT) PRODUCT

For n-vectors X and Y having entries $x_1, \ldots, x_n$ and $y_1, \ldots, y_n$, respectively, we define the *inner product* of X and Y as the number

$$\sum_{i=1}^{n} x_i y_i = x_1 y_1 + \cdots + x_n y_n \tag{1.2}$$

We use any of the notations $X \cdot Y$, $X^T Y$ or (X, Y) to indicate the inner product of X and Y. The inner product is also called the dot product.

MATRICES

An $m \times n$ *matrix* is a rectangular array of numbers containing m rows and n columns. For example,

$$A = \begin{bmatrix} a_{11} & \cdots & a_{1n} \\ & \cdots & \\ a_{m1} & \cdots & a_{mn} \end{bmatrix} \quad \text{and } B = \begin{bmatrix} b_{11} & \cdots & b_{1p} \\ & \cdots & \\ b_{n1} & \cdots & b_{np} \end{bmatrix}$$

denote, respectively, an $m \times n$ matrix and an $n \times p$ matrix. Here a_{ij} denotes the entry in the ith row and jth column of the matrix A. Note that a column vector is an $n \times 1$ matrix and a row vector is a $1 \times n$ matrix.

MATRIX PRODUCT

For an $m \times n$ matrix A and an $n \times p$ matrix B, the product AB is defined to be the $m \times p$ matrix C whose entries C_{ij} are equal to

$$C_{ij} = \sum_{k=1}^{n} a_{ik} b_{kj} \quad i = 1, \ldots, m \quad \text{and} \quad j = 1, \ldots, p \tag{1.3}$$

Note that the product AB is not defined if the number of columns of A is not equal to the number of rows of B. Note also that if A is $m \times n$ we can think of A as composed of m rows, each of which is a (row) m-vector; i.e.; A is composed of rows $R_1{}^T, \ldots, R_m{}^T$. Similarly, we can think of B as consisting of p columns, $C_1, \ldots, C_p$, each of which is a (column) n-vector. Then for $i = 1, \ldots, m$ and $j = 1, \ldots, p$, the (i, j) entry of the product matrix AB is equal to the dot product $R_i{}^T C_j$

$$AB = \begin{bmatrix} R_1^T C_1 & \cdots & R_1^T C_p \\ & \cdots & \\ R_m^T C_1 & \cdots & R_m^T C_p \end{bmatrix} \tag{1.4}$$

If A is $m \times n$ and B is $n \times m$, then the products AB and BA are both defined but produce results which are $m \times m$ and $n \times n$, respectively. Thus, if m is not equal to n, AB cannot equal BA. Even if m equals n, AB

is not necessarily equal to BA. The matrix product is not *commutative*.

IDENTITY MATRIX

Let I denote the $n \times n$ matrix whose entries I_{jk} are equal to 1 if j equals k and are equal to 0 if j is different from k. Then I is called the $n \times n$ *identity matrix* since for any $n \times n$ matrix A, we have $IA = AI = A$. Thus, I plays the role of the identity for matrix multiplication.

MATRIX NOTATION FOR SYSTEMS

The system of equations (1.1) can be expressed in matrix notation by writing $AX = B$; i.e.,

$$AX = \begin{bmatrix} a_{11} & \cdots & a_{1n} \\ & \cdots & \\ a_{m1} & \cdots & a_{mn} \end{bmatrix} \begin{bmatrix} x_1 \\ \vdots \\ x_n \end{bmatrix} = \begin{bmatrix} b_1 \\ \vdots \\ b_m \end{bmatrix} \tag{1.5}$$

We refer to A as the *coefficient matrix* for the system and to $\boldsymbol{X}$ and $\boldsymbol{B}$ as the *unknown vector* and the *data vector*, respectively.

Gaussian Elimination

AUGMENTED MATRIX

We shall now introduce the *Gaussian elimination* algorithm for solving the system (1.5) in the case $m = n$. We begin by forming the *augmented matrix* for the system (1.5). This is the $n \times n + 1$ matrix composed of the coefficient matrix A augmented by the data vector $\boldsymbol{B}$ as follows

$$[A|\,\boldsymbol{B}] = \begin{bmatrix} a_{11} & \cdots & a_{1n} & b_1 \\ & \cdots & & \vdots \\ a_{m1} & \cdots & a_{mn} & b_n \end{bmatrix}$$

EXAMPLE 1.2 AUGMENTED MATRIX

Consider the system of equations

$$\begin{aligned} x_1 + 2x_2 + 2x_3 &= 3 \\ 2x_1 + 5x_2 + 5x_3 &= 8 \\ -x_1 + 3x_3 &= -2 \end{aligned}$$

The augmented matrix for this system is

$$\begin{bmatrix} 1 & 2 & 2 & 3 \\ 2 & 5 & 5 & 8 \\ -1 & 0 & 3 & -2 \end{bmatrix}$$

Note that the augmented matrix contains all the information needed to form the original system of equations. Only extraneous symbols have been stripped away.

ROW OPERATIONS

Gaussian elimination consists of operating on the rows of the augmented matrix in a systematic way in order to bring the matrix into a form where the unknowns can be easily determined. The allowable row operations are

$E_{ij}(a)$ —add to row i of the augmented matrix (a times row j) to form a new ith row (for $a \neq 0$ and $i \neq j$),

$E_j(a)$ — multiply the jth row of the augmented matrix by the nonzero constant a to form a new jth row (for $a \neq 0$),

P_{ij} — exchange rows i and j of the augmented matrix.

These operations are referred to as *elementary row operations*. They do not alter the solution set of the system (1.5).

Theorem 1.1

Theorem 1.1. The solution set for a system of linear equations remains invariant under elementary row operations.

EXAMPLE 1.3 ROW REDUCTION TO UPPER TRIANGULAR FORM

Consider the system of equations from Example 1.2. The row operation $E_{12}(-2)$ has the following effect on the augmented matrix:

$$\begin{bmatrix} 1 & 2 & 2 & 3 \\ 2 & 5 & 5 & 8 \\ -1 & 0 & 3 & -2 \end{bmatrix} \text{—}E_{21}(-2)\text{—>} \begin{bmatrix} 1 & 2 & 2 & 3 \\ 0 & 1 & 1 & 2 \\ -1 & 0 & 3 & -2 \end{bmatrix}$$

We follow this with $E_{13}(1)$

$$\begin{bmatrix} 1 & 2 & 2 & 3 \\ 0 & 1 & 1 & 2 \\ -1 & 0 & 3 & -2 \end{bmatrix} \text{—}E_{31}(1)\text{—>} \begin{bmatrix} 1 & 2 & 2 & 3 \\ 0 & 1 & 1 & 2 \\ 0 & 2 & 5 & 1 \end{bmatrix}$$

and $E_{23}(-2)$

$$\begin{bmatrix} 1 & 2 & 2 & 3 \\ 0 & 1 & 1 & 2 \\ 0 & 2 & 5 & 1 \end{bmatrix} \text{—}E_{32}(-2)\text{—>} \begin{bmatrix} 1 & 2 & 2 & 3 \\ 0 & 1 & 1 & 2 \\ 0 & 0 & 3 & -3 \end{bmatrix}$$

The augmented matrix has been reduced to upper triangular form. A matrix is said to be in *upper triangular form* when all entries below the diagonal are zero. That is, A is in upper triangular form if $a_{ij} = 0$ for $i > j$.

BACK SUBSTITUTION

The final form of the augmented matrix in Example 1.3 is associated with the following system of equations $UX = B^*$; i.e.,

$$\begin{aligned} x_1 + 2x_2 + 2x_3 &= 3 \\ x_2 + x_3 &= 2 \\ 3x_3 &= -3 \end{aligned}$$

This system has the same solution set as the original system and since the coefficient matrix of the reduced system is upper triangular, we can solve for the unknowns by *back substitution*. That is,

$$3x_3 = -3 \text{ implies } x_3 = -1.$$

Substituting this result in the second equation of the reduced system leads to

$$x_2 + (-1) = 2 \text{ or } x_2 = 3.$$

Finally, substituting the values for x_3 and x_2 into the first equation yields

$$x_1 + 2(3) + 2(-1) = 3 \text{ or } x_1 = -1.$$

GAUSSIAN ELIMINATION

We can summarize the algorithm for solving the system $AX = B$ by Gaussian elimination as follows:

$$\left[A|\,B\right] \xrightarrow[\text{upper triangular form}]{\text{row reduction to}} \left[U|\,B^*\right] \xrightarrow[\text{substitution}]{\text{back}} X$$

The solved problems provide examples of systems requiring row exchanges for the reduction.

ECHELON FORM

The notion of an upper triangular matrix is not sufficiently precise for purposes of describing the possible outcomes of the Gaussian elimination algorithm. We say an upper triangular matrix is in *echelon form* if all the entries in a column that are below the first nonzero entry in each row are zero.

EXAMPLE 1.4 ECHELON FORM

$\begin{bmatrix} 1 & 2 & 2 \\ 0 & 0 & 1 \\ 0 & 0 & 3 \end{bmatrix}$ is upper triangular but is not in echelon form.

$\begin{bmatrix} 1 & 2 & 2 \\ 0 & 0 & 1 \\ 0 & 0 & 0 \end{bmatrix}$ is in echelon form.

ROW EQUIVALENCE AND RANK

If matrix A can be transformed into matrix B by a finite number of elementary row operations, we say that A and B are *row equivalent*. We define the *rank* of a matrix A to be the number of nontrivial rows in any echelon form upper triangular matrix that is row equivalent to A. A *trivial row* in a matrix is one in which every entry is a zero. Now we can state

Theorem 1.2

Theorem 1.2. Let A denote an $n \times n$ matrix and consider the associated linear system $AX = B$.

(a) If rank $A = n$, then the system is nonsingular and has a unique solution X for every data n-tuple B.

(b) If rank $A = \text{rank}[A|B] < n$, then the system is singular dependent; i.e., there are infinitely many solutions.

(c) If rank $A < \text{rank}[A|B]$, then the system is singular inconsistent; there is no solution.

EXAMPLE 1.5

(a) We apply Theorem 1.2 to the 2×2 system of Example 1.1(a)

$$A = \begin{bmatrix} 1 & 1 \\ 1 & -1 \end{bmatrix} \text{—}E_{21}(-1)\text{—>} \begin{bmatrix} 1 & 1 \\ 0 & -2 \end{bmatrix} = U, \ \text{rank}\, A = 2$$

We have rank $A = n = 2$. Then the theorem implies that $AX = B$ has a unique solution for every 2-tuple B.

(b) For the system of Example 1.1(b), we have

$$A = \begin{bmatrix} 1 & 1 \\ 2 & 2 \end{bmatrix} \text{—}E_{21}(-2)\text{—>} \begin{bmatrix} 1 & 1 \\ 0 & 0 \end{bmatrix} = U, \ \text{rank}\, A = 1$$

$$[A|B] = \begin{bmatrix} 1 & 1 & 4 \\ 2 & 2 & 8 \end{bmatrix} \text{—}E_{21}(-2)\text{—>} \begin{bmatrix} 1 & 1 & 4 \\ 0 & 0 & 0 \end{bmatrix} = [U|B^*], \ \text{rank}\, [A|B] = 1$$

Then rank $A = \text{rank}\, [A|B] = 1 < 2 = n$ and the theorem indicates that this system is singular dependent as we found previously by other means.

(c) For the third system in Example 1.1, we have rank $A = 1$ and

$$[A|B] = \begin{bmatrix} 1 & 1 & 4 \\ 2 & 2 & 0 \end{bmatrix} \text{—}E_{21}(-2)\text{—>} \begin{bmatrix} 1 & 1 & 4 \\ 0 & 0 & -8 \end{bmatrix} = [U|B^*], \ \text{rank}\, [A|B] = 2$$

Then the theorem states that this system is singular inconsistent, and there is no solution. The lines which are the graphs of the two equations of this system are parallel.

LU Decomposition

ELEMENTARY MATRICES AND ROW OPERATIONS

Each of the elementary row operations is equivalent to multiplication by an associated matrix called an *elementary matrix.*

Theorem 1.3

Theorem 1.3. Let I_n denote the $n \times n$ identity matrix and let E, F, G denote, respectively, the matrices obtained by applying to I_n the row operations $E_j(a)$, $E_{jk}(a)$, and P_{jk}, for $a \neq 0$ and $j \neq k$. Then, for $n \times n$ matrices A and B,

A —$E_j(a)$—> B if and only if $EA = B$

A —$E_{jk}(a)$—> B if and only if $FA = B$

A —P_{jk}—> B if and only if $GA = B$

EXAMPLE 1.6 ELEMENTARY MATRICES

Let I_3 denote the 3×3 identity. Then applying the row operations $E_1(a)$, $E_{32}(a)$, P_{23} to I_3 leads to

$$E = \begin{bmatrix} a & 0 & 0 \\ 0 & 1 & 0 \\ 0 & 0 & 1 \end{bmatrix}, F = \begin{bmatrix} 1 & 0 & 0 \\ 0 & 1 & 0 \\ 0 & a & 1 \end{bmatrix}, P = \begin{bmatrix} 1 & 0 & 0 \\ 0 & 0 & 1 \\ 0 & 1 & 0 \end{bmatrix}$$

Multiplying the matrix A by E, F, and P, respectively, we find

$$EA = \begin{bmatrix} aa_{11} & aa_{12} & aa_{13} \\ a_{21} & a_{22} & a_{23} \\ a_{31} & a_{32} & a_{33} \end{bmatrix}$$

$$FA = \begin{bmatrix} a_{11} & a_{12} & a_{13} \\ a_{21} & a_{22} & a_{23} \\ a_{31} + aa_{21} & a_{32} + aa_{22} & a_{33} + aa_{23} \end{bmatrix}$$

$$PA = \begin{bmatrix} a_{11} & a_{12} & a_{13} \\ a_{31} & a_{32} & a_{33} \\ a_{21} & a_{22} & a_{23} \end{bmatrix}$$

Note that the matrices E and F are lower triangular but P is not.

MATRIX INVERSE

For $n \times n$ matrices A and B, both products AB and BA are defined and produce an $n \times n$ result. In the special case that $AB = BA = I_n$, we say that B is the *matrix inverse* for A and we write $B = A^{-1}$. Note that only a square matrix may have an inverse but not every square matrix has an inverse. A square matrix with no inverse is said to be *singular* and a matrix that does have an inverse is said to be *invertible* or *nonsingular.*

Theorem 1.4 Inverse of the Product of Two Matrices

Theorem 1.4. Suppose A and B are invertible $n \times n$ matrices. Then the product AB is invertible and $(AB)^{-1} = B^{-1}A^{-1}$. In view of Theorem 1.3, every elementary matrix is invertible.

Theorem 1.5 Inverse of an Elementary Matrix

Theorem 1.5. Let I_n denote the $n \times n$ identity matrix and let E, F, P denote, respectively, the matrices obtained by applying to I_n the row operations $E_j(a)$, $E_{jk}(a)$, and P_{jk}, for $a \neq 0$ and $j \neq k$. Then E, F, and P are invertible with

$$I_n \text{—} E_j(\tfrac{1}{a}) \text{—>} E^{-1}; \quad I_n \text{—} E_{jk}(-a) \text{—>} F^{-1}; \quad \text{and} \quad I_n \text{—} P_{jk} \text{—>} P^{-1} = P$$

EXAMPLE 1.7 INVERSES OF ELEMENTARY MATRICES

Let E, F, and G be the elementary matrices of Example 1.6.

Then

$$E^{-1} = \begin{bmatrix} \frac{1}{a} & 0 & 0 \\ 0 & 1 & 0 \\ 0 & 0 & 1 \end{bmatrix}, \quad F^{-1} = \begin{bmatrix} 1 & 0 & 0 \\ 0 & 1 & 0 \\ 0 & -a & 1 \end{bmatrix}, \quad P^{-1} = \begin{bmatrix} 1 & 0 & 0 \\ 0 & 0 & 1 \\ 0 & 1 & 0 \end{bmatrix} = P$$

Note that the inverses for E and F are lower triangular and that P is its own inverse. We now illustrate the role played by elementary matrices in Gaussian elimination.

EXAMPLE 1.8 *LU* DECOMPOSITION

The following 3×3 matrix can be reduced to upper triangular form by applying (in order) the indicated row operations:

$$A = \begin{bmatrix} 1 & 2 & 2 \\ 2 & 5 & 5 \\ -1 & 0 & 3 \end{bmatrix} \text{—}E_{21}(-2),\ E_{31}(1),\ E_{32}(-2)\text{—>} \begin{bmatrix} 1 & 2 & 2 \\ 0 & 1 & 1 \\ 0 & 0 & 3 \end{bmatrix} = U$$

If we let E, F, and G denote the elementary matrices associated with the respective row operations, $E_{21}(-2)$, $E_{31}(1)$, and $E_{32}(-2)$, then $G(F(EA)) = U$. Thus, it follows that $A = E^{-1}(F^{-1}(G^{-1}U))$. Each of the matrices E, F, G is lower triangular as are each of the inverses. It is easy to check that the product of two lower triangular matrices is again lower triangular. Thus, the product $E^{-1}F^{-1}G^{-1}$ is a lower triangular matrix we shall denote by L. Then $A = LU$, where L is lower triangular and U is upper triangular. We refer to this as the *LU factorization* or the *LU decomposition* of A. Note further that, according to Theorem 1.5,

$$E^{-1} = \begin{bmatrix} 1 & 0 & 0 \\ 2 & 1 & 0 \\ 0 & 0 & 1 \end{bmatrix},\quad F^{-1} = \begin{bmatrix} 1 & 0 & 0 \\ 0 & 1 & 0 \\ -1 & 0 & 1 \end{bmatrix},\quad G^{-1} = \begin{bmatrix} 1 & 0 & 0 \\ 0 & 1 & 0 \\ 0 & 2 & 1 \end{bmatrix}$$

and, thus,

$$L = E^{-1}F^{-1}G^{-1} = \begin{bmatrix} 1 & 0 & 0 \\ 2 & 1 & 0 \\ -1 & 2 & 1 \end{bmatrix}$$

Observe that A is reduced to upper triangular form by the row operations

$$E_{21}(-2),\quad E_{31}(1),\quad E_{32}(-2)$$

and that L has off-diagonal entries

$$L_{21} = 2,\quad L_{31} = -1,\quad L_{32} = 2$$

Thus, in order to find the matrix L in the LU factorization of A, it is not necessary to compute the elementary matrices or their inverses. Instead, it is sufficient to keep track of the multipliers in the row operations used to reduce A to U. More generally, we have the next theorem.

Theorem 1.6

Theorem 1.6. Suppose $n \times n$ matrix A is reduced to upper triangular matrix U by row operations of the form $E_{jk}(a)$ and $E_j(a)$. Then $A = LU$ where the entries L_{jk} of the lower triangular matrix L are prescribed by:

1. Each row operation of the form $E_{jk}(a)$ leads to an off-diagonal entry $L_{jk} = -a$.
2. Each row operation of the form $E_j(a)$ leads to a diagonal entry $L_{jj} = 1/a$.

For simplicity we will avoid discussion of the effect of row exchanges on the factorization of A.

SOLUTION OF LINEAR SYSTEMS BY *LU* FACTORIZATION

Application of Gaussian elimination to the $n \times n$ system $AX = B$ reduces the system to an upper triangular systerm $UX = B^*$. By keeping track of the row operations employed, we obtain at the same time a lower triangular matrix L such that $LU = A$. Then $LUX = B$ or, since $UX = B^*$, we can write $LB^* = B$. Thus, the system $AX = B$ is seen to be equivalent to the two triangular systems

Forward Sweep: $LB^* = B$

Backward Sweep: $UX = B^*$

Solving $LB^* = B$ is accomplished by executing a *forward sweep*; that is, since the matrix L is lower triangular, we solve the first equation first, use the result in the second equation, etc. Then we solve the upper triangular system $UX = B^*$ by a *backward sweep*; i.e., by back substitution, as previously described. This strategy is particularly efficient if we are required to solve the system $AX = B$ for several different data vectors B since we do not have to repeat the time-consuming step of factorization but simply repeat the forward and backward sweeps with the same L and U matrices.

EXAMPLE 1.9 SOLVING *AX* = *B* BY *LU* FACTORIZATION

To solve

$$\begin{bmatrix} 1 & 2 & 2 \\ 2 & 5 & 5 \\ -1 & 0 & 3 \end{bmatrix} \begin{bmatrix} x \\ y \\ z \end{bmatrix} = \begin{bmatrix} 3 \\ 8 \\ -2 \end{bmatrix}$$

by LU factorization, recall from Example 1.7 that $A = LU$ where L and U are given in Example 1.7. We first execute a forward sweep, solving $LB^* = B$, i.e.,

$$\begin{bmatrix} 1 & 0 & 0 \\ 2 & 1 & 0 \\ -1 & 2 & 1 \end{bmatrix} \begin{bmatrix} b_1^* \\ b_2^* \\ b_3^* \end{bmatrix} = \begin{bmatrix} 3 \\ 8 \\ -2 \end{bmatrix}$$

This produces the following result:

$$b_1^* = 3$$

$$2 \cdot 3 + b_2^* = 8; \qquad b_2^* = 2$$

$$-1 \cdot 3 + 2 \cdot 2 + b_3^* = -2; \qquad b_3^* = -3$$

Then we execute the backward sweep to find X:

$$\begin{bmatrix} 1 & 2 & 2 \\ 0 & 1 & 1 \\ 0 & 0 & 3 \end{bmatrix} \begin{bmatrix} x \\ y \\ z \end{bmatrix} = \begin{bmatrix} 3 \\ 2 \\ -3 \end{bmatrix}; \qquad \begin{bmatrix} x \\ y \\ z \end{bmatrix} = \begin{bmatrix} -1 \\ 3 \\ -1 \end{bmatrix}$$

The solved problems illustrate some situations in which it is natural to solve $AX = B$ for a sequence of data vectors B. We summarize the LU decomposition algorithm for $AX = B$ in the following steps:

1. Reduce $[A|B]$ to upper triangular form $[U|B^*]$ by row operations.
2. Store multipliers from row operations to form the matrix L.
3. Execute a backward sweep to solve $UX = B^*$ for X.
4. To solve $AX = B$ for a new data vector B', execute a forward sweep to find B^* from $LB^* = B'$. Then repeat step 3.

LINEAR ALGEBRA

Vector Spaces

In order to understand the solution of systems of linear equations at more than an algorithmic level, it is necessary to introduce a number of abstract concepts. The proper context for considering questions relating to systems of linear equations is the setting of a *vector space*. In general, a vector space consists of:

- a collection of objects called *vectors*
- a second set of objects called *scalars*
- two operations, *vector addition* and *scalar multiplication* for combining scalars and vectors

It is customary to list a set of axioms describing how these objects and operations interact. For our purposes, it will be sufficient to proceed with less generality.

VECTORS AND SCALARS

We will take as our vectors n-tuples of real numbers, so-called n-vectors for which we have already introduced the notation X or X^{T} according to whether the vector is written as a column or a row. In this chapter we will use the real numbers as our scalars. In the next chapter it will be necessary to be slightly more general and to use the complex numbers for scalars.

VECTOR ADDITION, SCALAR MULTIPLICATION

We define *vector addition* by

$$X^{\mathrm{T}} + Y^{\mathrm{T}} = [x_1 + y_1, \ldots, x_n + y_n]$$

for arbitrary vectors X and Y. Scalar multiplication is defined as follows:

$$aX^{\mathrm{T}} = [ax_1, \ldots, ax_n]$$

for arbitrary scalar a and vector X.

EUCLIDEAN *n*-SPACE

By defining the vectors, scalars and operations as we have, we have presented a particular example of a vector space. This vector space is usually referred to as *Euclidean n-space* and is denoted by $\Re^n$. This is certainly not the only example of a vector space which is of importance in applied mathematics, but it is the only one we will consider in this chapter.

LINEAR COMBINATIONS, SUBSPACES

For scalars a and b and vectors X and Y, the combination

$$aX^{\mathsf{T}} + bY^{\mathsf{T}} = [ax_1 + by_1, \ldots, ax_n + by_n] = \mathbf{Z}$$

is also a vector. We refer to $\mathbf{Z}$ as a *linear combination* of X and Y. Certain sets of vectors have the property that they are closed under the operation of forming linear combinations. Such sets of vectors are called *subspaces*. That is, a set M of vectors is a subspace if and only if, for all scalars a and b and all vectors X and Y in M, the linear combination $aX + bY$ is also in M.

EXAMPLE 1.10 SUBSPACES

Null Space of a Matrix. Let A denote an $n \times n$ matrix. Then the set of all vectors X in $\Re^n$ with the property that $AX = \underline{0}$ forms a subspace called the *null space* of A. We denote this subspace by $N[A]$ and write the definition more briefly as

$$N[A] = \{X \text{ in } \Re^n\text{: } AX = \underline{0}\}$$

To see that $N[A]$ is a subspace, suppose that X and Y belong to $N[A]$; i.e., $AX = \underline{0}$ and $AY = \underline{0}$. For arbitrary scalars a and b, it follows from the definition of the product of a matrix times a vector that

$$A(aX + bY) = aAX + bAY = \underline{0} + \underline{0}.$$

Since $\mathbf{Z} = aX + bY$ satisfies $A\mathbf{Z} = \underline{0}$, $\mathbf{Z}$ belongs to $N[A]$ and $N[A]$ is closed under the operation of forming linear combinations.

Range of a Matrix. Note that for X in $\Re^n$, the product AX is also in $\Re^n$. The set of all vectors Y in $\Re^n$ that are of this form, $Y = AX$ for some X, form a subspace called the *range* of A. We denote this subspace by $\text{Rng}[A]$ and write

$$\text{Rng}[A] = \{Y \text{ in } \Re^n\text{: } Y = AX \text{ for some } X \text{ in } \Re^n\}$$

The proof that $\text{Rng}[A]$ is a subspace is found in the solved problems. The two subspaces $N[A]$ and Rng$[A]$ are subspaces defined by *membership rule*. The rule for membership in $N[A]$ is that AX must be the zero vector and the rule for membership in Rng$[A]$ is that Y must be of the form AX for some vector X. Thus, in order to decide if a given vector is in the subspace, it is sufficient to simply apply the membership rule to the vector. We consider now two examples of subspaces defined by another means.

Row Space of a Matrix. For $n \times n$ matrix A, we define the *row space* of A to be the collection of all possible linear combinations of the rows of A. The rows of A are (row) n-vectors and, hence, this is just the set of all vectors in $\Re^n$ that can be written as some linear combination of the rows of A. We denote this subspace by $RS[A]$ and note that since $RS[A]$ consists of *all possible* linear combinations of the rows of A, it must necessarily be closed under the operation of forming linear combinations. Thus, it is obvious that this is a subspace.

Column Space of a Matrix. The set of all possible linear combinations of the *columns* of the matrix A is, like the row space of A, a subspace of $\Re^n$. We denote this subspace by $CS[A]$ and refer to this as the *column space* of A.

Spanning Set. The row space and column space of a matrix A are both special cases of the more general concept of a subspace defined by a *spanning set*. Given any set $V_1, \ldots, V_p$ of vectors in $\Re^n$, we denote the collection of all the possible linear combinations of these vectors by span$[V_1, \ldots, V_p]$ and we refer to this subspace as the subspace *spanned by* the vectors $V_1, \ldots, V_p$. Thus, the row space of A is the subspace spanned by the rows of A and the column space of A is the subspace spanned by the columns of A. For $M = \text{span}[V_1, \ldots, V_p]$, it is obvious that M is indeed a subspace but it is not easy to determine whether an arbitrary vector in $\Re^n$ belongs to M. On the other hand, for a subspace defined by membership rule, it is necessary to prove that the set is, in fact, a subspace, but it is then easy to tell whether an arbitrary vector belongs to the subspace.

REPRESENTATIONS FOR *AX*

It will be convenient to take note of two particular ways of expressing the product of the $n \times n$ matrix A times the n-vector X. Let C_i denote the ith column of A and R_i denote the ith row of A. First we have

$$\begin{bmatrix} a_{11} & \cdots & a_{1n} \\ & \cdots & \\ a_{n1} & \cdots & a_{nn} \end{bmatrix} \begin{bmatrix} x_1 \\ \vdots \\ x_n \end{bmatrix} = \begin{bmatrix} a_{11}x_1 + a_{12}x_2 + \cdots + a_{1n}x_n \\ \vdots \\ a_{n1}x_1 + a_{n2}x_2 + \cdots + a_{nn}x_n \end{bmatrix}$$

$$= \begin{bmatrix} a_{11}x_1 \\ \vdots \\ a_{n1}x_1 \end{bmatrix} + \cdots + \begin{bmatrix} a_{1n}x_n \\ \vdots \\ a_{nn}x_n \end{bmatrix}$$

$$= x_1 \begin{bmatrix} a_{11} \\ \vdots \\ a_{n1} \end{bmatrix} + \cdots + x_n \begin{bmatrix} a_{1n} \\ \vdots \\ a_{nn} \end{bmatrix} = x_1C_1 + \cdots + x_nC_n;$$

i.e.,

$$AX = x_1C_1 + \cdots + x_nC_n \tag{1.6}$$

where C_k denotes the kth column of the matrix A. Note also that

$$Ax = \begin{bmatrix} a_{11}x_1 + a_{12}x_2 + \cdots + a_{1n}x_n \\ \vdots \\ a_{n1}x_1 + a_{n2}x_2 + \cdots + a_{nn}x_n \end{bmatrix}$$

i.e.,

$$AX = \begin{bmatrix} R_1^{\mathsf{T}}X \\ R_n^{\mathsf{T}}X \end{bmatrix} \tag{1.7}$$

The representations (1.6) and (1.7) will be useful in establishing certain properties of the four subspaces $N[A]$, Rng$[A]$, $RS[A]$, and $CS[A]$ related to A.

Theorem 1.7

Theorem 1.7. For any $m \times n$ matrix A, $CS[A] = \text{Rng}\,[A]$.

LINEAR DEPENDENCE

The vectors $V_1, \ldots, V_p$ are said to be *linearly dependent* if and only if there exists scalars $a_1, \ldots, a_p$, not all zero, such that

$$a_1V_1 + \cdots + a_pV_p = \underline{0}.$$

If the vectors $V_1, \ldots, V_p$ are not linearly dependent, they are said to be *linearly independent*. Clearly, the vectors $V_1, \ldots, V_p$ are linearly independent if and only if $a_1V_1 + \cdots + a_pV_p = \underline{0}$ implies that $a_1 = \cdots = a_p = 0$.

Theorem 1.8 Test for Linear Dependence

Theorem 1.8. Let A denote the $p \times n$ matrix whose rows are the row vectors $V_1^{\mathsf{T}}, \ldots, V_p^{\mathsf{T}}$. Then the vectors $V_1, \ldots, V_p$ are linearly dependent if and only if rank $A < p$.

EXAMPLE 1.11 LINEAR DEPENDENCE

(a) Let E_k denote the n-vector whose kth component is a 1 and all other components are zero; i.e.,

$$E_1^{\mathsf{T}} = [1, 0, \ldots, 0],\ E_2^{\mathsf{T}} = [0, 1, 0, \ldots, 0], \ldots, E_n^{\mathsf{T}} = [0, 0, \ldots, 1]$$

The $n \times n$ matrix whose kth row is E_k^{T} for $k = 1$ to n is just I_n and, thus, has rank equal to n. Then, by Theorem 1.8, these vectors are linearly independent. Note that the vectors $E_1, \ldots, E_n$ span all of $\Re^n$ since for an arbitrary vector $X^{\mathsf{T}} = [x_1, \ldots, x_n]$ we have $X = x_1E_1 + \cdots + x_nE_n$; i.e., X can be written as a linear combination of the E_k's.

(b) Suppose the $n \times n$ matrix A has rank $p < n$. That is, A is row equivalent to an upper triangular matrix U having p nontrivial rows. Then, by Theorem 1.8, the nontrivial rows of U are linearly independent, whereas the rows of A are linearly dependent. Note that since the rows of U were obtained by performing elementary row operations on the rows of A, the rows of U are linear combinations of the rows of A. Then if $A_1^{\mathsf{T}}, \ldots, A_n^{\mathsf{T}}$ denote the rows of A and if $U_1^{\mathsf{T}}, \ldots, U_p^{\mathsf{T}}$ denote the nontrivial rows of U, it follows that

$$RS[A] = \text{span}[A_1, \ldots, A_n] = \text{span}[U_1^{\mathsf{T}}, \ldots, U_p^{\mathsf{T}}] = RS[U]$$

BASES

The vectors $V_1, \ldots, V_p$ are said to form a *basis* for the subspace M if M is spanned by the vectors and if, in addition, the vectors are linearly independent. A subspace may have many different bases, but all of the bases have certain properties in common.

Theorem 1.9 Properties of Bases

Theorem 1.9. Let $V_1, \ldots, V_p$ be a basis for the subspace M. Then every vector X in M can be written uniquely as a linear combination of the V_k's; i.e., there exists a unique set of scalars $x_1, \ldots, x_p$ such that $X = x_1V_1 + \cdots + x_pV_p$. Moreover, every basis for M must contain precisely p vectors.

DIMENSION

If $V_1, \ldots, V_p$ is a basis for the subspace M, then we define the *dimension* of M to equal p. Since every basis for M contains the same number of vectors, the dimension of M is well defined (does not depend on which basis we choose).

EXAMPLE 1.12 BASIS AND DIMENSION

(a) The vectors $E_1, \ldots, E_n$ form a basis for $\Re^n$ and it follows that the dimension of $\Re^n$ equals n. This basis is called the *standard basis* for

$\Re^n$. Any other choice on n linearly independent vectors in $\Re^n$ is also a basis for $\Re^n$. In particular, we can show that if $V_1, \ldots, V_p$ are linearly independent vectors in $\Re^n$ with $p < n$, then there is a basis for $\Re^n$ consisting of $V_1, \ldots, V_p$ and $n - p$ additional vectors.

(b) Suppose the $n \times n$ matrix A has rank equal to p. This means A is row equivalent to an upper triangular matrix U having p nontrivial rows. Then the p nontrivial rows of U form a basis for the row space of U and it follows that the row space of U has dimension equal to p. But we have seen in the previous example that $RS[A] = RS[U]$. Thus,

$$\text{rank}\, A = \text{dimension of}\, RS[A] = \dim RS[A] \tag{1.8}$$

NORM, UNIT VECTOR

In (1.2), we defined the inner product of vectors X and Y. Then we may define the *norm* of a vector X to be the scalar quantity

$$\|X\| = (X, X)^{1/2}$$

Then $\|X\| \geq 0$ for every X and $\|X\| = 0$ if and only if $X = \underline{0}$. For $n = 2, 3$, the norm of a vector X in $\Re^n$ coincides with the notion of the length of the directed line segment X. A vector X for which $\|X\| = 1$ is called a *unit vector.*

ORTHOGONALITY

Vectors X and Y are said to be *orthogonal* if $(X, Y) = 0$. For $n = 2, 3$, two vectors in $\Re^n$ are orthogonal if and only if they are perpendicular as line segments. If the vectors $\{X_1, \ldots, X_p\}$ form a basis for subspace M in $\Re^n$, then a procedure known as the Gram-Schmidt orthogonalization process generates a set of mutually orthogonal unit vectors $\{Y_1, \ldots, Y_p\}$ which also span M. The set $\{Y_1, \ldots, Y_p\}$ is called an *orthonormal basis* for M. Thus, from any basis, we can generate an orthonormal basis spanning the same subspace (see Problem 1.9).

ORTHOGONAL COMPLEMENT

For any collection of vectors M in $\Re^n$, the *orthogonal complement* of M consists of all vectors X such that $(X, Y) = 0$ for all Y in M. The orthogonal complement of M is denoted by $M^\perp$. Although the orthogonal complement is defined for any collection M of vectors in $\Re^n$, we will only be concerned with the case in which M is a subspace in $\Re^n$.

EXAMPLE 1.13 ORTHOGONAL COMPLEMENTS

(a) Let M denote the subspace of $\Re^3$ that is spanned by the single vector $E_1 = [1, 0, 0]^T$; i.e., M contains all the vectors lying along the x_1 axis. Then $M^\perp$ consists of all the vectors that are perpendicular to the x_1 axis, namely, all the vectors in the x_2x_3 plane.

(b) Let A denote an $m \times n$ matrix whose rows are vectors in $\Re^n$ denoted by $R_1, \ldots, R_m$. Then it follows from (1.7) that

$$X \text{ in } \Re^n \text{ satisfies } AX = \underline{0} \text{ if and only if } (R_j, X) = 0 \text{ for } j = 1, \ldots, m.$$

But if X is orthogonal to each of the vectors $R_1, \ldots, R_m$, then X is in the orthogonal complement of the row space of A which is spanned by the rows of A. Thus, every X in the null space of A is in the orthogonal complement of the row space of A.

Conversely, if X is orthogonal to every row of A, then it follows from (1.7) that X satisfies $AX = \underline{0}$. Thus, every X that is orthogonal to the row space of A lies in the null space of A. We have proved $RS[A]^{\perp} = N[A]$.

Theorem 1.10 Properties of the Orthogonal Complement

Theorem 1.10. For M any subspace in $\Re^n$ with orthogonal complement denoted by $M^{\perp}$:

1. $M^{\perp}$ is a subspace in $\Re^n$.
2. $(M^{\perp})^{\perp} = M$.
3. The only vector belonging to M and to $M^{\perp}$ is the zero vector.
4. Each X in $\Re^n$ can be written uniquely as a sum $X = Y + Z$ for Y in M and Z in $M^{\perp}$.
5. If $\dim M = p$, then $\dim M^{\perp} = n - p$.

RANK AND NULLITY OF A MATRIX

It follows from Example 1.12(b) and part 5 of Theorem 1.10 that, for an $m \times n$ matrix A,

$$\dim N[A] = n - \dim RS[A] = n - \operatorname{rank} A$$

If the rank of A is equal to r, then $p = n - r$ is called the *nullity* of A. If the rank of A equals n, then the nullity equals zero. For a matrix of nullity zero, the only vector satisfying $AX = \underline{0}$ is the zero vector.

Theorem 1.11 Fundamental Theorem for Matrices

Theorem 1.11. For any $m \times n$ matrix A, $\dim RS[A] = \dim CS[A]$.

TRANSPOSE OF A MATRIX

We define the transpose of the $m \times n$ matrix A to be the $n \times m$ matrix whose rows are the columns of A. If the transpose is denoted by A^{T}, then

$$RS[A^{\mathsf{T}}] = CS[A] \quad \text{and} \quad CS[A^{\mathsf{T}}] = RS[A]$$

For any $m \times n$ matrix A, we have the following properties of A^{T}:

1. $(A^{\mathsf{T}})^{\mathsf{T}} = A$.
2. If the product AB is defined, $(AB)^{\mathsf{T}} = B^{\mathsf{T}}A^{\mathsf{T}}$.
3. For any $\boldsymbol{X}$ in $\Re^n$ and every $\boldsymbol{B}$ in $\Re^m$, $(A\boldsymbol{X}, \boldsymbol{B}) = (\boldsymbol{X}, A^{\mathsf{T}}\boldsymbol{B})$.

Any matrix with the property that $A = A^{\mathsf{T}}$ is said to be *symmetric*.

We refer to the subspaces $N[A]$, Rng$[A]$, $RS[A]$, and $CS[A]$ as the fundamental subspaces for the matrix A. These subspaces are related as follows.

Theorem 1.12 Fundamental Subspaces of A

Theorem 1.12. For an $m \times n$ matrix A with rank equal to r, we have:

1. $N[A] = RS[A]^{\perp}$.
2. $CS[A] = \text{Rng}\,[A]$.
3. $\dim RS[A] = \dim CS[A] = r$; $\dim N[A] = n - r = p$.
4. $N[A^{\mathsf{T}}] = RS[A^{\mathsf{T}}]^{\perp} = CS[A]^{\perp} = (\text{Rng}\,[A])^{\perp}$.
5. $RS[A] = CS[A^{\mathsf{T}}] = \text{Rng}\,[A^{\mathsf{T}}]$.
6. $\dim RS[A^{\mathsf{T}}] = \dim CS[A^{\mathsf{T}}] = r$; $\dim N[A^{\mathsf{T}}] = m - r = q$.

Now we can state the fundamental theorem on solvability of linear systems.

Theorem 1.13 Fundamental Theorem for Linear Systems

Theorem 1.13. Let A denote an $m \times n$ matrix with rank $A = r > 0$ and consider the linear system $A\boldsymbol{X} = \boldsymbol{B}$.

Solvability. If $\dim N[A^{\mathsf{T}}] = m - r$ equals zero, then at least one solution $\boldsymbol{X}$ exists for every $\boldsymbol{B}$ in $\Re^m$. If $\dim N[A] = m - r$ is positive, then there is no solution $\boldsymbol{X}$ unless the data vector $\boldsymbol{B}$ in $\Re^m$ is orthogonal to every solution of the system $A^{\mathsf{T}}\boldsymbol{Y} = \underline{\mathbf{0}}$; i.e., if $m - r = q > 0$ and $[\boldsymbol{Y}_1, \ldots, \boldsymbol{Y}_q]$ is a basis for $N[A^{\mathsf{T}}]$, then $A\boldsymbol{X} = \boldsymbol{B}$ has no solution unless $\boldsymbol{B} \bullet \boldsymbol{Y}_j = 0$ for $1 \leq j \leq q$.

Uniqueness. If $\dim N[A] = n - r$ equals zero, then $A\boldsymbol{X} = \boldsymbol{B}$ has at most one solution $\boldsymbol{X}$. If $\dim N[A] = n - r$ is positive, then, for any solution $\boldsymbol{X}$, the sum $\boldsymbol{X} + \mathbf{Z}$ is also a solution for any vector $\mathbf{Z}$ from $N[A]$; i.e., if $n - r = p > 0$ and $[\mathbf{Z}_1, \ldots, \mathbf{Z}_p]$ is a basis for $N[A]$, then

$$X = V + C_1 Z_1 + \cdots + C_p Z_p$$

is a solution of $A\boldsymbol{X} = \boldsymbol{B}$ for any choice of the scalars $C_1, \ldots, C_p$ and any $\boldsymbol{V}$ such that $A\boldsymbol{V} = \boldsymbol{B}$.

SOLVED PROBLEMS

Mathematical Models Leading to Systems of Linear Equations

PROBLEM 1.1 2-DIMENSIONAL STEADY STATE DIFFUSION

Consider a diffusion cell composed of two flat square plates separated by a distance that is small compared to the sidelength of the plates. Suppose the space between the plates is filled with fluid in which a contaminant can diffuse. Suppose further that the East and West edges of this square sandwich are sealed against the passage of fluid or contaminant, but that the North and South edges of the cell are in contact with contaminant sources of prescribed but position dependent levels of concentration. This contaminant is free to diffuse into the cell, and after all transient behavior has died out, it assumes some equilibrium distribution within the cell. Derive a mathematical model for the equilibrium distribution of contaminant and show that the model consists of a system of algebraic equations in which the unknowns are the contaminant concentrations at various positions in the cell.

SOLUTION 1.1

Let the sidelength of the square cell be denoted by L and suppose that the cell is then subdivided into n^2 square subcells each of sidelength $W = L/n$. Then the contaminant distribution is in a state of equilibrium if the contaminant flow into any subcell equals the contaminant outflow. We can express this as an equation in terms of contaminant flux, denoted by F. The flux through the East face of the subcell will be denoted F_E with similar notation for the flux through the other three faces. If we adopt the convention that horizontal flux is positive if it moves East to West and a vertical flux moving South to North is positive, then the equilibrium condition may be stated in the form

$$\text{inflow} - \text{outflow} = F_E - F_W + F_S - F_N = 0 \tag{1}$$

This equation must be satisfied in each subcell. Contaminant flows from regions where concentration is high to areas of lower concentration. A precise statement of this observation is known as Fick's law which expresses the flux F through any face in terms of the contaminant concentrations on the two sides of the face. In particular,

$$F = -D\frac{u_p - u_m}{w} \tag{2}$$

where u_p and u_m denote the concentrations on the "plus" and "minus" sides of the face, respectively. For a horizontal flux, the plus side is to the right of the face, whereas the upper side of the face is the plus side for a vertical flux. Here w is the width of every subcell and D is a positive constant whose value depends on the materials involved.

In order to apply equations (1) and (2), suppose $n = 4$, so the diffusion cell is subdivided into 16 square subcells. We number these cells from 1 to 16 as seen in Figure 1.1. In addition, we label cells along the North and South edges of the diffusion cell with numbers 17 to 24. Then the concentrations $u_1, \dots, u_{16}$ in cells 1 to 16 are the unknown equilibrium concentrations and the concentrations $u_{17}, \dots, u_{24}$ are the specified concentrations of contaminant along the upper and lower boundaries of the diffusion cell. The East and West edges of the diffusion cell are sealed so that the fluxes through those faces are zero.

Using (2) or the sealed edge boundary conditions, we find the fluxes in cell 1 to be

$$F_E = 0 \quad F_W = \frac{-D(u_2 - u_1)}{w} \quad F_S = \frac{-D(u_2 - u_6)}{w} \quad F_N = \frac{-D(u_{17} - u_1)}{w}$$

Similarly, in cell 16 they are

$$F_E = \frac{-D(u_{16} - u_{15})}{w} \quad F_W = 0 \quad F_S = \frac{-D(u_{16} - u_{24})}{w} \quad F_N = \frac{-D(u_{12} - u_{16})}{w}$$

Using these expressions for the fluxes in (1), we obtain the following equations for the unknowns $u_1, \dots, u_{16}$ in terms of the data $u_{17}, \dots, u_{24}$:

cell 1: $3u_1 - u_2 - u_5 = u_{17}$

cell 2: $-u_1 + 4u_2 - u_3 - u_6 = u_{18}$

cell 3: $-u_2 + 4u_3 - u_4 - u_7 = u_{19}$

cell 4: $-u_3 + 3u_4 - u_8 = u_{20}$

etc.

This leads to the system of equations $\boldsymbol{AX} = \boldsymbol{B}$ where A denotes the 16×16 coefficient matrix composed of sixteen 4×4 blocks as follows:

$$\begin{bmatrix} A_1 & -I & 0 & 0 \\ -I & A_1 & -I & 0 \\ 0 & -I & A_1 & -I \\ 0 & 0 & -I & A_1 \end{bmatrix}$$

where I denotes the 4×4 identity matrix, 0 denotes the 4×4 matrix of zeroes, and

$$A_1 = \begin{bmatrix} 3 & -1 & 0 & 0 \\ -1 & 4 & -1 & 0 \\ 0 & -1 & 4 & -1 \\ 0 & 0 & -1 & 3 \end{bmatrix}$$

The data vector $\boldsymbol{B}$ is equal to $[u_{17}, u_{18}, u_{19}, u_{20}, 0, 0, 0, 0, 0, 0, 0, 0, u_{21}, u_{22}, u_{23}, u_{24}]^T$. The coefficient matrix A consists of three diagonal strips of 4×4 blocks and is called a *block tridiagonal* matrix. This system can be solved by the algorithm described in Problem 1.4.

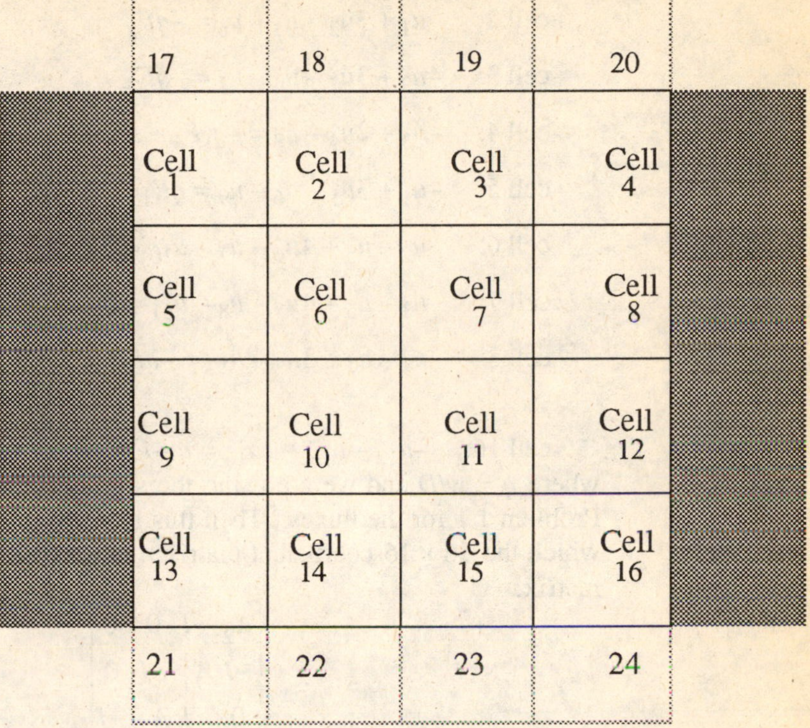

Figure 1.1
Concentration Specified on N and S Faces
E and W Faces Sealed

PROBLEM 1.2

Derive the system of equations that models the 2-dimensional diffusion system of the previous problem in the case that boundary fluxes are specified on all four edges of the diffusion cell.

SOLUTION 1.2

We shall continue to suppose the diffusion cell is divided into 16 subcells as in the previous problem. Then the concentrations $u_1, \ldots, u_{16}$ in these subcells are the unknowns in the system we will derive. Now let $F_1, \ldots, F_4$ denote the fluxes across the faces on the North side of cells 1 through 4. These fluxes are given and will form part of the data in our system of equations. Similarly, let $F_5, \ldots, F_8$ denote the given fluxes through the East faces of the cells along the East edge of the array of subcells. Figure 1.2 shows the array of subcells and the given fluxes.

Using equations (1) and (2) from the previous problem for each of the 16 cells in the array, we find

cell 1: $2u_1 - u_2 - u_5 = q(-F_1 + F_{16})$

cell 2: $-u_1 + 3u_2 - u_3 - u_6 = -qF_2$

cell 3: $-u_2 + 3u_3 - u_4 - u_7 = -qF_3$

cell 4: $-u_3 + 2u_4 - u_8 = -q(F_4 + F_5)$

cell 5: $-u_1 + 3u_5 - u_6 - u_9 = qF_{15}$

cell 6: $-u_2 - u_5 + 4u_6 - u_7 - u_{10} = 0$

cell 7: $-u_3 - u_6 + 4u_7 - u_8 - u_{11} = 0$

cell 8: $-u_4 - u_7 + 3u_8 - u_{12} = -qF_6$

⋮

cell 16: $-u_{12} - u_{15} + 2u_{16} = q(-F_8 + F_9)$

where $q = w/D$ and we are using the sign convention of equation (1) in Problem 1.1 for the fluxes. Then this is a system of the form $\boldsymbol{AX} = \boldsymbol{B}$ in which the 16×16 coefficient matrix A is the following block tridiagonal matrix:

$$\begin{bmatrix} A_2 & -I & 0 & 0 \\ -I & A_1 & -I & 0 \\ 0 & -I & A_1 & -I \\ 0 & 0 & -I & A_2 \end{bmatrix}$$

where A_1 is as in the previous problem and

$$A_2 = \begin{bmatrix} 2 & -1 & 0 & 0 \\ -1 & 3 & -1 & 0 \\ 0 & -1 & 3 & -1 \\ 0 & 0 & -1 & 2 \end{bmatrix}$$

The data vector in this problem is

$$\boldsymbol{B} = q[F_{16} - F_1, -F_2, -F_3, -F_4 - F_5, F_{15}, 0, 0, -F_6, F_{14}, 0, 0, -F_7, F_{12} + F_{13}, F_{11}, F_{10}, F_9 - F_8]^T$$

Note that this matrix A has the property that the sum of the entries in each row is zero. We will consider this system again in Problem 1.15.

PROBLEM 1.3 1-DIMENSIONAL TIME DEPENDENT DIFFUSION

Consider a thin straight tube filled with a fluid in which a contaminant can diffuse. Assume that the initial distribution of contaminant in the tube is given and that the ends of the tube are in contact with reservoirs in which the concentration of contaminant varies with time in a (given) prescribed way. Then show that this physical system can be modelled by a linear system of algebraic equations that must be solved repeatedly with the same coefficient matrix.

SOLUTION 1.3

We suppose that the walls of the tube are impervious to any flow and that the width of the tube is small compared to its length. Then the flow of contaminant can be assumed to be 1-dimensional; i.e., the concentration of contaminant varies only with position *along* the tube and with time. Let the length of the tube be denoted by L and suppose that the tube is divided into N equal "cells." Then each cell has length $w = L/N$ and we assume the tube has uniform cross-sectional area A so this is the cross-sectional area of each cell as well. Let T denote some suitable increment

	Flux 1	Flux 2	Flux 3	Flux 4	
Flux 16	Cell 1	Cell 2	Cell 3	Cell 4	Flux 5
Flux 15	Cell 5	Cell 6	Cell 7	Cell 8	Flux 6
Flux 14	Cell 9	Cell 10	Cell 11	Cell 12	Flux 7
Flux 13	Cell 13	Cell 14	Cell 15	Cell 16	Flux 8
	Flux 12	Flux 11	Flux 10	Flux 9	

Figure 1.2
Flux Specified On All External Faces.

of time and, for $j = 0, 1, \ldots$, let $t_j = jT$. Then we define $u_{i,j}$ to be the concentration of contaminant in cell number i at the time t_j for $i = 1, \ldots, N$ and $j = 0, 1, \ldots$. We are assuming that the ends of the tube are each in contact with reservoirs in which the concentration of contaminant varies with time in a specified way. Thus, we label these given concentrations at the left and right ends of the tube as $u_{0,j}$ and $u_{N+1,j}$, respectively, and treat these quantities as given data in the problem. The *amount* of contaminant in cell i at time t_j then is equal to $u_{i,j}\, A_w$ and we can state the following balance law for the contaminant in cell number i during the time interval from $t_{j\,i}$ to t_j

$$A_w(u_{i,j} - u_{i,j-1}) = (\text{inflow} - \text{outflow}) \text{ in cell number } i \text{ during } (t_{j-i}, t_j) \quad (1)$$

The inflow and outflow can be expressed in terms of the concentrations by means of Fick's law:

$$\text{inflow across left face of cell } i = -ADT\frac{u_{i,j} - u_{i-1,j}}{w}$$

$$\text{outflow across right face of cell } i = -ADT\frac{u_{i+1,j} - u_{i,j}}{w} \quad (2)$$

Then using these equations in the balance law, we find

$$u_{i,j} - u_{i,j-1} = r(u_{i+1,j} - 2u_{i,j} + u_{i-1,j}) \text{ for } i = 1, \ldots, N \text{ and } j = 1, 2, \ldots \quad (3)$$

where $r = DT/w^2$. Since the initial concentrations for each cell are known; i.e., $u_{i,0}$ is given for $i = 1, \ldots, N$, this leads to the following system of equations for the unknown concentrations $u_{1,1}, \ldots, u_{N,1}$ at time t_1:

$$(1 + 2r)u_{1,1} - ru_{2,1} = ru_{0,1} + u_{1,0}$$
$$-ru_{1,1} + (1 + 2r)u_{2,1} - ru_{3,1} = ru_{2,0}$$
$$-ru_{2,1} + (1 + 2r)u_{3,1} - ru_{4,1} = ru_{3,0}$$
$$\vdots$$
$$-ru_{N-1,1} + (1 + 2r)u_{N,1} = ru_{N+1,1} + u_{N,0}.$$

This is a system of the form $AU_1 = \boldsymbol{B}$ where U_1 contains the N unknown concentrations at time t_1 and A denotes the following $N \times N$ matrix

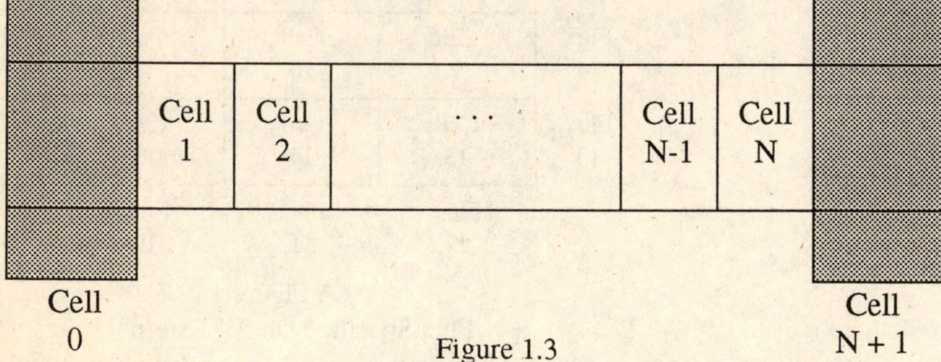

Figure 1.3

$$\begin{bmatrix} 1+2r & -r & 0 & \ldots & 0 \\ -r & 1+2r & -r & \ldots & 0 \\ & & & & -r \\ 0 & & (-r) & & 1+2r \end{bmatrix}$$

The data vector is composed of the sum $B_0 = U_0 + rC_1$, where

$$C_1 = [u_{0,1}, 0, \ldots, 0, u_{N+1,1}]^T$$

$$U_0 = [u_{1,0}, \ldots, u_{N,0}]^T$$

Note that U_0 represents the initial state of the system, whereas C_1 constitutes the input from the boundary of the system during the first time interval. If we solve this system to obtain U_1, then we can solve the system $AU_2 = B_1$ for the unknown concentrations at time t_2 where A denotes the same coefficient matrix and $B_1 = rC_2 + U_1$. Thus, the solution vector from one time level forms a part of the data vector at the next time level as we solve the system repeatedly with same coefficient matrix (see Problem 1.7).

Gaussian Elimination

PROBLEM 1.4

Write a BASIC computer program to carry out the steps of the Gaussian elimination algorithm on the following system of six equations in six unknowns:

$$\begin{bmatrix} A & -I \\ -I & A \end{bmatrix} X = [1, 1, 1, 0, 0, 0]^T$$

where I denotes the 3×3 identity and A denotes the following 3×3 block

$$A = \begin{bmatrix} 4 & -1 & 0 \\ -1 & 4 & -1 \\ 0 & -1 & 4 \end{bmatrix}$$

This is the system of equations that must be solved to find the equilibrium concentrations of contaminant for the diffusion cell pictured in Figure 1.4. The data vector given here is consistent with given concentrations of 1 in the cells 7, 8, and 9 and concentrations of 0 in cells 10 through 16.

SOLUTION 1.4

The following BASIC program executes the Gaussian elimination algorithm in its simplest form (i.e., with no row interchanges):

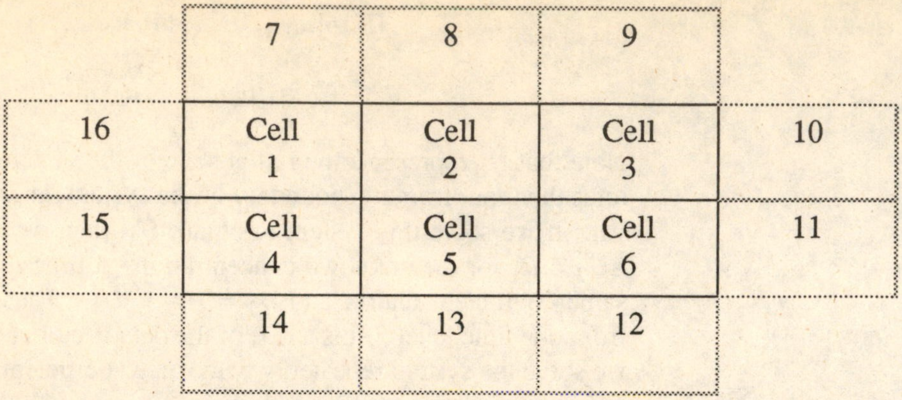

	7	8	9	
16	Cell 1	Cell 2	Cell 3	10
15	Cell 4	Cell 5	Cell 6	11
	14	13	12	

Figure 1.4

```
REM GAUSSIAN ELIMINATION TO SOLVE AX = B
REM
DIM A(50,50), B(50), C(50)
REM
READ N
FOR I = 1 TO N
    FOR J = 1 TO N
       READ A(I,J)
    NEXT J
NEXT I
FOR I = 1 TO N
    READ B(I)
NEXT I
REM          READ IN MATRIX A AND DATA VECTOR B
DATA 6
DATA 4,-1,0,-1,0,0
DATA -1,4,-1,0,-1,0
DATA 0, -1,4,0,0,-1
DATA -1,0,0,4,-1,0
DATA 0,-1,0,-1,4,-1
DATA 0,0,-1,0,-1,4
DATA 1,1,1,0,0,0
REM          PRINT MATRIX A AND DATA VECTOR B
REM          FORM AUGMENTED MATRIX
FOR I = 1 TO N
    A(I,N+1) = B(I)
```

```
NEXT I
FOR I = 1 TO N-1
IF ABS(A(I,I))<1.0 E-06 THEN
PRINT "MATRIX A MAY BE SINGULAR"
STOP
ELSE
   FOR K = I+1 TO N
      R = A(K,I)/A(I,I)          calculate multiplier
         FOR J = 1 TO N+1
            A(K,J) = A(K,J) - R*A(I,J)  Row operation Eki(R)
         NEXT J
   NEXT K
END IF
NEXT I
IF ABS(A(N,N))<1.0E-06 THEN
PRINT "MATRIX A MAY BE SINGULAR"
STOP
ELSE
C(N) = A(N,N+1)/A(N,N)
FOR I = N-1 TO 1, STEP -1        Back sweep
SUM = 0                          Save solution vector in C
   FOR J = I+1 TO N
      SUM = SUM + A(I,J)*C(J)
   NEXT J
C(I) = (A(I,NA=1)-SUM)/A(I,I)
NEXT I
END IF
REM        PRINT SOLUTION VECTOR
END
```

Applying this program to the 6×6 system given above leads to

$$X = [.416, .509, .416, .155, .205, .155]^T$$

In general, for given concentrations of u_7 through u_{16} in cells 7 through 16, the data vector for the system shown in Figure 1.4 is given by

$$B = [u_7 + u_{16}, u_8, u_9 + u_{10}, u_{14} + u_{15}, u_{13}, u_{11} + u_{12}]^T$$

PROBLEM 1.5

Write a subroutine in BASIC to factor an $n \times n$ matrix A into a product LU of a lower triangular times an upper triangular matrix.

SOLUTION 1.5

The following is a BASIC subroutine to factor an $N \times N$ matrix A into a product LU where L and U are, respectively, lower and upper triangular $N \times N$ matrices. This is accomplished by reducing A to upper triangular form U by means of row operations and storing the multipliers in the sub-diagonal entries of L as described in the text. The subroutine assumes no row exchanges will be necessary and hence is not fully general.

```
SUBROUTINE FACTOR(A, L, U)
REM     the matrices A, L and U are dimensioned N×N in the main
REM     program
FOR I = 1 TO N
   IF ABS(A(I,I))<1.0E-06 THEN
      PRINT "A MAY BE SINGULAR"
      STOP
   ELSE
      L(I,I) = 1
   FOR K = I + 1 TO N
      R = A(K,I)/A(I,I)            : Compute multiplier r
      L(K,I) = R                   : Save in L
         FOR J = 1 TO N
            A(K,J) = A(K, J) – R*A(I,J)  : Row Operation E_ki(r)
            U(K,J) = A(K,J)
         NEXT J
   NEXT K
   END IF
NEXT I
L(N,N) = 1
IF ABS(A(N,N))<1.0E-06 THEN
   PRINT "A MAY BE SINGULAR"
   STOP
END IF
RETURN
```

PROBLEM 1.6

Write a subroutine in BASIC to carry out the forward and backward sweeps in solving $LUX = B$.

SOLUTION 1.6

The system $AX = B$ is equivalent to the two systems $LB* = B$ and $UX = B*$ if $A = LU$. Solving the lower triangular system $LB* = B$ amounts to a forward sweep and solving $UX = B*$ is accomplished by a back substitution. Here is a subroutine to carry out these two steps once A has been factored:

```
SUBROUTINE SOLV(L, U, B)
REM  L and U are dimensioned N (rows) by N + 1 (columns) in the
     main program
```

```
REM  B and C are dimensioned as N vectors in the main program
REM  the data vector B is stored in the N + 1 column of L
REM  the vector B* (called C) is placed in the N + 1 column of U
REM  solution is returned in the vector B
REM
C(1) = L(1,N+1)                : begin the forward sweep to solve LC = B
FOR I = 2 TO N
    C(I) = L(I,N+1)
    FOR J = 1 TO I - 1
        C(I) = C(I) - L(I,J)*C(J)
    NEXT J
NEXT I
REM                            :load C into N + 1 column of
FOR J = 1 TO N U
    U(J,N+1) = C(J)
NEXT J
REM                            : begin backward sweep to solve UB = C
B(N) = U(N,N+1)/U(N,N)
FOR I = N - 1 TO 1  STEP -1
    B(I) = U(I,N+1)
    FOR J = I+1 TO N
        B(I) = B(I) - U(I,I)* B(J)
    NEXT J
    B(I) = B(I)/U(I,I)         : solution is now stored in vector B
NEXT I
RETURN
```

PROBLEM 1.7

Use the subroutines of the previous two problems to solve for the time dependent concentrations for the 1-dimensional diffusion system described in Problem 1.3. Assume there are 20 cells in the tube ($N = 20$) and that the initial concentration in cells 1 through 20 is zero. Suppose further that controlled concentration in cell 0 is maintained at the value 1 while cell 21 at the other end of the tube is kept at concentration zero during the experiment. We will suppose that the parameters D, T, and L in the problem are such that $r = DT/w^2 = DTN^2/L^2 = .5$ and compute the value of $u_{i,j}$ for $i = 1$ to 20 and $j = 1$ to 100.

SOLUTION 1.7

In Problem 1.3 we showed that the vector U_j containing the concentrations in the N cells at the time level t_j could be obtained by solving the system $AU_j = B_{j-1}$ where A denotes the following $N \times N$ matrix

$$\begin{bmatrix} 1+2r & -r & 0 & \cdots & 0 \\ -r & 1+2r & -r & \cdots & 0 \\ & & \vdots & & \\ & & & & -r \\ 0 & & & -r & 1+2r \end{bmatrix}$$

and the data vector is composed of the sum $B_{j-1} = U_{j-1} + rC_j$ where U_{j-1} is the vector of concentrations from the previous time level and

$$C_j = [u_{0,j}, 0, \ldots, 0, u_{N+1,j}]^T = \text{boundary input}$$

We are supposing here that $N = 20$ and $u_{0,j} = 1$ and $u_{21,j} = 0$ for all j. In addition, we assume that the initial concentration vector is the zero vector; i.e., $U_0 = \underline{0}$. Then the following is a BASIC program to solve the system $AU_j = B_{j-1}$ for $j = 1, \ldots, 100$ with $r = .5$.

```
REM  Program to factor A into LU then solve by forward and backward
     sweeps
DIM A(20,20), L(20,20), U(20,20), B(20), C(20)
REM                       : generate the matrix A and the initial data vector B
R = .5
FOR I = 1 TO 19
    A(I,I+1) = -R
    A(I,I) = 1 + 2* R
    A(I+1,I) = -R
    B(I) = 0
NEXT I
GOSUB FACTOR(A,L,U)       : Factor A into LU
REM
FOR J = 1 TO 100
    B(1) = B(1) + 1*R     : enter boundary concentration into data vector
REM
FOR I = 1 TO 20
    L(I,21) = B(I)        : Load data vector into column 21 of L matrix
NEXT I
REM
GOSUB SOLV(L,U,B)         :execute forward and backward sweeps
REM
REM                       : generate output (See Figure 1.5)
NEXT J
END
```

Note that the time-consuming step in the solution algorithm is the factoring of the matrix A and this is done just once. Then solving repeatedly with updated data vectors requires only the forward and backward sweeps to be repeated. We have chosen to generate the output graphically, plotting $u_{i,j}$ versus i for selected values of j as shown in Figure 1.5. The curves obtained are referred to as concentration *profiles*. Note that the profiles indicate that the concentration distribution tends, with increasing time, toward a linear distribution of contaminant across the tube, varying from concentration equal to 1 at the left end down to concentration 0 at the right end of the tube.

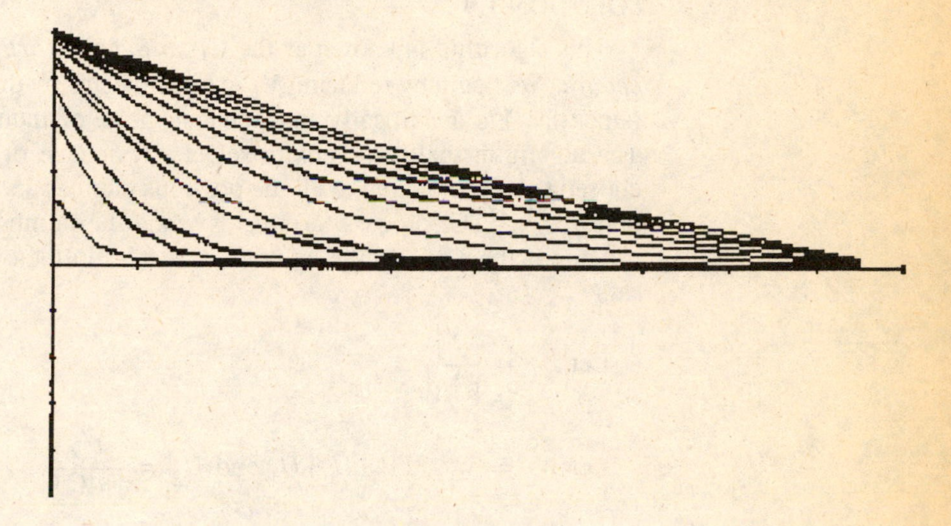

Figure 1.5
Selected Profiles for Problem 1.7

Linear Algebra

PROBLEM 1.8

Show that for any $m \times n$ matrix A, Rng $[A]$ is a subspace of $\Re^m$.

SOLUTION 1.8

For an $m \times n$ matrix A, the product AX is defined for $\boldsymbol{X}$ in $\Re^n$ and produces a vector $\boldsymbol{B}$ in $\Re^m$. We define the set Rng $[A]$ to be the set of all vectors in $\boldsymbol{B}$ in $\Re^m$ that are equal to AX for some vector $\boldsymbol{X}$ in $\Re^n$. To show that Rng $[A]$ is a subspace, we must show that for any two vectors $\boldsymbol{B}$ and $\boldsymbol{B}'$ in Rng $[A]$, the linear combination $a\boldsymbol{B} + b\boldsymbol{B}'$ is again in Rng $[A]$ for all scalars a and b. But if $\boldsymbol{B}$ and $\boldsymbol{B}'$ are in Rng $[A]$, then $\boldsymbol{B} = AX$ and $\boldsymbol{B}' = AX'$ for some X and X' in $\Re^n$. Then it follows from the definition of matrix multiplication that

$$A(aX + bX') = A(aX) + A(bX') = aA(X) + bA(X') = a\boldsymbol{B} + b\boldsymbol{B}'$$

Thus, $aB + bB' = AZ$ for $Z = aX + bX'$ in $\Re^n$ and we have proved that the linear combination is again in Rng $[A]$.

PROBLEM 1.9 GRAM-SCHMIDT ORTHOGONALIZATION

Let $V_1, \ldots, V_m$ be m linearly independent but not necessarily orthogonal vectors in $\Re^n$. Then show how to generate, from these vectors, an orthogonal set of unit vectors $U_1, \ldots, U_m$.

SOLUTION 1.9

This algorithm is known as the *Gram-Schmidt orthogonalization procedure*. We begin by reducing V_1 to a unit vector by dividing by its length (since the V's are linearly independent, none of them has zero length). The algorithm then proceeds to select a sequence of new vectors, each chosen to be orthogonal to all the previous choices and each of which is a unit vector. The process terminates when the number of vectors in the orthonormal set equals the number of vectors in the original independent set.

$$\text{Let } U_1 = \frac{V_1}{\|V_1\|}.$$

$$\text{Let } W_2 = V_2 - (V_2, U_1)\,U_1 \text{ and } U_2 = \frac{W_2}{\|W_2\|}.$$

$$\text{Let } W_3 = V_3 - (V_3, U_2)\,U_2 - (V_3, U_1)\,U_1 \text{ and } U_3 = \frac{W_3}{\|W_3\|}.$$

$$\vdots$$

$$\text{Let } W_m = V_m - \sum_{j=1}^{m-1} (V_m, U_j)\,U_j \text{ and } U_m = \frac{W_m}{\|W_m\|}.$$

The vectors U_i generated by this algorithm are clearly unit vectors, and to check that they are mutually orthogonal is a simple exercise. To see that the algorithm does not terminate before we obtain m vectors, suppose that $W_j = \underline{0}$ for some $j \le m$. This means $W_j = V_j + c_{j-1} + \cdots + c_1 V_1 = \underline{0}$. But the V's are linearly independent and thus the linear combination cannot vanish. This proves that the orthogonalization procedure will not terminate until m vectors have been obtained.

PROBLEM 1.10

Let A denote an $m \times n$ matrix. Then show that:

(a) For each X in $\Re^n$, AX belongs to $\Re^m$.
(b) $CS[A] = \text{Rng }[A]$.
(c) $\Re^n = N[A] + \text{Rng }[A^T]$; i.e., each X in $\Re^n$ can be written uniquely as a

sum $\boldsymbol{X} = \boldsymbol{U} + \boldsymbol{V}$ where $\boldsymbol{U}$ is in $N[A]$ and $\boldsymbol{V}$ belongs to Rng $[A^{\mathrm{T}}]$.

(d) $\Re^m = N[A^{\mathrm{T}}] + \mathrm{Rng}\, A$; i.e., each $\boldsymbol{B}$ in $\Re^m$ can be written uniquely as a sum $\boldsymbol{B} = \boldsymbol{Y} + \boldsymbol{Z}$ where $\boldsymbol{Y}$ is in $N[A^{\mathrm{T}}]$ and $\boldsymbol{Z}$ belongs to Rng $[A]$.

SOLUTION 1.10

Since A is $m \times n$ and each X in $\Re^n$ is $n \times 1$, it follows from the definition of matrix multiplication that AX is $m \times 1$; i.e., AX belongs to $\Re^m$. We write $A{:}\Re^n \to \Re^m$ to indicate that A carries vectors in $\Re^n$ into vectors in $\Re^m$.

It follows from (1.6) that any vector of the form AX can be written as a linear combination of the columns of A. Conversely, any vector that is a linear combination of the columns of A is of the form AX where the jth entry of X equals the jth coefficient in the linear combination. This proves that Rng $[A] = CS[A]$. Since this is true for any matrix, it is also true that Rng $[A^{\mathrm{T}}] = CS[A^{\mathrm{T}}]$.

To show (c), recall from Example 1.12(b) that $N[A] = RS[A]^{\perp}$. Also, by definition, $RS[A] = CS[A^{\mathrm{T}}]$. Since (b) implies $CS[A^{\mathrm{T}}] = \mathrm{Rng}\ [A^{\mathrm{T}}]$, we have $N[A] = (\mathrm{Rng}\ [A^{\mathrm{T}}])^{\perp}$ and then (c) follows from Theorem 1.10, part 4. Now (d) is obtained by replacing A by A^{T} in (c) and recalling that $(A^{\mathrm{T}})^{\mathrm{T}} = A$.

PROBLEM 1.11

Describe the solution set for the system $AX = B$ in the case that A is an $m \times n$ matrix with rank $A = m < n$; i.e., the number of equations is fewer than the number of unknowns and the system is therefore *underdetermined.*

SOLUTION 1.11

We are assuming that the rank of A is equal to m, the number of rows of A. Thus, the rows of A are linearly independent. Then, by Theorem 1.11, it follows that

$$\dim N[A] = n - m = p > 0 \quad \text{and} \quad \dim N[A^{\mathrm{T}}] = m - m = q = 0$$

i.e., there are p linearly independent vectors in $N[A]$ and $\underline{0}$ is the only vector in $N[A^{\mathrm{T}}]$. Now $N[A^{\mathrm{T}}] = \underline{0}$ together with part (d) of the previous problem implies that Rng $[A]$ is the whole space $\Re^m$. This means that, for every $\boldsymbol{B}$ in $\Re^m$, there exists X in $\Re^n$ such that $AX = \boldsymbol{B}$, which is to say, the underdetermined system is always solvable.

Since dim $N[A] = p$, there exists a basis for $N[A]$ containing p vectors, $\boldsymbol{Z}_1, \ldots, \boldsymbol{Z}_p$, and if $\boldsymbol{U}$ is any vector in $\Re^n$ satisfying $A\boldsymbol{U} = \boldsymbol{B}$, then

$$X = \boldsymbol{U} + c_1\boldsymbol{Z}_1 + \cdots + c_p\boldsymbol{Z}_p \tag{1}$$

also solves $AX = \boldsymbol{B}$ for any choice of the scalars $c_1, \ldots, c_p$. Thus, the solution of the underdetermined system is not unique. We refer to (1) as the *general form* of the solution for the underdetermined system and say the solution contains p *degrees of freedom*.

PROBLEM 1.12

Find the general solution of the underdetermined system

$$\begin{bmatrix} 1 & 1 & 1 \\ 0 & 1 & 2 \end{bmatrix} \begin{bmatrix} x_1 \\ x_2 \\ x_3 \end{bmatrix} = \begin{bmatrix} 4 \\ 5 \end{bmatrix}$$

SOLUTION 1.12

It is readily apparent that the rank of the 2×3 coefficient matrix A is 2. Thus, since A maps $\Re^3$ into $\Re^2$, we find

$$\dim N[A] = n - r = 3 - 2 = 1 \quad \text{and} \quad \dim N[A^T] = m - r = 2 - 2 = 0.$$

Since $\dim N[A^T] = 0$, the range of A is all of $\Re^2$ and the system is solvable for all vectors $\boldsymbol{B}$ in $\Re^2$, including $[4, 5]^T$. To find the general solution, we solve the second equation for x_2 to get $x_2 = 5 - 2x_3$. Then, substituting this into the first equation leads to

$$x_1 + 5 - 2x_3 + x_3 = 4;$$

i.e.,
$$x_1 = -1 + x_3.$$

Then for any choice of the scalar x_3, $\boldsymbol{X} = [-1 + x_3, 5 - 2x_3, x_3]^T$ solves both equations. If we write this solution in the form

$$\boldsymbol{X} = [-1, 5, 0]^T + [x_3, -2x_3, x_3]^T = [-1, 5, 0]^T + x_3[1, -2, 1]^T, \tag{1}$$

then we see that $\boldsymbol{X} = \boldsymbol{U} + c\boldsymbol{Z}_1$ where $\boldsymbol{U} = [-1, 5, 0]^T$ satisfies $A\boldsymbol{U} = \boldsymbol{B}$ and $\boldsymbol{Z}_1 = [1, -2, 1]^T$ is in $N[A]$. Since $N[A]$ is 1-dimensional, $\boldsymbol{Z}_1$ is a basis for $N[A]$ and (1) is a general solution for the underdetermined system. Since the scalar x_3 can have any value whatever, the solution is seen to have one degree of freedom.

PROBLEM 1.13

Suppose the $m \times n$ $(m < n)$ matrix A has rank m. Show that out of all the solutions to the underdetermined system $A\boldsymbol{X} = \boldsymbol{B}$, the solution of minimum norm is given by

$$\boldsymbol{X} = A^T\boldsymbol{W} \quad \text{where} \quad AA^T\boldsymbol{W} = \boldsymbol{B}. \tag{1}$$

Find the solution of minimum norm for the system

$$\begin{aligned} x_1 + x_2 - x_3 - x_4 &= 4 \\ x_2 \qquad + x_4 &= 2 \end{aligned} \tag{2}$$

SOLUTION 1.13

The general solution of $\boldsymbol{AX} = \boldsymbol{B}$ is of the form $\boldsymbol{U} + \boldsymbol{Z}$ where $\boldsymbol{U}$ is any solution of $\boldsymbol{AU} = \boldsymbol{B}$ and $\boldsymbol{Z}$ belongs to $N[A]$. Then the solution of minimum length is the solution vector such that $\boldsymbol{Z} = \underline{0}$ and $\boldsymbol{U}$ is orthogonal to $N[A]$; i.e., the solution of minimum norm lies in the orthogonal complement of $N[A]$. According to Theorem 1.11 $N[A]^{\perp} = RS[A] = \text{Rng}\,[A^{\mathsf{T}}]$ and, thus, the solution of minimum norm is the solution of $\boldsymbol{AX} = \boldsymbol{B}$ which lies in Rng $[A^{\mathsf{T}}]$. But $\boldsymbol{U}$ is in Rng $[A^{\mathsf{T}}]$ if and only if $\boldsymbol{U} = A^{\mathsf{T}}\boldsymbol{W}$ for some $\boldsymbol{W}$ in $\Re^m$ and, thus, the solution of minimum norm satisfies

$$\boldsymbol{U} = A^{\mathsf{T}}\boldsymbol{W} \text{ and } A\boldsymbol{U} = \boldsymbol{B}$$

That is,

$$\boldsymbol{U} = A^{\mathsf{T}}\boldsymbol{W} \text{ where } AA^{\mathsf{T}}\boldsymbol{W} = \boldsymbol{B}.$$

Applying this result in the case of the system (2), we first compute AA^{T}

$$\begin{bmatrix} 1 & 1 & -1 & -1 \\ 0 & 1 & 0 & 1 \end{bmatrix} \begin{bmatrix} 1 & 0 \\ 1 & 1 \\ -1 & 0 \\ -1 & 1 \end{bmatrix} = \begin{bmatrix} 4 & 0 \\ 0 & 2 \end{bmatrix}$$

Then the system

$$AA^{\mathsf{T}}\boldsymbol{W} = \begin{bmatrix} 4 & 0 \\ 0 & 2 \end{bmatrix} \begin{bmatrix} w_1 \\ w_2 \end{bmatrix} = \begin{bmatrix} 4 \\ 2 \end{bmatrix}$$

has the unique solution $w_1 = w_2 = 1$. The minimum norm solution to the original system (2) is equal to $\boldsymbol{X} = A^{\mathsf{T}}\boldsymbol{W}$; i.e.,

$$\begin{bmatrix} 1 & 0 \\ 1 & 1 \\ -1 & 0 \\ -1 & 1 \end{bmatrix} \begin{bmatrix} 1 \\ 1 \end{bmatrix} = \begin{bmatrix} 1 \\ 2 \\ -1 \\ 0 \end{bmatrix}$$

PROBLEM 1.14

Describe the solution set for the system of equations $\boldsymbol{AX} = \boldsymbol{B}$ in the case that A is an $m \times n$ matrix with $m > n = \text{rank}\,A$; i.e., the number of equations exceeds the number of unknowns and, thus, the system is *overdetermined.*

SOLUTION 1.14

We assume here that the rank of A equals n, the number of columns of A. Thus, the columns of A are independent and from Theorem 1.11, we have

$$\dim N[A] = n - \text{rank}\,A = 0 \text{ and } \dim N[A^{\mathsf{T}}] = m - n = q > 0$$

Since $N[A]$ has dimension zero, the solution for $AX = B$ is unique if it exists. However, since $q = \dim N[A^T]$ is positive, the range of A is not all of $\Re^m$ and the solution fails to exist unless B belongs to the range of A; in view of part (d) of Problem 1.10, the solution fails to exist unless B is orthogonal to $N[A^T]$. Every basis of $N[A^T]$ contains q vectors and if $V_1, \ldots, V_q$ denotes any basis for $N[A^T]$, then $AX = B$ is solvable if and only if $B \cdot V_k = 0$ for $k = -1, \ldots, q$. We will see in later problems how an overdetermined system can arise naturally and how such a system can be "solved."

PROBLEM 1.15

Describe the solution set for the system of equations $AX = B$ in the case that A is an $n \times n$ matrix with rank $A < n$. In particular, consider the 16×16 system from Problem 1.2 and give physical interpretations of the results.

SOLUTION 1.15

For any matrix A, we have from Theorem 1.10 that $\dim RS[A] = \dim CS[A] = r$ where r denotes the rank of A. Then, for a square matrix, it follows that

$$\dim N[A] = \dim N[A^T] = n - r = p$$

If $n = r$, then $p = 0$ and $N[A] = N[A^T] = \underline{0}$; i.e., the solution for $AX = B$ is unique and exists for every B. We are assuming here that $p > 0$. Then there are p vectors in every basis $Z_1, \ldots, Z_p$ for $N[A]$ and each basis $V_1, \ldots, V_p$ for $N[A^T]$. No solution for $AX = B$ exists unless B satisfies $B \cdot V_k = 0$ for $k = 1, \ldots, p$ and in this case the general solution is of the form $X = U + c_1 Z_1 + \cdots + c_p Z_p$ where U is any solution of $AX = B$ and $c_1, \ldots, c_p$ are arbitrary constants.

For the 16×16 matrix A from Problem 1.2 it is not hard to check that the sum of the entries in each row of A equals zero. Then it follows that $AZ = \underline{0}$ for the vector Z whose 16 entries are all equal to 1; i.e., Z belongs to $N[A]$. Then $\dim N[A] \geq 1$. In fact, using row reduction, it is possible to show that rank $A = 15$, so $\dim N[A] = 1$. Now this matrix A is symmetric ($A = A^T$) and so $N[A] = N[A^T]$ and Z is a basis for $N[A] = N[A^T]$. By the remarks of the previous paragraph we see no solution for $AX = B$ unless B satisfies $B \cdot Z = 0$. The data vector B is given in Problem 1.2, and, since every entry of Z is 1, we find

$$B \cdot Z = (F_9 + \cdots + F_{16}) - (F_1 + \cdots + F_8) = \text{inflow} - \text{outflow} = 0.$$

Thus, it is seen that the condition on the specified fluxes that must be satisfied if a solution to the system is to exist is that the net flux across the boundary of the diffusion cell equals zero. But this is precisely the condition that *must* be satisfied if the diffusion cell is to remain in a state of equilibrium. If this condition is violated, then there will be a net increase or a net decrease in the amount of contaminant in the cell and no equilibrium solution is possible.

If the compatibility condition on the fluxes is satisfied, then the general solution of the system is of the form $\boldsymbol{X} = \boldsymbol{U} + c\boldsymbol{Z}$ where $\boldsymbol{U}$ is any solution of $A\boldsymbol{X} = \boldsymbol{B}$ and c is any constant. This means that any equilibrium distribution $\boldsymbol{U}$ can be altered by changing the concentration of contaminant in every cell by the same constant amount c and the result will still be an equilibrium distribution of contaminant in the diffusion cell. In other words, specifying the fluxes over the entire boundary of a diffusion cell determines the *relative* concentrations inside the cell, but it does not determine the *absolute* concentrations.

PROBLEM 1.16

For $\boldsymbol{X} = [x_1, \ldots, x_n]^{\mathsf{T}}$, consider the expression

$$Q(x_1, \ldots, x_n) = \boldsymbol{X}^{\mathsf{T}} A\boldsymbol{X} - 2\boldsymbol{X}^{\mathsf{T}}\boldsymbol{B} + c \tag{1}$$

where A denotes an $n \times n$ matrix, $\boldsymbol{B}$ denotes a given vector in $\Re^n$, and c is a given scalar. The $Q = Q(\boldsymbol{X})$ is called a *quadratic form* in $\boldsymbol{X}$. Show that if $A = A^{\mathsf{T}}$, then **grad** $Q = 2(A\boldsymbol{X} - \boldsymbol{B})$.

SOLUTION 1.16

Let ∂_j denote differentiation with respect to x_j, $j = 1, \ldots, n$. Then

$$\partial_j Q = \partial_j \boldsymbol{X}^{\mathsf{T}} A\boldsymbol{X} + \boldsymbol{X}^{\mathsf{T}} \partial_j A\boldsymbol{X} - 2\partial_j \boldsymbol{X}^{\mathsf{T}}\boldsymbol{B} + \partial_j c \tag{2}$$

and, for each $j = 1, \ldots, n$, we have

$$\partial_j \boldsymbol{X}^{\mathsf{T}} = \partial_j (x_1 \boldsymbol{E}_1{}^{\mathsf{T}} + \cdots + x_n \boldsymbol{E}_n{}^{\mathsf{T}}) = \boldsymbol{E}_j{}^{\mathsf{T}} \quad \text{[see Example 1.10(a)]}$$

$$\partial_j A\boldsymbol{X} = \partial_j (x_1 \boldsymbol{C}_1 + \cdots + x_n \boldsymbol{C}_n) = \boldsymbol{C}_j \quad \text{[from (1.6)]}$$

$$\partial_j c = 0 \quad (c \text{ is a constant})$$

Here $\boldsymbol{C}_j$ denotes the jth column of the matrix A, and since $A = A^{\mathsf{T}}$, it follows that $\boldsymbol{C}_j = \boldsymbol{R}_j$ where $\boldsymbol{R}_j$ denotes the jth row of A. Then we can write (2) as

$$\partial_j Q = \boldsymbol{E}_j{}^{\mathsf{T}} A\boldsymbol{X} + \boldsymbol{X}^{\mathsf{T}}\boldsymbol{R}_j - 2\boldsymbol{E}_j{}^{\mathsf{T}}\boldsymbol{B} + 0 \tag{3}$$

But it follows from (1.7) that $A\boldsymbol{X} = (\boldsymbol{X}_1{}^{\mathsf{T}}\boldsymbol{R}_1)\boldsymbol{E}_1 + \cdots + (\boldsymbol{X}_n{}^{\mathsf{T}}\boldsymbol{R}_n)\boldsymbol{E}_n$ and, thus, $\boldsymbol{X}^{\mathsf{T}}\boldsymbol{R}_j = \boldsymbol{E}_j{}^{\mathsf{T}} A\boldsymbol{X}$ for $j = 1, \ldots, n$. Then (3) becomes

$$\partial_j Q = 2\boldsymbol{E}_j{}^{\mathsf{T}} A\boldsymbol{X} - 2\boldsymbol{E}_j{}^{\mathsf{T}}\boldsymbol{B} = 2\boldsymbol{E}_j{}^{\mathsf{T}}(A\boldsymbol{X} - \boldsymbol{B}) \tag{4}$$

and since $\partial_{jQ} = E_j^{\mathsf{T}}$ **grad** Q, it follows from (4) that **grad** $Q = 2(AX - B)$.

PROBLEM 1.17

An n by n matrix A is said to be positive definite if $X^{\mathsf{T}}AX > 0$ for all nonzero vectors X in $\Re^n$. Let $Q(X)$ be the quadratic form (1) of the previous problem for A symmetric and positive definite. Then show that the following statements are equivalent:

(a) X minimizes $Q(X)$ over $\Re^n$.

(b) X satisfies $AX = B$.

SOLUTION 1.17

To show that (a) implies (b) note that at a global minimum of the function $Q = Q(x_1, \ldots, x_n)$, we have **grad** $Q = \underline{0}$. Then if $X = [x_1, \ldots, x_n]^{\mathsf{T}}$ minimizes Q over $\Re^n$, **grad** Q vanishes at X and it follows from the result of the previous problem that $AX = B$. Thus (a) implies (b).

Now if $AX = B$, then for arbitrary Z in $\Re^n$ we have

$$
\begin{aligned}
Q(Z) - Q(X) &= Z^{\mathsf{T}}AZ - 2Z^{\mathsf{T}}B - X^{\mathsf{T}}AX + 2X^{\mathsf{T}}B \\
&= Z^{\mathsf{T}}AZ - 2Z^{\mathsf{T}}AX - X^{\mathsf{T}}AX + 2X^{\mathsf{T}}AX \\
&= Z^{\mathsf{T}}AZ - 2Z^{\mathsf{T}}AX + X^{\mathsf{T}}AX
\end{aligned}
$$

Note that

$$
\begin{aligned}
(Z - X)^{\mathsf{T}}A(Z - X) &= Z^{\mathsf{T}}AZ - X^{\mathsf{T}}AZ - Z^{\mathsf{T}}AX + X^{\mathsf{T}}AX \\
&= Z^{\mathsf{T}}AZ - Z^{\mathsf{T}}A^{\mathsf{T}}X - Z^{\mathsf{T}}AX + X^{\mathsf{T}}AX
\end{aligned}
$$

If $A = A^{\mathsf{T}}$ (A is symmetric), then $Q(Z) - Q(X) = (Z - X)^{\mathsf{T}}A(Z - X)$ and, for A positive definite, $(Z - X)^{\mathsf{T}}A(Z - X) > 0$ for Z not equal to X. Thus, for A symmetric and positive definite, $Q(Z) - Q(X) > 0$ for Z not equal to X. But this is just the statement the X minimizes Q over $\Re^n$; hence (b) implies (a).

PROBLEM 1.18 LEAST SQUARES SOLUTION OF AN OVERDETERMINED SYSTEM

Let A denote an $m \times n$ matrix of rank n with $m > n$. Then X in $\Re^n$ is said to be a *least squares solution* to the overdetermined system $AX = B$ if X minimizes the *squared error*:

$$E(X) = \|AX - B\|^2 \tag{1}$$

Show that the following are equivalent:

(a) X is the unique least squares solution of the overdetermined system $AX = B$.

(b) X solves the nonsingular system of equations $A^TAX = A^TB$.
The system in (b) is referred to as the *normal equations* for the least squares solution to $AX = B$.

SOLUTION 1.18

If A is an $m \times n$ matrix of rank n with $m > n$, then the system $AX = B$ is overdetermined and, as seen in Problem 1.14, no solution exists in general. In such a case, we can seek instead a vector X which minimizes the squared error $E(X)$ defined in (1). Clearly, $E(X) \geq 0$ for all X and if $E(X) = 0$, then X solves $AX = B$. Note that

$$\begin{aligned} E(X) &= (AX - B)^T(AX - B) \\ &= (AX)^TAX - (AX)^TB - B^TAX + B^TB \\ &= X^TA^TAX - 2X^TA^TB + B^TB \end{aligned}$$

Evidently $E(X)$ is a quadratic form whose terms are the $n \times n$ matrix A^TA, n-vector A^TB, and scalar B^TB. $E(X)$ is analogous to $Q(X)$ defined in Problem 1.16 with terms A, B and c. Note that $(A^TA)^T = A^TA^{TT} = A^TA$; i.e., A^TA is symmetric. In addition, A^TA is positive definite. To see this write

$$X^TA^TAX = (AX)^TAX = \|AX\|^2 \geq 0 \tag{2}$$

and recall that since rank $A = n$, dim $N[A] = 0$. Then $AX = \underline{0}$ if and only if $X = \underline{0}$ and (2) is a strict inequality for nonzero vectors X. Since A^TA is symmetric and positive definite, we may apply the results of the previous problem to infer that X is the unique minimum for the quadratic functional $E(X)$ if and only if X solves $A^TAX = A^TB$. But this implies that (a) is equivalent to (b).

PROBLEM 1.19 LINEAR REGRESSION

Suppose we have collected m pairs of experimental data (x_i, y_i), $i = 1, \ldots, m$, to support the hypothesis that x and y are related by a linear equation of the form $y = ax + b$. Show that this leads to an overdetermined system of m equations for two unknowns and find the least squares solution for the system.

SOLUTION 1.19

We are seeking numbers a and b such that

$$\begin{gathered} ax_1 + b = y_1 \\ \cdot \\ \cdot \\ \cdot \end{gathered}$$

$$ax_m + b = y_m$$

In matrix notation, this system becomes

$$\begin{bmatrix} x_1 & 1 \\ \vdots & \vdots \\ x_m & 1 \end{bmatrix} \begin{bmatrix} a \\ b \end{bmatrix} = \begin{bmatrix} y_1 \\ \vdots \\ y_m \end{bmatrix}; \quad \text{i.e., } AX = Y$$

Here A denotes the $m \times 2$ coefficient matrix for the overdetermined system. As long as the x_i's are not all the same, the rank of A equals two and $A^{\mathrm{T}}A$ is a symmetric positive definite matrix. The nonsingular system of normal equations contains the ingredients

$$A^{\mathrm{T}}A = \begin{bmatrix} \sum x_i^2 & \sum x_i \\ \sum x_i & m \end{bmatrix}, \quad A^{\mathrm{T}}Y = \begin{bmatrix} \sum x_i y_i \\ \sum y_i \end{bmatrix}$$

The unique solution $X = [a, b]^{\mathrm{T}}$ of this system minimizes the square error

$$E(a, b) = \sum_{i=1}^{m} (ax_i + b - y_i)^2$$

For example, suppose x and y are related by $y = 2x + 1$ and that in trying to extablish this fact, we record the following data

x	0	1	2	3	4
y	1.1	2.9	4.8	7.2	9

This leads to the following normal equations

$$\begin{bmatrix} 30 & 10 \\ 10 & 5 \end{bmatrix} \begin{bmatrix} a \\ b \end{bmatrix} = \begin{bmatrix} 70.1 \\ 25.0 \end{bmatrix}$$

whose solution is $X = [2.01, .98]^{\mathrm{T}}$. Then the straight line $y = 2.01x + .98$ is referred to as the *least squares straight line fit* for this data. Note that while none of the data points actually falls on the line, the sum of the squares of the distances of the points from the line is minimal.

PROBLEM 1.20 POLYNOMIAL REGRESSION

Suppose we have collected m pairs of experimental data (x_i, y_i), $i = 1, \ldots, m$, to support the hypothesis that x and y are related by a polynomial equation of the form $y = ax^2 + bx + c$. Show that this leads to an overdetermined system of m equations for three unknowns and find the associated normal system for the least squares solution.

SOLUTION 1.20

In this case, we are looking for numbers a, b, and c such that

$$ax_1{}^2 + bx_1 + c = y_1$$

$$\vdots$$

$$ax_m{}^2 + bx_m + c = y_m$$

or

$$\begin{bmatrix} x_1^2 & x_1 & 1 \\ \vdots & \vdots & \\ x_m^2 & x_m & 1 \end{bmatrix} \begin{bmatrix} a \\ b \\ c \end{bmatrix} = \begin{bmatrix} y_1 \\ \vdots \\ y_m \end{bmatrix}; \text{ i.e., } AX = Y$$

With experimental error in the data, this overdetermined system will generally have no solution. Instead, we solve the associated system of normal equations $A^TAX = A^TY$ where

$$A^TA = \begin{bmatrix} \sum x_i^4 & \sum x_i^3 & \sum x_i^2 \\ \sum x_i^3 & \sum x_i^2 & \sum x_i \\ \sum x_i^2 & \sum x_i & m \end{bmatrix}, \quad A^TY = \begin{bmatrix} \sum x_i^2 y_i \\ \sum x_i y_i \\ \sum y_i \end{bmatrix}$$

As long as the x_i's are not all identical, the rank of A is 3 and the 3×3 matrix A^TA is then symmetric positive definite and the normal equations are nonsingular.

*The Gaussian elimination algorithm can be used to find the solution of a nonsingular system of linear equations. The algorithm consists of applying so-called elementary row operations to the augmented matrix of the system to reduce this matrix to upper triangular form. The solution set of the reduced system is identical to that of the original system but the solution is readily obtained from the reduced system by back substitution. By keeping track of and properly storing the multipliers in the row operations used in the reduction procedure, we can, with very little additional effort, obtain the LU factorization of the coefficient matrix A. This is especially useful if we are required to solve AX = **B** repeatedly with the same A but different data vectors **B**. This is just what is done in modeling evolution of a time dependent system as illustrated in Problem 1.7.*

In order to be able to discuss the solution of linear systems with precision and efficiency we introduce abstract concepts from linear algebra. We define vectors to be n-tuples of real numbers and define what is meant by linear combinations of vectors. We are particularly interested in special collections of vectors, called subspaces, which are closed under the operation of forming linear combinations. Associated with any $m \times n$ matrix A, there are four subspaces with relevance to the solution of the system $AX = \mathbf{B}$:

N[A]	*the null space of A*	*subspaces of* $\Re^n$
RS[A]	*the row space of A*	
Rng[A]	*the range of A*	*subspaces of* $\Re^m$
CS[A]	*the column space of A*	

Associated with the $n \times m$ matrix A^T, whose rows are just the columns of A, we have similar subspaces $N[A^T]$ and $RS[A^T]$ in $\Re^m$ and Rng $[A^T]$ and $CS[A^T]$ in $\Re^n$. These subspaces are related as follows:

$RS[A] = CS[A^T] = Rng[A^T]$ and $N[A]$ is their orthogonal complement
$Rng[A] = CS[A] = RS[A^T]$ and $N[A^T]$ is their orthogonal complement

The six subspaces $RS[A] = CS[A] = Rng\ [A]$ and $Rng\ [A] = CS[A] = RS[A^T]$ all have the same dimension, equal to r, the rank of the matrix A. Then the dimensions of $N[A]$ and $N[A^T]$ are equal to $n - r$ and $m - r$, respectively. If A is a square matrix, then these subspaces have the same dimension (although they are not the same subspace unless $A = A^T$).

We may think of the $m \times n$ matrix A as a transformation which (by multiplication) transforms vectors in $\Re^n$ into vectors in $\Re^m$. Then we may also think of $\Re^n$ as being split into the mutually orthogonal subspaces $N[A]$ and Rng $[A^T]$ and of $\Re^m$ as similarly split into Rng $[A]$ and $N[A^T]$. Then A carries all of the vectors in $N[A]$ to zero while all the vectors in Rng $[A^T]$ are carried onto the vectors of Rng $[A]$. The vectors in $N[A^T]$ are the vectors in $\Re^m$ that are not accessible to the transformation A in the sense that no vector in $N[A^T]$ is the image of any vector in $\Re^n$. This motivates the statement: The solution of the system $AX = \mathbf{B}$ exists for all $\mathbf{B}$ in $\Re^m$ if and only if $N[A^T] = 0$ and is unique if and only if $N[A] = 0$. A more detailed description of the structure of the solution for $AX = \mathbf{B}$ is the content of Theorem 1.12.

We can make further use of this splitting of $\Re^n$ and $\Re^m$ to characterize the solution of minimum norm for the underdetermined system $AX = \mathbf{B}$. The solution of minimum norm contains no component in $N[A]$, hence it lies in Rng $[A^T]$; i.e., $X = A^T W$ where $AA^T W = \mathbf{B}$. Similarly, the least squares solution of thme overdetermined system $AX = \mathbf{B}$ is found to be the unique solution of the normal equations $A^T AX = A^T \mathbf{B}$.

2

Algebraic Eigenvalue Problems

There are two principal problems motivating the study of linear algebra. The previous chapter was devoted to the first of these problems; the solution of systems of simultaneous linear equations. In this chapter we consider the other major problem in linear algebra: the algebraic eigenvalue problem.

It is easy to see that for $X = 0$ *the* $n \times n$ *matrix equation* $AX = \lambda X$ *is satisfied for any choice of the scalar* λ. *What is not so obvious is that for certain values of* λ *called eigenvalues, there exist nonzero vectors* X *which satisfy* $AX = \lambda X$. *We define eigenvalues in terms of the matrix determinant and show how to compute the eigenvalues and eigenvectors of small matrices. For large matrices, it is generally necessary to resort to numerical methods to compute eigenvalues and eigenvectors. There are many sophisticated algorithms for solving the matrix eigenvalue problem but these lie outside the scope of this text. Problems 2.5 and 2.6 illustrate two exceptional cases where it is possible to compute the eigenvalues and eigenvectors of a large matrix by analytic means.*

There are certain general properties of eigenvalues and eigenvectors that can be deduced from certain other properties of the matrix A. For example, if A has the property of symmetry – if $A = A^{T}$ *– then the eigenvalues of A must be real and eigenvectors corresponding to distinct eigenvalues are necessarily orthogonal. A number of results of this type are discussed. Applications of the results of this chapter are found in later chapters, particularly those dealing with systems of ordinary differential equations.*

DETERMINANTS

The *determinant* of an $n \times n$ matrix $A = (a_{jk})$ is defined to equal

$$\det A = a_{11}a_{22} - a_{12}a_{21} \qquad \text{if } n = 2$$

$$\det A = \sum_{j=1}^{n} (-1)^{j+k} a_{jk} M_{jk} \qquad \text{if } n > 2$$

$$= \sum_{k=1}^{n} (-1)^{j+k} a_{jk} M_{jk} \qquad \text{if } n > 2$$

Here M_{jk} denotes the determinant of the $n-1 \times n-1$ matrix obtained by deleting from A the jth row and kth column. Determinants are not defined for nonsquare matrices.

EXAMPLE 2.1

Consider the matrix

$$A = \begin{bmatrix} 1 & 2 & 4 \\ 1 & 3 & 9 \\ 1 & 4 & 16 \end{bmatrix}$$

Then

$$\det A = 1 \det \begin{bmatrix} 3 & 9 \\ 4 & 16 \end{bmatrix} - 1 \det \begin{bmatrix} 2 & 4 \\ 4 & 16 \end{bmatrix} + 1 \det \begin{bmatrix} 2 & 4 \\ 3 & 9 \end{bmatrix} = 12 - 16 + 6 = 2$$

Alternatively,

$$\det A = 1 \det \begin{bmatrix} 3 & 9 \\ 4 & 16 \end{bmatrix} - 2 \det \begin{bmatrix} 1 & 9 \\ 1 & 16 \end{bmatrix} + 4 \det \begin{bmatrix} 1 & 3 \\ 1 & 4 \end{bmatrix} = 12 - 14 + 4 = 2$$

PROPERTIES OF DETERMINANTS

It follows at once from our definition that the determinant of any triangular square matrix is equal to the product of the diagonal entries. In addition, if $E_{jk}(a)$, $E_j(a)$, P_{jk} denote the elementary row operations introduced in the previous chapter, then for any square matrix A

$\det E_{jk}(a)\, A = \det A$ for a not zero
$\det E_j(a)\, A = a \det A$ for a not zero
$\det P_{jk}\, A = -\det A$ for j different from k

Finally, for $n \times n$ matrices A and B, we can show

$$\det A = 0 \qquad \text{if and only if rank } A < n$$
$$\det AB = \det A \det B$$
$$\det A^{\mathrm{T}} = \det A$$

Theorem 2.1

Theorem 2.1. The $n \times n$ system $AX = \mathbf{0}$ has nonzero solutions if and only if $\det A = 0$.

EIGENVALUES AND EIGENVECTORS

Let A denote an $n \times n$ matrix and consider the problem of finding a scalar λ and a vector X such that $AX = \lambda X$. Clearly, $X = \mathbf{0}$ is a solution for all choices of λ but for certain scalar values λ, called eigenvalues, there exist nonzero solutions X for this system. A nonzero vector X such that $AX = \lambda X$ is said to be an *eigenvector* for A corresponding to the *eigenvalue* λ. Note that if $AX = \lambda X$, then $A(aX) = \lambda(aX)$ for all scalars a and it follows that if X is an eigenvector for A corresponding to the eigenvalue λ, then the same may be said about the vector aX for any nonzero scalar a. Thus, an eigenvector is determined only up to a multiplicative constant.

$AX = \lambda X$ is equivalent to $(A - \lambda I)X = \mathbf{0}$ and, by Theorem 2.1, this latter equation has nonzero solutions X if and only if $\det(A - \lambda I) = 0$. Thus, we form the following definitions:

λ is an *eigenvalue* of the $n \times n$ matrix A if and only if $\det(A - \lambda I) = 0$; i.e., λ is a root of the nth degree polynomial equation $\det(A - \lambda I) = 0$.

X is an *eigenvector* for A corresponding to eigenvalue λ if and only if X belongs to $N[A - \lambda I]$.

Note that if A is singular, then $\lambda = 0$ is an eigenvalue for A and every vector in $N[A]$ is an eigenvector for A corresponding to $\lambda = 0$. If A is nonsingular, then all eigenvalues of A are nonzero.

Theorem 2.2

Theorem 2.2. Every square matrix A has at least one eigenvalue λ and at least one corresponding eigenvector X.

EXAMPLE 2.2 EIGENVALUES AND EIGENVECTORS

(a) Consider the matrix

$$A_1 = \begin{bmatrix} 0 & 0 & 2 \\ 0 & 0 & 0 \\ 2 & 0 & 0 \end{bmatrix}$$

Then

$$\det(A_1 - \lambda I) = \lambda^3 - 4\lambda = \lambda(\lambda^2 - 4)$$

and the eigenvalues are easily seen to be $\lambda = 0, -2, 2$. To compute the eigenvector associated with $\lambda = 0$, note that $A_1 - 0I = A_1$ has rank 2 and, thus, dim $N[A_1] = 1$. To find a basis for $N[A_1]$, we solve the singular dependent system $A_1 \mathbf{X} = \mathbf{0}$. The equations clearly imply $x_1 = x_3 = 0$ and x_2 is arbitrary. Then the eigenvector associated with $\lambda = 0$ is of the form $\mathbf{X}(0) = [0, x_2, 0]^\top = x_2[0, 1, 0]^\top$; i.e., $A_1\mathbf{X}(0) = \mathbf{0}$ for all choices of the scalar x_2. This is in keeping with what we said earlier about eigenvectors being determined only up to a multiplicative constant.

To find the eigenvector associated with $\lambda = 2$, we note that $A_1 - 2I$ has rank 2 and so dim $N[A_1 - 2I] = 1$. Then

$$(A_1 - 2I)\mathbf{X} = \begin{bmatrix} -2x_1 + 2x_3 \\ -2x_2 \\ 2x_1 - 2x_3 \end{bmatrix} = \mathbf{0}$$

implies $x_1 = x_3$ and $x_2 = 0$. Thus, for each nonzero choice of the scalar x_1, the vector $\mathbf{X}(2) = [x_1, 0, x_1]^\top = x_1[1, 0, 1]^\top$ is an eigenvector for A_1 corresponding to $\lambda = 2$. In the same way we find that for each choice of the scalar x_1, the vector $\mathbf{X}(-2) = [x_1, 0, -x_1]^\top = x_1[1, 0, -1]^\top$ is an eigenvector for A_1 corresponding to $\lambda = -2$.

(b) Next consider the matrix

$$A_2 = \begin{bmatrix} 0 & 0 & 2 \\ 0 & 0 & 0 \\ -2 & 0 & 0 \end{bmatrix}$$

Here we find $\det(A_2 - \lambda I) = \lambda^3 + 4\lambda = \lambda(\lambda^2 + 4)$, and this third degree polynomial has roots, $\lambda = 0, 2i, -2i$. Here i denotes the imaginary unit $\sqrt{-1}$ and we see that even in a matrix with all real entries it is common to find complex eigenvalues. In the usual way we find the associated eigenvectors:

$$\begin{aligned} \mathbf{X}(0) &= [0, x_2, 0]^\top &&= x_2[0, 1, 0]^\top && \text{for } \lambda = 0 \\ \mathbf{X}(2i) &= [x_1, 0, ix_1]^\top &&= x_1[1, 0, i]^\top && \text{for } \lambda = 2i \\ \mathbf{X}(-2i) &= [x_1, 0, -ix_1]^\top &&= x_1[1, 0, -i]^\top && \text{for } \lambda = -2i \end{aligned}$$

(c) Let A_3 equal the $n \times n$ identity matrix. Then $A_3 - \lambda I$ is diagonal and $\det(A_3 - \lambda I) = (\lambda - 1)^n$, so A_3 has the single eigenvalue $\lambda = 1$

(repeated n times). We say that the eigenvalue $\lambda = 1$ has algebraic multiplicity equal to n. In general, the *algebraic multiplicity* of an eigenvalue refers to the number of times it is repeated as a root of the eigenvalue equation $\det(A - \lambda I) = 0$.

Here the $n \times n$ matrix $A_3 - I$ is just the $n \times n$ zero matrix having rank equal to 0. Thus, $\dim N[A_3 - I] = n - 0 = n$. We say that $\lambda = 1$ has a geometric multiplicity equal to n. In general, the *geometric multiplicity* of the eigenvalue λ equals the dimension of $N[A - \lambda I]$. We can show that the geometric multiplicity of an eigenvalue is less than or equal to its algebraic multiplicity with strict inequality in some examples. Here every vector is an eigenvector of $A_3 = I$, and we are free to choose the vectors $\underline{E}_1, \ldots, \underline{E}_n$ of the standard basis as a basis for $N[A_3 - I]$, the subspace of eigenvectors of A_3. Often we refer to $N[A - \lambda I]$, the subspace of eigenvectors of A corresponding to the eigenvalue λ, as the *eigenspace* of λ. Note that with regard to eigenvalues, Theorem 2.2 could state (with more precision) that if we count multiplicities, then each $n \times n$ matrix has n eigenvalues.

(d) Let A_4 denote the following 2 by 2 matrix:

$$A_4 = \begin{bmatrix} 0 & 1 \\ 0 & 0 \end{bmatrix}$$

It is easy to show that this matrix has eigenvalues $\lambda = 0, 0$. Since $A_4 - 0I = A_4$ has rank equal to 1, it follows that $\dim N[A_4 - 0I] = 1$, so there is only a single independent eigenvector corresponding to the eigenvalue $\lambda = 0$. Thus, $\lambda = 0$ is an eigenvalue of algebraic multiplicity 2 and geometric multiplicity 1.

COMPLEX EIGENVALUES AND EIGENVECTORS

Example 2.2(c) makes it clear that even a matrix with all real entries may have complex eigenvalues with corresponding eigenvectors containing complex entries. Thus, to treat the topic of eigenvalues and eigenvectors in its widest sense, we would have to work in the vector space of complex n-tuples over the field of complex scalars. For the applications we have in mind, this will not be necessary and so we confine our attention for the most part to matrices whose eigenvalues and eigenvectors are all real. We point out, however, the following facts about complex eigenvalues and eigenvectors:

- If the real matrix A has complex eigenvalues, then these must occur in complex conjugate pairs λ, λ^*.
- If the real matrix A has complex eigenvalue λ with the corresponding eigenvector X, then X must have complex entries and the eigenvector corresponding to the conjugate eigenvalue λ^* is the vector X^* whose

entries are the complex conjugates of the entries of X.

The real $n \times n$ matrix A is said to be skew symmetric if $A = -A^{\mathsf{T}}$. In this case, the eigenvalues of A are pure imaginary (i.e., all have real part equal to zero).

Properties of Eigenvalues and Eigenvectors

There are certain properties of the eigenvalues and eigenvectors of a matrix A that can be anticipated from the properties of A.

Theorem 2.3 Symmetric Matrices

Theorem 2.3. The real $n \times n$ matrix A is said to *symmetric* if $A = A^{\mathsf{T}}$. In this case it follows that:

1. Every eigenvalue of A is real.
2. Eigenvectors corresponding to distinct eigenvalues are orthogonal; i.e., $\lambda \neq \mu$ implies $N[A - \lambda I]$ is orthogonal to $N[A - \mu I]$.
3. If λ is an eigenvalue of algebraic multiplicity k, then $\dim N[A - \lambda I] = k$.
4. A has an orthonormal set of eigenvectors which form a basis for $\Re^n$.
5. If Q denotes the $n \times n$ matrix whose columns are the orthonormal family of eigenvectors of A, then $A = QDQ^{\mathsf{T}}$, where D denotes the diagonal matrix whose diagonal entries are the eigenvalues of A.

ORTHOGONAL MATRICES

Any real matrix Q whose columns form an orthonormal set can be seen to have the property that $QQ^{\mathsf{T}} = Q^{\mathsf{T}}Q = I$; i.e., $Q^{-1} = Q^{\mathsf{T}}$. Such matrices are said to be *orthogonal* matrices.

Theorem 2.4

Theorem 2.4. If the real $n \times n$ matrix A has n distinct eigenvalues, then the n corresponding eigenvectors are linearly independent. In particular, they form a basis for $\Re^n$. If P denotes the $n \times n$ matrix whose columns are independent eigenvectors of A then P is invertible and $A = PDP^{-1}$ where D denotes the $n \times n$ diagonal matrix whose diagonal entries are the eigenvalues of A.

SOLVED PROBLEMS

Determinants

PROBLEM 2.1

Compute the determinant of each of the following matrices:

$$\text{(a)} \begin{bmatrix} 2 & -1 & 0 & 0 \\ -1 & 2 & -1 & 0 \\ 0 & -1 & 2 & -1 \\ 0 & 0 & -1 & 2 \end{bmatrix}, \quad \text{(b)} \begin{bmatrix} 1 & -1 & 0 & 0 \\ -1 & 2 & -1 & 0 \\ 0 & -1 & 2 & -1 \\ 0 & 0 & -1 & 2 \end{bmatrix}, \quad \text{(c)} \begin{bmatrix} 1 & -1 & 0 & 0 \\ -1 & 2 & -1 & 0 \\ 0 & -1 & 2 & -1 \\ 0 & 0 & -1 & 1 \end{bmatrix}$$

SOLUTION 2.1

Recall that the determinant is left unchanged by row operations of the form $E_{jk}(a)$ for a not zero. Then for the first of the three matrices we find that the row operations $E_{21}(1/2)$, $E_{32}(2/3)$, and $E_{43}(3/4)$, applied in succession, eliminate the –1's on the subdiagonal and reduce the matrix to upper triangular form:

$$\begin{bmatrix} 2 & -1 & 0 & 0 \\ -1 & 2 & -1 & 0 \\ 0 & -1 & 2 & -1 \\ 0 & 0 & -1 & 2 \end{bmatrix} \text{— } E_{21}(1/2)\text{: } E_{32}(2/3)\text{: } E_{43}(3/4) \longrightarrow \begin{bmatrix} 2 & -1 & 0 & 0 \\ 0 & \frac{3}{2} & -1 & 0 \\ 0 & 0 & \frac{4}{3} & -1 \\ 0 & 0 & 0 & \frac{5}{4} \end{bmatrix}$$

Then $\det A = 2(3/2)(4/3)(5/4) = 5$. For the second matrix, we apply the row operations $E_{21}(1)$, $E_{32}(1)$, $E_{43}(1)$ to reduce the matrix to upper triangular form:

$$\begin{bmatrix} 1 & -1 & 0 & 0 \\ -1 & 2 & -1 & 0 \\ 0 & -1 & 2 & -1 \\ 0 & 0 & -1 & 2 \end{bmatrix} \text{— } E_{21}(1)\text{: } E_{32}(1)\text{: } E_{43}(1) \longrightarrow \begin{bmatrix} 1 & -1 & 0 & 0 \\ 0 & 1 & -1 & 0 \\ 0 & 0 & 1 & -1 \\ 0 & 0 & 0 & 1 \end{bmatrix}$$

In this case, the determinant is clearly equal to 1. Finally, the row operations $E_{21}(1)$: $E_{32}(1)$: $E_{43}(1)$ reduce the third matrix to the upper triangular matrix

$$\begin{bmatrix} 1 & -1 & 0 & 0 \\ 0 & 1 & -1 & 0 \\ 0 & 0 & 1 & -1 \\ 0 & 0 & 0 & 0 \end{bmatrix}$$

for which the determinant is clearly zero. The vanishing of this determinant could have been anticipated by noting that the elements in each row of this part (c) matrix sum to zero. Then it follows that the vector $[1, 1, 1, 1]^T$ belongs to the null space of this matrix and so the rank is less than 4. But an $n \times n$ matrix of rank less than n must have determinant equal to zero.

PROBLEM 2.2 CRAMER'S RULE

Solve the 2 by 2 system $a_{11}x_1 + a_{12}x_2 = b_1$, $a_{21}x_1 + a_{22}x_2 = b_2$ and show that the solution can be expressed in the form

$$x_1 = \frac{\det A_1}{\det A}, \quad x_2 = \frac{\det A_2}{\det A} \quad \text{for } \det A \neq 0, \tag{1}$$

where $\det A$ denotes the determinant of the 2 by 2 coefficient matrix and A_1 (A_2) denotes the 2 by 2 matrix obtained by replacing the first (second) column of A by the data vector $[b_1, b_2]^T$.

SOLUTION 2.2

If we multiply the first equation by a_{22} and the second equation by $-a_{12}$ and add the resulting equations, we find

$$(a_{11}a_{22} - a_{12}a_{21})x_1 = a_{22}b_1 - a_{12}b_2;$$

i.e.,

$$(\det A)x_1 = \det A.$$

If $\det A$ is nonzero, then this is equivalent to the first result in (1). Similarly, multiplying the second equation by a_{11} and adding this to $-a_{21}$ times the first leads to $(\det A)x_2 = \det A_2$. The result (1) is known as Cramer's rule for the solution of nonsingular systems of equations. More generally, for an $n \times n$ system of equations we have

$$x_m = \frac{\det A_m}{\det A}, \qquad m = 1, \ldots, m$$

where A denotes the $n \times n$ coefficient matrix and A_m denotes the $n \times n$ matrix obtained by replacing the mth column of A by the data vector. Cramer's rule is not practical for large systems since the difficulty of evaluating a determinant increases very rapidly with increasing n.

Eigenvalues and Eigenvectors

PROBLEM 2.3

Find the eigenvalues and eigenvectors for the matrix

$$A = \begin{bmatrix} 2 & -1 & 0 \\ -1 & 2 & -1 \\ 0 & -1 & 2 \end{bmatrix}.$$

SOLUTION 2.3

Note that this matrix is real and symmetric and, thus, we anticipate real eigenvalues and a full set of orthogonal eigenvectors. We compute

$$\det(A-\lambda I) = (\lambda-2)\left[(\lambda-2)^2-1\right]-1(\lambda-2) = (\lambda-2)(\lambda^2-4\lambda+2)$$

The roots of this third degree polynomial are found to be $\lambda = 2-\sqrt{2}, 2, 2+\sqrt{2}$. The corresponding eigenvectors are found by solving $(A-\lambda I)X = \mathbf{0}$ with each of the three values for λ.

$$\text{for } \lambda = 2: \ \lambda I - A = \begin{bmatrix} 0 & 1 & 0 \\ 1 & 0 & 1 \\ 0 & 1 & 0 \end{bmatrix} \text{---}P_{12}{:}E_{32}(-1)\text{--->} \begin{bmatrix} 1 & 0 & 1 \\ 0 & 1 & 0 \\ 0 & 0 & 0 \end{bmatrix}, \qquad \begin{matrix} x_1 = -x_3 \\ x_2 = 0 \end{matrix}$$

Thus $X(2) = x_3[-1, 0, 1]^T$

$$\text{for } \lambda = 2+\sqrt{2}: \ \lambda I - A = \begin{bmatrix} \sqrt{2} & 1 & 0 \\ 1 & \sqrt{2} & 1 \\ 0 & 1 & \sqrt{2} \end{bmatrix}$$

Then the row operations $E_2(\sqrt{2})$: $E_{21}(-1)$: $E_{32}(-1)$: $E_{12}(-1)$ applied in succession reduce this matrix to

$$\begin{bmatrix} \sqrt{2} & 0 & -\sqrt{2} \\ 0 & 1 & \sqrt{2} \\ 0 & 0 & 0 \end{bmatrix}, \ x_1 = x_3, \ x_2 = -\sqrt{2}x_3; X(2+\sqrt{2}) = x_3 \begin{bmatrix} 1 \\ -\sqrt{2} \\ 1 \end{bmatrix}$$

Similarly, we find $X(2-\sqrt{2}) = x_3[1, \sqrt{2}, 1]^T$. Note that the eigenvectors are mutually orthogonal as Theorem 2.3 predicts. If we choose x_3 to have the value $1/\sqrt{2}$, then each of the eigenvectors becomes a unit vector.

PROBLEM 2.4

Find the eigenvalues and eigenvectors of the matrix

$$A = \begin{bmatrix} 1 & 2 & 0 & 0 \\ 3 & 6 & 0 & 0 \\ 0 & 0 & 1 & 1 \\ 0 & 0 & 0 & 1 \end{bmatrix}.$$

SOLUTION 2.4

The matrix A is not symmetric, so Theorem 2.3 cannot be applied in order to anticipate that the eigenvalues are real. However, note that the second row of A is a multiple of the first row. Then rank $A < 4$ and since A is singular, zero must be one of the eigenvalues for A. To find all the eigenvalues, we compute

$$\det(A-\lambda I) = (1-\lambda)^2 \det\begin{bmatrix} 1-\lambda & 2 \\ 3 & 6-\lambda \end{bmatrix} = (1-\lambda)^2(7\lambda-\lambda^2)$$

from which it follows at once that the eigenvalues are $\lambda = 0, 1, 1, 7$. Since the eigenvalues are not distinct, Theorem 2.4 does not apply, and it remains to be seen whether A has four linearly independent eigenvectors. To find the independent eigenvectors for the eigenvalue λ, we must find a basis for $N[A-\lambda I]$. In particular, for $\lambda = 0$ we note that A is row equivalent to an upper triangular matrix U of rank 3. Then we find the general solution for $UX = \mathbf{0}$; that is,

$$U = \begin{bmatrix} 1 & 2 & 0 & 0 \\ 0 & 0 & 0 & 0 \\ 0 & 0 & 1 & 1 \\ 0 & 0 & 0 & 1 \end{bmatrix}, \qquad \begin{aligned} x_1 + 2x_2 &= 0 \\ x_3 &= 0 \\ x_4 &= 0 \end{aligned} \qquad X(0) = x_2\begin{bmatrix} -2 \\ 1 \\ 0 \\ 0 \end{bmatrix}$$

Similarly, $A - 7I$ is easily seen to be row equivalent to an upper triangular matrix U from which we can find the eigenvector $X(7)$,

$$U = \begin{bmatrix} 3 & -1 & 0 & 0 \\ 0 & 0 & 0 & 0 \\ 0 & 0 & -6 & 1 \\ 0 & 0 & 0 & -6 \end{bmatrix} \qquad \begin{aligned} 3x_1 - x_2 &= 0 \\ x_3 &= 0 \\ x_4 &= 0 \end{aligned} \qquad X(7) = x_1\begin{bmatrix} 1 \\ 3 \\ 0 \\ 0 \end{bmatrix}$$

Corresponding to the repeated eigenvalue $\lambda = 1$, we find $A - I$ to be row equivalent to an upper triangular matrix U of rank 3. Then $\dim N[A-I] = 1$ and it follows that there is just one independent eigenvector corresponding to $\lambda = 1$.

$$U = \begin{bmatrix} 3 & 0 & 0 & 0 \\ 0 & 2 & 0 & 0 \\ 0 & 0 & 0 & 1 \\ 0 & 0 & 0 & 0 \end{bmatrix} \qquad \begin{aligned} x_1 &= 0 \\ x_2 &= 0 \\ x_4 &= 0 \end{aligned} \qquad X(1) = x_3\begin{bmatrix} 0 \\ 0 \\ 1 \\ 0 \end{bmatrix}$$

The eigenvectors $\mathbf{X}(0)$, $\mathbf{X}(7)$, and $\mathbf{X}(1)$ span a 3-dimensional subspace in $\Re^4$.

PROBLEM 2.5

Find the eigenvalues and eigenvectors for the $n \times n$ matrix

$$A = \begin{bmatrix} 2 & -1 & 0 & \cdots & 0 \\ -1 & 2 & -1 & \cdots & 0 \\ & & \vdots & & -1 \\ 0 & & & -1 & 2 \end{bmatrix}$$

The matrix A has 2's on the diagonal and -1's on the super- and subdiagonal with all other entries equal to zero. Such a matrix is said to be *tridiagonal.*

SOLUTION 2.5

A solution of this problem is only possible because of the high degree of structure present in this matrix A. In general, finding the eigenvalues and eigenvectors for an $n \times n$ matrix is possible only when n and all the matrix entries are specified and, even then, numerical methods must be used if n is very large. Note that when $n = 3$, this problem reduces to Problem 2.3.

We begin by writing out the n equations associated with the system

$$(A - \lambda I)X = \mathbf{0}, \tag{1}$$

i.e.,

$$\begin{aligned} (2 - \lambda)x_1 - \quad x_2 \quad &= 0 \\ -x_1 + (2 - \lambda)x_2 - x_3 &= 0 \\ \vdots \qquad & \\ -x_{n-1} + (2 - \lambda)x_n &= 0 \end{aligned} \tag{2}$$

This system can be written more concisely as follows:

$$-x_{m-1} + (2 - \lambda)x_m - x_{m+1} = 0 \quad \text{for } m = 1, \ldots, n$$

where (3)

$$x_0 = x_{n+1} = 0$$

Then the systems (1), (2), and (3) are equivalent in the sense that a solution λ, $[x_1, \ldots, x_n]^T$ for one is a solution for all.

We will solve (3) by supposing the entries x_m are of the form r^m for some nonzero value r. Substituting this into (3) and dividing by $-r^{m-1}$ leads to

$$r^2 - (2 - \lambda)r + 1 = 0 \tag{4}$$

This is a quadratic equation in r with roots r_1 and r_2 which satisfy

$$r_1 r_2 = 1 \quad \text{and} \quad r_1 + r_2 = 2 - \lambda \tag{5}$$

i.e., $\quad (r - r_1)(r - r_2) = r^2 - (r_1 + r_2)r + r_1 r_2.$

Then it follows that the general solution for x_m is of the form

$$x_m = C_1 r_1{}^m + C_2 r_2{}^m \tag{6}$$

for arbitrary constants C_1 and C_2. Now substituting (6) into the conditions $x_0 = 0$ and $x_{n+1} = 0$ leads to

$$x_0 = C_1 + C_2 = 0 \quad \text{and} \quad x_{n+1} = C_1 r_1{}^{n+1} + C_2 r_2{}^{n+1} = 0 \tag{7}$$

The first equation in (7) implies that $C_1 = -C_2$, and when this is used in the second equation in (7) we obtain

$$C_2(r_2{}^{n+1} - r_1{}^{n+1}) = 0$$

$$\text{i.e.,} \quad \left(\frac{r_2}{r_1}\right)^{n+1} = 1 = e^{i2\pi N}, \quad N = \text{integer} \tag{8}$$

Here we have used the fact that for any integer N, $e^{i2\pi N} = 1$. Now (8), together with the fact that $r_1 r_2 = 1$, implies

$$r_1 = e^{iN\pi/(n+1)} \quad \text{and} \quad r_2 = e^{-iN\pi/(n+1)} \tag{9}$$

Using this in (6) leads to the result that the mth entry of the eigenvector $\boldsymbol{X}$ corresponding to the solution (9) is given by

$$x_m = C_2(r_2{}^m - r_1{}^m) = 2C_2 \text{Sin}\frac{mN\pi}{n+1} \qquad m = 1, \ldots, n \tag{10}$$

According to the second equation in (5), these are the entries of the eigenvector $\boldsymbol{X}(\lambda)$ corresponding to the eigenvalue $\lambda = 2 - r_1 - r_2$; i.e.,

$$\lambda = 2 - e^{iN\pi/(n+1)} - e^{-iN\pi/(n+1)} = 2(1 - \text{Cos}\frac{N\pi}{n+1}). \tag{11}$$

Note that these eigenvalues are real, as they should be, since A is symmetric. Of course, A is an $n \times n$ matrix and can, therefore, have only n eigenvalues and n corresponding mutually orthogonal eigenvectors. Thus, while (10) and (11) are valid solutions to (3) for every integer value of N, the periodicity of the sine and cosine functions lead to the result that the only values of N which produce distinct values of λ with corresponding nonzero vectors $\boldsymbol{X}$ are the n values $N = 1, \ldots, n$. Thus, the eigenvalues and eigenvectors for A are as follows:

$$\lambda_N = 2(1 - \text{Cos}\frac{N\pi}{n+1}) \quad \text{with} \quad X(\lambda_N) = \begin{bmatrix} \text{Sin}\frac{N\pi}{n+1} \\ \vdots \\ \text{Sin}\frac{nN\pi}{n+1} \end{bmatrix}, \quad N = 1, \ldots, n$$

PROBLEM 2.6

Find the eigenvalues and eigenvectors for the $n \times n$ matrix

$$A = \begin{bmatrix} 1 & -1 & 0 & \cdots & 0 \\ -1 & 2 & -1 & \cdots & 0 \\ & \vdots & & & -1 \\ 0 & & & -1 & 2 \end{bmatrix}$$

SOLUTION 2.6

In this case, the eigenvalue problem $(A - \lambda I)X = \mathbf{0}$ can be shown to be equivalent to the problem of finding $\lambda, x_1, \ldots, x_n$ satisfying

$$-x_{m-1} + (2 - \lambda)x_m - x_{m+1} = 0, \quad m = 1, \ldots, n$$

where $\quad x_1 = x_0 \quad$ and $\quad x_{n+1} = 0. \qquad (1)$

Then the assumption that $x_m = r^m$, $m = 0, 1, \ldots, n + 1$, for a nonzero number r reduces (1) to the quadratic equation

$$r^2 - (2 - \lambda)r + 1 = 0.$$

As in the previous problem this equation has roots r_1 and r_2 which satisfy

$$r_1 r_2 = 1 \quad \text{and} \quad r_1 + r_2 = 2 - \lambda \qquad (2)$$

Then the general solution for x_m is of the form

$$x_m = C_1 r_1{}^m + C_2 r_2{}^m \qquad (3)$$

for arbitrary constants C_1 and C_2. Using (3) in the conditions $x_1 = x_0$ and $x_{n+1} = 0$ leads to

$$C_1(r_1 - 1) + C_2(r_2 - 1) = 0$$

$$C_1 r_1{}^{n+1} + C_2 r_2{}^{n+1} = 0$$

i.e.,

$$\frac{r_2 - 1}{r_1 - 1} = -\frac{C_1}{C_2} = \frac{r_2{}^{n+1}}{r_1{}^{n+1}} \qquad (4)$$

Now we use the fact that $r_1 r_2 = 1$ in (4) to obtain

$$\frac{(r_2 - 1)}{(r_1 - 1)} = \frac{(r_2 - r_1 r_2)}{(r_1 - 1)} = -r_2$$

$$\left(\frac{r_2}{r_1}\right)^{n+1} = (r_2{}^2)^{n+1}$$

and, thus,

$$r_2{}^{2n+1} = -1 = e^{i\pi + i2N\pi} \text{ for } N = \text{integer.} \tag{5}$$

Now (5) implies

$$r_2 = e^{i\pi(2N+1)/(2n+1)} \quad \text{and} \quad r_1 = r_2{}^{-1} = e^{-i\pi(2N+1)/(2n+1)} \tag{6}$$

If we make use of $C_2 = -C_1\left(\frac{r_1}{r_2}\right)^{n+1}$ we find

$$x_m = C_1(r_1{}^m - r_1{}^{n+1}r_2{}^{m-n-1}) = -C_1r_1{}^{n+1}(r_2{}^{m-n-1} - r_1{}^{m-n-1}).$$

Then the general solution for x_m can be expressed in the form

$$x_m = B\text{Sin}\frac{(n+1-m)\pi(2N+1)}{2n+1}, \quad m = 1, \ldots, n \tag{7}$$

for arbitrary values of B. For the eigenvalue λ we have from (2)

$$\lambda = 2 - r_1 - r_2 = 2 - 2\text{Cos}\frac{(2N+1)\pi}{2n+1} \tag{8}$$

As in the previous problem, (7) and (8) provide solutions to (1) for all integer values of N. However, because of the periodicity of the sine and cosine functions, the only values of N that lead to distinct values of λ and independent eigenvectors are the n values $N = 0, \ldots, n-1$. Then the eigenvalues and eigenvectors for A are given for $N = 0, 1, \ldots, n-1$ by

$$\lambda_N = 2 - 2\text{Cos}\frac{(2N+1)\pi}{2n+1}, \quad X(\lambda_N) = \begin{bmatrix} \text{Sin}\dfrac{n\pi(2N+1)}{2n+1} \\ \vdots \\ \text{Sin}\dfrac{\pi(2N+1)}{2n+1} \end{bmatrix}$$

Properties of Eigenvalues and Eigenvectors

PROBLEM 2.7

Show that if the $n \times n$ real matrix A is symmetric, then the eigenvalues of A are all real.

SOLUTION 2.7

Suppose that λ is an eigenvalue of the $n \times n$ real symmetric matrix A with corresponding eigenvector X. Then

$$(AX)^{\mathsf{T}}X^* = (\lambda X)^{\mathsf{T}}X^* = \lambda(X^{\mathsf{T}}X^*)$$

and

$$X^T(AX)^* = X^T(\lambda X)^* = \lambda^*(X^T X^*)$$

where λ^* and X^* denote the complex conjugates of λ and X, respectively. Now we use the fact that for a symmetric matrix $(AX)^T X^* = X^T(AX)^*$ and it follows that

$$\lambda^*(X^T X^*) = \lambda(X^T X^*)$$

i.e.,

$$(\lambda^* - \lambda)(X^T X^*) = 0.$$

For a nonzero vector X, the scalar quantity $X^T X^*$ is different from zero and, thus, $\lambda^* = \lambda$, the eigenvalue is real. Since the eigenvalues of the real symmetric matrix A are real, we can assume that the eigenvectors are real as well.

PROBLEM 2.8

Suppose X and Y are eigenvectors of the real symmetric matrix A corresponding to distinct eigenvalues λ and μ. Show that X and Y are necessarily orthogonal.

SOLUTION 2.8

If $AX = \lambda X$ and $AY = \mu Y$ for $\lambda \neq \mu$, then λ, μ, X, and Y are all real with

$$(AX)^T Y = (\lambda X)^T Y = \lambda(X^T Y)$$

and

$$X^T(AY) = X^T(\mu Y) = \mu(X^T Y).$$

Since A is symmetric, $(AX)^T Y = X^T(AY)$ and, thus, $(\lambda - \mu)(X^T Y) = 0$. We have assumed $\lambda \neq \mu$ and it follows that $X^T Y = 0$; i.e., X is orthogonal to Y. Since this must hold for any X in $N[A - \lambda I]$ and Y in $N[A - \mu I]$, we have shown that the subspace $N[A - \lambda I]$ is orthogonal to $N[A - \mu I]$.

PROBLEM 2.9

Show that if the matrix A has distinct eigenvalues, then the corresponding set of eigenvectors is linearly independent.

SOLUTION 2.9

For simplicity, let us suppose that A is 2 by 2 with distinct eigenvalues λ and μ and corresponding eigenvectors X and Y. Suppose that for scalars a and b

$$aX + bY = 0 \tag{1}$$

Then

$$A(aX + bY) = aAX + bAY = a\lambda X + b\mu Y = 0 \tag{2}$$

But (1) implies

$$a\lambda X + b\lambda Y = 0 \tag{3}$$

and subtracting (3) from (2) leads to the conclusion $b(\mu - \lambda)Y = \mathbf{0}$. Since μ is distinct from λ and Y is an eigenvector and thus not the zero vector, it follows that $b = 0$. Then if $b = 0$, (1) implies a must also vanish. Thus, the only way that (1) can hold is to have $a = b = 0$. Then X and Y must be linearly independent. In much the same way, we can prove the result for general n.

PROBLEM 2.10

Prove that the following statements are equivalent:

1. The n by n matrix A has n linearly independent eigenvectors.
2. There exists an n by n invertible matrix P such that $A = PDP^{-1}$ where D denotes an $n \times n$ diagonal matrix.

When 1 holds, the matrix P in statement 2 is the matrix whose columns are the n eigenvectors of A and the matrix D has the eigenvalues of A for its diagonal entries.

SOLUTION 2.10

Suppose first that statement 2 holds. Then $AP = PD$. But note that

$$PD = \begin{bmatrix} p_{11} & \cdots & p_{1n} \\ p_{n1} & \cdots & p_{nn} \end{bmatrix} \begin{bmatrix} d_1 & & \\ & \ddots & \\ & & d_n \end{bmatrix} = \begin{bmatrix} d_1 p_{11} & \cdots & d_n p_{1n} \\ d_1 p_{n1} & \cdots & d_n p_{nn} \end{bmatrix}$$

$$= [d_1 P_1, \ldots, d_n P_n] \quad \text{for } P_m = \text{the } m\text{th column of } P$$

and

$$AP = A[P_1, \ldots, P_n] = [AP_1, \ldots, AP_n]$$

Since P is invertible, the columns of P are linearly independent and, clearly, $AP = PD$ implies that $AP_m = d_m P_m$ for $m = 1, \ldots, n$. Thus, statement 2 implies statement 1.

To show statement 1 implies statement 2, suppose A has n linearly independent eigenvectors and let P denote the $n \times n$ matrix whose columns are these eigenvectors. Then P is invertible (this follows from Theorem 1.10). If we denote the eigenvalues of A by $d_1, \ldots, d_n$ then

$$AP = A[P_1, \ldots, P_n] = [AP_1, \ldots, AP_n] = [d_1 P_1, \ldots, d_n P_n] = PD$$

and since P is invertible, $A = PDP^{-1}$. Thus, statement 1 implies statement 2.

If Q is the $n \times n$ matrix whose columns are the normalized eigenvectors of the $n \times n$ real symmetric matrix A, then $Q^{-1} = Q^{\mathrm{T}}$ and $A = QDQ^{\mathrm{T}}$ where D denotes the diagonal matrix whose diagonal entries are the eigenvalues of A.

PROBLEM 2.11

Suppose A is a 4×4 matrix with eigenvalue λ of algebraic multiplicity 4 and geometric multiplicity 1. Then show that there is an invertible matrix P with the property that $A = PJP^{-1}$, where J is the following 4×4

matrix

$$\begin{bmatrix} \lambda & 1 & 0 & 0 \\ 0 & \lambda & 1 & 0 \\ 0 & 0 & \lambda & 1 \\ 0 & 0 & 0 & \lambda \end{bmatrix}$$

The matrix J is called the *Jordan canonical form* for A.

SOLUTION 2.11

If λ has algebraic multiplicity 4 and geometrical multiplicity 1, then dim $N[A - \lambda I]$ is equal to 1. Let $\boldsymbol{X}$ denote a vector in $N[A - \lambda I]$; i.e., $A\boldsymbol{X} = \lambda\boldsymbol{X}$. Then there are no eigenvectors for A that are independent of $\boldsymbol{X}$. Now define vectors $\boldsymbol{Y}_1$, $\boldsymbol{Y}_2$ and $\boldsymbol{Y}_3$ by the conditions

$$\begin{aligned} (A - \lambda I)\boldsymbol{Y}_1 &= \boldsymbol{X}; \quad \text{i.e.,} \quad (A - \lambda I)^2\boldsymbol{Y}_1 = \boldsymbol{0} \\ (A - \lambda I)\boldsymbol{Y}_2 &= \boldsymbol{Y}_1; \quad \text{i.e.,} \quad (A - \lambda I)^3\boldsymbol{Y}_2 = \boldsymbol{0} \\ (A - \lambda I)\boldsymbol{Y}_3 &= \boldsymbol{Y}_2; \quad \text{i.e.,} \quad (A - \lambda I)^4\boldsymbol{Y}_3 = \boldsymbol{0} \end{aligned}$$

We refer to the vectors $\boldsymbol{Y}_m$, $m = 1, 2, 3$ as *generalized eigenvectors* for A with the eigenvalue λ. We can show easily that the vectors $\boldsymbol{X}, \boldsymbol{Y}_1, \boldsymbol{Y}_2, \boldsymbol{Y}_3$ form a linearly independent set. Thus, the matrix P whose columns are these four vectors is invertible. Let the rows of the inverse matrix P_2 be denoted by $R_1^{\mathsf{T}}, \ldots, R_4^{\mathsf{T}}$. Then since $PP^{-1} = I$, it follows that the inner product $R_j^{\mathsf{T}} P_k$ equals 1 if j equals k and is zero otherwise. Here, for convenience, we let P_k denote the kth column of P; i.e., $P_1 = \boldsymbol{X}$, $P_2 = \boldsymbol{Y}_1$, etc.

Now

$$\begin{aligned} AP = A[\boldsymbol{X}, \boldsymbol{Y}_1, \boldsymbol{Y}_2, \boldsymbol{Y}_3] &= [A\boldsymbol{X}, A\boldsymbol{Y}_1, A\boldsymbol{Y}_2, A\boldsymbol{Y}_3] \\ &= [\lambda\boldsymbol{X}, \boldsymbol{X} + \lambda\boldsymbol{Y}_1, \boldsymbol{Y}_1 + \lambda\boldsymbol{Y}_2, \boldsymbol{Y}_2 + \lambda\boldsymbol{Y}_3] \\ &= [\lambda\boldsymbol{X}, \lambda\boldsymbol{Y}_1, \lambda\boldsymbol{Y}_2, \lambda\boldsymbol{Y}_3] + [0, \boldsymbol{X}, \boldsymbol{Y}_1, \boldsymbol{Y}_2] \end{aligned}$$

and, thus,

$$\begin{aligned} P^{-1}AP &= P^{-1}\lambda P + [R_1\ R_2\ R_3\ R_4]^{\mathsf{T}}[0\ P_1\ P_2\ P_3] \\ &= \begin{bmatrix} \lambda & 0 & 0 & 0 \\ 0 & \lambda & 0 & 0 \\ 0 & 0 & \lambda & 0 \\ 0 & 0 & 0 & \lambda \end{bmatrix} + \begin{bmatrix} 0 & 1 & 0 & 0 \\ 0 & 0 & 1 & 0 \\ 0 & 0 & 0 & 1 \\ 0 & 0 & 0 & 0 \end{bmatrix} \end{aligned}$$

i.e., $P^{-1}AP = J$.

PROBLEM 2.12

Suppose that the $n \times n$ matrix A is *diagonalizable*; i.e., there exists an invertible $n \times n$ matrix P and a diagonal matrix D such that $A = PDP^{-1}$. Then show that:

(a) For every positive integer n, $A^n = PD^nP^{-1}$.

(b) For each real t, $e^{tA} = Pe^{tD}P^{-1}$, where e^{tD} denotes the diagonal matrix whose diagonal entries are e^{tdj} for $j = 1, \ldots, n$.

SOLUTION 2.12

To see that a) holds note that

$$A = PDP^{-1}$$

$$A^2 = PDP^{-1}PDP^{-1} = PD^2P^{-1}$$

$$A^3 = PDP^{-1}PD^2P^{-1} = PD^3P^{-1}, \text{ etc.}$$

where we have used that $P^{-1}P = I$. To show (b), write (formally)

$$e^{tA} = I + tA + \frac{t^2}{2}A^2 + \frac{t^3}{3!}A^3 + \cdots \tag{1}$$

If we make use of (a), then this can be written in the equivalent form

$$e^{tA} = PIP^{-1} + PtDP^{-1} + P\frac{t^2}{2}D^2P^{-1} + P\frac{t^3}{3!}D^3P^{-1} + \cdots$$

$$= P\left(I + tD + \frac{t}{2}D^2 + \frac{t^3}{3!}D^3 + \cdots\right)P^{-1}$$

Now for each integer m, $(\frac{t^m}{m!})D^m$ is a diagonal matrix, all of whose

entries are of the form $(\frac{t^m}{m!})\,d_j^m$. Summing over all non-negative integer values of m,

$$\sum_{m=0}^{\infty} \frac{t^m}{m!} d_j^m = e^{tdj}, \quad j = 1, \ldots, n \tag{1}$$

and it follows that the matrix series

$$I + tD + \frac{t}{2}D^2 + \frac{t^3}{3!}D^3 + \cdots$$

can be viewed as a single matrix whose off-diagonal terms are all zero and whose diagonal terms are each an infinite series of the form (1). This produces a diagonal matrix whose diagonal entries are of the form e^{td_j} for $j = 1, \ldots, n$. If we denote this matrix by e^{tD}, then it follows that

$$e^{tA} = Pe^{tD}P^{-1}.$$

We can show that the series expansion (1) is a valid definition for e^{tA}, even if A is not diagonalizable.

The eigenvalues of the $n \times n$ matrix A are the roots of the nth degree polynomial equation $\det(A - \lambda I) = 0$. The eigenvectors of A corresponding to the eigenvalue λ are the nonzero vectors in $N[A - \lambda I]$. If the eigenvalue λ is repeated m times as a root of the eigenvalue equation we say λ has algebraic multiplicity m. Counting multiplicities, every $n \times n$ matrix has n eigenvalues. If $N[A - \lambda I]$ has dimension k, then we say λ has geometric multiplicity k. The geometric multiplicity of any eigenvalue is at least 1 and is less than or equal to its algebraic multiplicity.

A real matrix may have complex eigenvalues but in this case they must occur in complex conjugate pairs. The eigenvectors corresponding to complex eigenvalues of a real matrix also occur in complex conjugate pairs.

A real antisymmetric $n \times n$ matrix (i.e., $A = -A^T$) has all imaginary eigenvalues (since they occur in complex conjugate pairs, if n is odd, then zero is an eigenvalue). The eigenvalues of any real symmetric matrix (i.e., $A = A^T$) are all real. In this chapter we have not discussed matrices having complex entries.

Any $n \times n$ matrix having n linearly independent eigenvectors can be diagonalized; i.e., there exists an invertible $n \times n$ matrix P such that $A = PDP^{-1}$ where D is an $n \times n$ diagonal matrix whose entries are the eigenvalues of A. Here P denotes an $n \times n$ matrix whose columns are independent eigenvectors of A. Cases in which A has n linearly independent eigenvectors include:

(a) If the real $n \times n$ matrix A is symmetric, then A has n mutually orthogonal eigenvectors.

(b) If A has n distinct eigenvalues, then A has n linearly independent eigenvectors

The eigenvalues of small matrices can be found by solving the eigenvalue equation. For large matrices it is generally necessary to resort to numerical methods. Occasionally, for large matrices with a high degree of structure, it is possible to find the eigenvalues and eigenvectors by analytic means.

3

Multivariable Calculus

In this chapter we briefly survey some of the principal results in the calculus of scalar valued and vector valued functions of several variables. We begin by discussing the directional derivative as one measure of the rate of change of a scalar field in a given direction. This notion leads naturally to the related subject of partial derivative and to the concept of the gradient of a scalar function. We also discuss the mean value theorem and its generalization to higher order derivatives, called Taylor's theorem.

Just as the gradient can be used in various ways to provide information about the variation of a scalar field, the variation of a vector field can be described in terms of the divergence and curl of the field. All of these measures of variation can be expressed in terms of the vector differential operator, (called the "del" operator), defined by $\nabla = \partial_x \boldsymbol{i} + \partial_y \boldsymbol{j} + \partial_z \boldsymbol{k}$.

Integrals on intervals in $\Re^1$ *and* $\Re^2$ *generalize to line and surface integrals. Without dwelling on the technical aspects of evaluating such integrals we present some of the qualitative results which are generalizations to higher dimensions of the fundamental theorem of calculus. These generalizations take many forms including Green's theorem, Gauss' divergence theorem, Green's identities, and Stokes' theorem. Each of these results is used extensively in the solved problems for deriving some of the equations of mathematical physics, including the Navier-Stokes equations of fluid dynamics and the Maxwell equations for electromagnetic fields.*

MULTIVARIABLE DIFFERENTIAL CALCULUS

Functions of Several Variables

DEPENDENT AND INDEPENDENT VARIABLES

We will begin by considering real valued functions of several real variables. For example, $z = f(x, y)$, $w = F(x, y, z)$, and $y = G(x_1, x_2, \ldots, x_n)$ denote functions of two, three, and n variables, respectively. We speak of (x, y), (x, y, z), and $(x_1, x_2, \ldots, x_n)$ as the *independent variables* or *arguments* for the functions f, F, and G respectively. Similarly, z, w, and y are referred to as the *dependent variables* or *values* of these functions.

DOMAIN AND RANGE

A real valued function of two real variables can be defined as a rule that assigns to each point (x, y) in a subset D of the plane, a unique real number $f(x, y)$. The set D in the plane is called the *domain* of the function f, and the set of values $z = f(x, y)$ generated as (x,y) varies over D is called the *range* of the function.

SCALAR FIELDS AND VECTOR FIELDS

More generally, we can define a real valued function of n real variables to be a rule that assigns to each n-tuple $(x_1, x_2, \ldots, x_n)$ in a subset D of n-space, a unique real number $G(x_1, x_2, \ldots, x_n)$. Then the set of n-tuples D is the domain of the function G, and as $(x_1, x_2, \ldots, x_n)$ ranges over D, the set of values that is assumed by G is the range. The rule, usually given by a formula or set of formulas, is called a *scalar valued function* since it assigns a scalar value to each point in D. We shall also use the term *scalar field* to refer to scalar valued functions. A function which assigns a vector value to each point of D will be referred to as a *vector valued function* or *vector field*. For example,

$$V(x, y, z) = v_1(x, y, z)\boldsymbol{i} + v_2(x, y, z)\boldsymbol{j} + v_3(x, y, z)\boldsymbol{k}$$

denotes a vector field in $\Re^3$.

CONTINUITY

Let $y = G(X)$ denote a real valued function of n real variables. Then we say that G is *continuous* at the point $\boldsymbol{P}$ in the domain D of G if and only if for each $\varepsilon > 0$ there exists a $\delta > 0$ such that $|G(\boldsymbol{X}) - G(\boldsymbol{P})| < \varepsilon$ for $\boldsymbol{X}$ in any ball of radius δ centered at $\boldsymbol{P}$. If G is continuous at every point $\boldsymbol{P}$ in D, then we say G belongs to the class $C(D)$ of functions that are continuous on D.

Variation of a Scalar Field

THE DIRECTIONAL DERIVATIVE

If p, q, r are real numbers such that $p^2 + q^2 + r^2 = 1$, then the directed line segment from the origin to the point (p, q, r) makes angles α, β, and γ with the x, y, and z axes, respectively, where $\text{Cos } \alpha = p$, $\text{Cos } \beta = q$ and $\text{Cos } \gamma = r$. Then

$$\boldsymbol{u} = \text{Cos } \alpha \boldsymbol{i} + \text{Cos } \beta \boldsymbol{j} + \text{Cos } \gamma \boldsymbol{k}$$

is a *unit vector* (vector of length one) with direction numbers p, q, and r. If $f = f(x, y, z)$ has domain D in $\Re^3$, if $\boldsymbol{P} = (a, b, c)$ is in D, and if the following limit exists, we define

$$\nabla_u f(a, b, c) = \lim_{h \to 0} \frac{f(a + h \text{ Cos } \alpha, b + h \text{ Cos } \beta, c + h \text{ Cos } \gamma) - f(a, b, c)}{h}$$

as the *directional derivative of* $f(x, y, z)$ *in the direction* $\boldsymbol{u} = (\text{Cos } \alpha, \text{Cos } \beta, \text{Cos } \gamma)$ at the point $\boldsymbol{P}$. If $\nabla_u f$ is continuous at $\boldsymbol{P}$ for every direction $\boldsymbol{u}$, then $\nabla_u f$ is said to be continuous at $\boldsymbol{P}$ and if $\nabla_u f$ is continuous at every point $\boldsymbol{P}$ in D, then we say that f belongs to the class $C^1(D)$ of functions that are continuously differentiable on D.

PARTIAL DERIVATIVES

In the special case that $\boldsymbol{u}$ is directed along the x_m axis, (i.e., $\boldsymbol{u} = \boldsymbol{E}_m$ for some m), then $\nabla_u f$ is referred to as the *partial derivative* of f with respect to x_m. For example, if $n = 3$ and $f = f(x, y, z)$ then

for $\boldsymbol{u} = (1, 0, 0)$:

$$\nabla_u f = \frac{\partial f}{\partial x} = \partial_x f = f_x$$

= partial derivative of f with respect to x

for $\boldsymbol{u} = (0, 1, 0)$:

$$\nabla_u f = \frac{\partial f}{\partial y} = \partial_y f = f_y$$

= partial derivative of f with respect to y

for $\boldsymbol{u} = (0, 0, 1)$:

$$\nabla_u f = \frac{\partial f}{\partial z} = \partial_z f = f_z$$

= partial derivative of f with respect to z

We will use the notations $\frac{\partial f}{\partial x} = \partial_x f = f_x$ for partial derivatives interchangeably.

THE GRADIENT OF A SCALAR FIELD

Let $f = f(x, y, z)$ denote a scalar field. Then we introduce the notation

$$\mathbf{grad}\, f(x, y, z) = \nabla f(x, y, z) = (\partial_x f, \partial_y f, \partial_z f) = \partial_x f\mathbf{i} + \partial_y f\mathbf{j} + \partial_z f\mathbf{k}$$

and we refer to the vector field **grad** f as the *gradient* of f. We will use the notations **grad** $f(x, y, z) = \nabla f(x, y, z)(x, y, z)$ for the gradient interchangeably. We collect several important properties of the gradient of a scalar field. For simplicity, we state these results for a function $f = f(x, y, z)$ of three variables.

Theorem 3.1

Theorem 3.1. Let $f = f(x, y, z)$ denote a C^1 scalar field with domain D in $\Re^3$ and define the vector differential operator

$$\nabla = \frac{\partial}{\partial x} i + \frac{\partial}{\partial y} j + \frac{\partial}{\partial z} k$$

Then ∇ applied to f produces the vector field **grad** $f = \nabla f$ with the properties:

(a) The directional derivative $\nabla_u f$ can be expressed in terms of the gradient of f: $\nabla_u f = \mathbf{grad}\, f \cdot \boldsymbol{u}$. Then at each point $\boldsymbol{P}$ in D, f is increasing most rapidly in the direction **grad** $f(\boldsymbol{P})$ and decreases most rapidly in the direction, $-$**grad** $f(\boldsymbol{P})$.

(b) At each point $\boldsymbol{P}$ inside D where f has a relative maximum/minimum, **grad** $f(\boldsymbol{P}) = \mathbf{0}$.

(c) Let S denote a level surface of f; i.e., $S = \{(x, y, z): f(x, y, z) = \text{constant}\}$. Then at each point $\boldsymbol{P}$ on S, the vector **grad** $f(\boldsymbol{P})$ is normal to S.

THE MEAN VALUE THEOREM

We will have need of the following multivariable version of the mean value theorem.

Theorem 3.2

Theorem 3.2. Suppose $f \in C^1(D)$ for D in $\Re^2$ and that for a point (a, b) in D there exists a $\rho > 0$ such that the disc $(x-a)^2 + (y-b)^2 \le \rho^2$ is contained in D. Then for each h, k such that $h^2 + k^2 < \rho^2$, there exists a pair of numbers λ_1, λ_2 with $0 < \lambda_1, \lambda_2 < 1$ and

$$f(a+h, b+k) = f(a, b) + f_x(a + \lambda_1 h, b)\, h + f_y(a+h, b+\lambda_2 k)\, k.$$

Alternatively, there exists a single number ϑ, $0 < \vartheta < 1$, such that

$$f(a+h, b+k) = f(a,b) + f_x(a+\vartheta h, b+\vartheta k)\,h + f_y(a+\vartheta h, b+\vartheta k)\,k$$

There are theorems similar to Theorem 3.2 for functions of more than two variables.

PARTIAL DERIVATIVES OF HIGHER ORDER

Applying differentiation operators repeatedly leads to derivatives of order higher than one. A function $f = f(x_1, \ldots, x_n)$ of n variables that is continuous, together with all of its derivatives of order less than or equal to m, is said to belong to the class $C^m(D)$, where D denotes the domain of f. For a function f in $C^m(D)$, the order of differentiation in mixed partial derivatives of order less than or equal to m is not important. For example, if f is in $C^2(D)$, then $f_{xy}(\boldsymbol{P})$ and $f_{yx}(\boldsymbol{P})$ are each continuous in D and are, therefore, equal at each point $\boldsymbol{P}$ in D.

TAYLOR'S THEOREM

Taylor's theorem is an extension of the mean value theorem to derivatives of higher order. We have the following version of Taylor's theorem for functions of several variables. For simplicity, we will state the theorem for functions of two variables.

Theorem 3.3 Taylor's Theorem in Two Variables

Theorem 3.3. Suppose $f = f(x, y)$ belongs to $C^{m+1}(D)$. Suppose also the point $\boldsymbol{P} = (a, b)$ in D is such that for some $r > 0$ there is a disc of radius r centered at $\boldsymbol{P}$ contained in D. Then for all (x, y) in the disc

$$\begin{aligned} f(x, y) &= f(a, b) + f_x(a, b)h + f_y(a, b)k \\ &\qquad + \frac{1}{2}\left(h^2 f_{xx}(a, b) + 2hk f_{xy}(a, b) + k^2 f_{yy}(a, b)\right) + \cdots \\ &= \sum_{j=0}^{m} \frac{1}{j!}(h\partial_x + k\partial_y)^j f(a, b) + R_m \end{aligned}$$

where

$h = x - a$, $k = y - b$, and for some λ, $0 < \lambda < 1$,

$$R_m = \frac{1}{(m+1)!}(h\partial_x + k\partial_y)^{m+1} f(a + \lambda h, b + \lambda k)$$

For each positive integer j, the expression $(h\partial_x + k\partial_y)^j$ is intended to be expanded formally by the binomial theorem.

Variation of a Vector Field

The vector operator ∇ can be applied to the scalar field f to produce the vector field **grad** f. Theorem 3.1 shows various ways that **grad** f provides information about the variation in the scalar field f. Similarly the ∇ operator can be used to express the variation in a vector field.

Divergence of a Vector Field

For vector field $\boldsymbol{V} = \boldsymbol{V}(x, y, z)$, the inner product

$$\begin{aligned}\operatorname{div} \boldsymbol{V} &= \nabla \bullet \boldsymbol{V} \\ &= \left(\frac{\partial}{\partial x}\boldsymbol{i} + \frac{\partial}{\partial y}\boldsymbol{j} + \frac{\partial}{\partial z}\boldsymbol{k}\right) \bullet \left(v_1(x, y, z)\boldsymbol{i} + v_2(x, y, z)\boldsymbol{j} + v_3(x, y, z)\boldsymbol{k}\right) \\ &= \frac{\partial v_1}{\partial x} + \frac{\partial v_2}{\partial y} + \frac{\partial v_3}{\partial z}\end{aligned}$$

is a scalar field called the *divergence* of $\boldsymbol{V}(x, y, z)$. It is denoted by div $\boldsymbol{V}$ or $\nabla \bullet \boldsymbol{V}$.

Curl of a Vector Field For a vector field $\boldsymbol{V} = \boldsymbol{V}(x, y, z)$ the vector product

$$\nabla \times \boldsymbol{V}(x, y, z) = \operatorname{curl} \boldsymbol{V} = \det \begin{bmatrix} \boldsymbol{i} & \boldsymbol{j} & \boldsymbol{k} \\ \frac{\partial}{\partial x} & \frac{\partial}{\partial y} & \frac{\partial}{\partial z} \\ v_1 & v_2 & v_3 \end{bmatrix}$$

$$= \boldsymbol{i}\left(\frac{\partial v_3}{\partial y} - \frac{\partial v_2}{\partial z}\right) - \boldsymbol{j}\left(\frac{\partial v_3}{\partial x} - \frac{\partial v_1}{\partial z}\right) + \boldsymbol{k}\left(\frac{\partial v_2}{\partial x} - \frac{\partial v_1}{\partial y}\right)$$

is a vector field called the *curl* of $\boldsymbol{V}(x, y, z)$. It is denoted by curl $\boldsymbol{V}$ or by $\nabla \times \boldsymbol{V}$.

VECTOR IDENTITIES

The gradient, divergence, and curl operators may be combined in various ways. If we suppose that $F = F(x, y, z)$ is a C^2 scalar field and $\boldsymbol{V} = \boldsymbol{V}(x, y, z)$ is a C^2 vector field, then **grad** F and curl $\boldsymbol{V}$ are vector fields, whereas div $\boldsymbol{V}$ is a scalar field. Then the following combinations of vector differential operators are defined:

$$\textbf{grad}(\operatorname{div} \boldsymbol{V}) = \nabla(\nabla \bullet \boldsymbol{V}),$$

$$\operatorname{div}(\operatorname{curl} \boldsymbol{V}) = \nabla \bullet (\nabla \times \boldsymbol{V}),$$

$$\operatorname{curl}(\operatorname{curl} \boldsymbol{V}) = \nabla \times (\nabla \times \boldsymbol{V}),$$
$$\operatorname{curl}(\textbf{grad}\, F) = \nabla \times (\nabla F),$$

$$\operatorname{div}(\mathbf{grad}\ F) = \nabla \cdot \nabla F \quad (\text{also denoted by } \nabla^2 F)$$

We have the following results regarding these combinations of operators:

Theorem 3.4

Theorem 3.4. Let $F = F(x, y, z)$ denote a C^2 scalar field and let $\boldsymbol{V} = \boldsymbol{V}(x, y, z)$ be a C^2 vector field. Then

1. $\operatorname{div}(\mathbf{grad}\ F) = \nabla^2 F = F_{xx} + F_{yy} + F_{zz}$.
2. $\operatorname{curl}(\operatorname{curl} \boldsymbol{V}) = \mathbf{grad}(\operatorname{div} \boldsymbol{V}) - (\nabla \cdot \nabla)\boldsymbol{V}$.
3. $\operatorname{div}(\operatorname{curl} \boldsymbol{V}) = 0$.
4. $\operatorname{curl}(\mathbf{grad}\ F) = \mathbf{0}$.

MULTIVARIABLE INTEGRAL CALCULUS

We suppose that the reader is familiar with the topic of multiple integrals of scalar valued functions of several variables. Under rather general conditions, a multiple integral can be shown to equal corresponding sequences of single integrals which are then evaluated by means of the fundamental theorem of calculus. Thus, we focus here on line and surface integrals and various identities relating them. We will not discuss the technicalities of evaluating such integrals but are interested rather in using the integral identities in a qualitative way. For example, these identities will be used repeatedly in the solved problems in deriving mathematical models for various physical systems.

Line Integrals

PARAMETRIC DESCRIPTION OF A CURVE IN SPACE

Let C denote the set of points $\{P(t) \in \Re^3\colon\ P(t) = (x(t), y(t), z(t)), a \le t \le b\}$. Then we say that C is a *curve* in $\Re^3$ described *parametrically* in terms of the parameter t. As t varies from a to b, $\boldsymbol{P}(t)$ traces out the points of C. The functions $x(t)$, $y(t)$, $z(t)$ in the parametric description are not unique. In fact, for any given curve C, there are infinitely many different parameterizations.

ARCS

If $\boldsymbol{P}(a)$ and $\boldsymbol{P}(b)$ are distinct points, then C is said to be an *arc*, and if there are no values t_1, t_2, $a \le t_1 < t_2 \le b$, such that $\boldsymbol{P}(t_1) = \boldsymbol{P}(t_2)$, then we say that C is a *simple arc*; i.e., C does not cross over itself at any point.

SIMPLE CLOSED CURVES

If $P(a) = P(b)$, then we say that C is a *closed curve*; i.e., the initial and final points of C coincide. A closed curve for which $P(t_1) \neq P(t_2)$ for $a < t_1 < t_2 < b$ is called a *simple closed curve*.

SMOOTH CURVES

If the functions $x(t)$, $y(t)$, and $z(t)$ belong to $C^1[a, b]$, we say that C is a *smooth curve*. We will always assume that there is no t in $[a, b]$ where the derivatives $x'(t)$, $y'(t)$, $z'(t)$ vanish simultaneously. This ensures that C has a continuously turning tangent line at each point.

PIECEWISE SMOOTH CURVES

More generally, we say that C is *piecewise smooth* if there exists a partition $\{a = t_0, t_1, \ldots, t_n = b\}$ for $[a, b]$ such that x, y, z are C^1 on $[t_{k-1}, t_k]$ for $k = 1, \ldots, n$. Then each of the subarcs $C_k = \{P(t) \in \Re^3: P(t) = (x(t), y(t), z(t)), t_{k-1} \le t \le t_k\}$ is a smooth arc.

ORIENTATION

Once a parameterization has been chosen for a curve $C = \{P(t) \in \Re^3: P(t) = (x(t), y(t), z(t)), a \le t \le b\}$, then an *orientation* is induced on C. We say that C is traced in the *positive* sense from $P(a)$ to $P(b)$ and is traced in the *negative* sense from $P(b)$ to $P(a)$. We refer to $P(a)$ and $P(b)$ respectively as the *initial* endpoint and *final* endpoint for C.

If C is a simple closed curve with $P(a) = P(b)$, then C may be traversed in either the clockwise or the counterclockwise sense. Either of these may be designated as the positive sense.

LINE INTEGRALS OF SCALAR AND VECTOR FIELDS

An integral of a scalar or vector field along a space curve is a generalization of the notion of integration over a closed bounded interval of the real axis. Such integrals are called line integrals and may be reduced to standard integrals over intervals on the real axis by means of a parametric description of C.

Theorem 3.5

Theorem 3.5. Let $C = \{P(t) \in \Re : P(t) = (x(t), y(t), z(t)), a \le t \le b\}$ be a piecewise smooth curve and let the scalar field $F = F(x, y, z)$ and the vector field $V = V(x, y, z)$ be continuous at each point of C. Then

$$\int_C F dx = \int_a^b F(x(t), y(t), z(t))\, x'(t)\, dt = \int_a^b F(t)\, x'(t)\, dt$$

$$\int_C F dy = \int_a^b F(t)\, y'(t)\, dt, \quad \int_C F dz = \int_a^b F(t)\, z'(t)\, dt,$$

$$\int_C F ds = \int_a^b F(x(t), y(t), z(t)) \sqrt{x'(t)^2 + y'(t)^2 + z'(t)^2}\, dt$$

$$\int_C \boldsymbol{V} \cdot d\boldsymbol{r} = \int_C v_1 dx + v_2 dy + v_3 dz$$

$$= \int_C \boldsymbol{V}(x(t), y(t), z(t)) \cdot \boldsymbol{P}'(t) = \int_C \boldsymbol{V} \cdot \boldsymbol{T} ds$$

where

$$d\boldsymbol{r} = dx\, \boldsymbol{i} + dy\, \boldsymbol{j} + dz\, \boldsymbol{k} = \boldsymbol{P}'(t)\, dt = \boldsymbol{T}(t) \frac{ds}{dt} dt.$$

In particular, each of the line integrals listed above exists.

WORK

If the vector field $\boldsymbol{V} = \boldsymbol{V}(x, y, z)$ is interpreted as a force field then the integral

$$W = \int_C \boldsymbol{V} \cdot d\boldsymbol{r}$$

is interpreted as the *work* done by the field as a particle moves along the path C. For $\boldsymbol{V} = v_1(x, y, z)\boldsymbol{i} + v_2(x, y, z)\boldsymbol{j} + v_3(x, y, z)\boldsymbol{k}$, we have

$$W = \int_C v_1 dx + v_2 dy + v_3 dz$$

Then the work can be evaluated by means of Theorem 3.5.

CIRCULATION OF A VECTOR FIELD

In the case of a piecewise smooth, simple closed curve C and a smooth vector field $\boldsymbol{V} = v_1(x, y, z)\boldsymbol{i} + v_2(x, y, z)\boldsymbol{j} + v_3(x, y, z)\boldsymbol{k}$, we define the *circulation* of $\boldsymbol{V}$ around C to be

$$\Gamma[V; C] = \int_C \boldsymbol{V} \cdot \boldsymbol{T} ds.$$

Then

$$\Gamma[V; C] = \int_C v_1 dx + v_2 dy + v_3 dz.$$

Thus, work and circulation are different interpretations of the same line integral (although circulation is defined only for *closed* curves C).

GREEN'S THEOREM

Green's theorem is an integral identity applying to line integrals in the plane. For Ω, a closed bounded domain in $\Re^2$ having a piecewise smooth boundary Γ with the usual orientation, and for a plane smooth vector field $V(x, y) = v_1(x,y)i + v_2(x,y)j$, we have

$$\int_{\Gamma} v_1(x, y)\,dx + v_2(x, y)\,dy = \iint_{\Omega} (\partial_x v_2(x, y) - \partial_y v_1(x, y))\,dxdy$$

PROPERTIES OF VECTOR FIELDS

There are various properties attributable to vector fields and under certain conditions, these properties are equivalent to one another. We will discuss these properties for vector fields in $\Re^3$, but first we must describe a special class of domains where the results hold.

SIMPLY CONNECTED DOMAINS

Recall that a set Ω in $\Re^3$ is called a *domain* if any two points in Ω can be joined by a polygonal path containing only points of Ω and if, in addition, Ω is open. A set is *open* if for each point P of Ω there is a neighborhood of P that is inside Ω. We say that the domain Ω is *simply connected* if for every simple closed curve C in Ω, there is a surface in Ω having C as its boundary. More plainly, a simply connected domain is a domain without "holes" bored completely through it. For example, a sphere is a simply connected domain and the annular region between two concentric spheres is simply connected, but the annular region between two infinitely long coaxial cylinders is not simply connected.

PATH INDEPENDENT VECTOR FIELDS

A continuous vector field $V(x, y, z)$ on domain Ω in $\Re^3$ is said to be *path independent* if for any two curves C_1 and C_2 in Ω having the same initial endpoint and the same final endpoint, we have

$$\int_{C_1} V \cdot dr = \int_{C_2} V \cdot dr$$

CONSERVATIVE VECTOR FIELDS

A continuous vector field $V(x, y, z)$ on domain Ω in $\Re^3$ is said to be *conservative* if for any simple closed curve C in Ω we have

$$\int_{C} V \bullet dr = 0$$

POTENTIAL FIELDS

A continuous vector field $V(x, y, z)$ on domain Ω in $\Re^3$ is said to be a *potential field* if $V(x, y, z) = \textbf{grad}\, F$ for some scalar field $F = F(x, y, z)$ in $C^1(\Omega)$.

IRROTATIONAL FIELDS

A C^1 vector field $V(x, y, z)$ on Ω in $\Re$ is said to be *irrotational* if curl $\mathbf{V} = \mathbf{0}$ throughout the domain Ω.

Theorem 3.6

Theorem 3.6. For $V(x, y, z)$ a continuous vector field in domain Ω in $\Re^3$, the following properties are equivalent:

1. V is path independent in Ω.
2. V is conservative in Ω.
3. V is a potential field in Ω.

If V is C^1 and Ω is simply connected, then these conditions are equivalent to

4. V is irrotational.

In addition, for Ω a simply connected domain in $\Re^3$ and $\mathbf{V} \in C^1(\Omega)$

curl $\mathbf{V} = \underline{\mathbf{0}}$ in Ω if and only if $\mathbf{V} =$ **grad** F for $F \in C^2(\Omega)$
div $\mathbf{V} = 0$ in Ω if and only if $\mathbf{V} =$ curl $\mathbf{G}$ for $\mathbf{G} \in C^2(\Omega)$

Surface Integrals

REPRESENTATIONS OF SURFACES

Whereas a curve in $\Re^3$ is a set of points having a 1-dimensional character, a surface is a point set where the points in the set have two degrees of freedom. There are several ways to represent a surface:

Explicit Representation S is the set of points $\{(x, y, z)\}$ where

$z = f(x,y)$ for smooth function f with domain U_{xy} in $\Re^2$

or

$y = g(x, z)$ for smooth function g with domain V_{xz} in $\Re^2$

or

$x = h(y, z)$ for smooth function h with domain W_{yz} in $\Re^2$

Implicit Representation S is the set of points $\{(x, y, z)\colon\ F(x, y, z) = 0\}$ for F a smooth function on domain Ω in $\Re^3$.

Parametric Representation S is the set of points $\{(x, y, z)\}$ where

$$x = x(u, v)$$
$$y = y(u, v) \quad \text{for } a \le u \le b,\ c \le v \le d,$$
$$z = z(u, v)$$

for x, y, z smooth functions on the rectangle $[a, b] \times [c, d]$.

OUTWARD NORMAL TO A BOUNDARY SURFACE

A surface S having a well defined continuously varying normal vector N at each point is said to be a *smooth surface*. If S is the union of finitely many nonoverlapping smooth surfaces, then we say S is *piecewise smooth*. The surface of a cube is piecewise smooth but not smooth. If S is the boundary surface for a solid region Ω in $\Re^3$, then S is a *closed surface*, and if S is smooth, then S has a well defined unit normal vector $\mathbf{N}$ at each point $\mathbf{P} = (x_0, y_0, z_0) \in S$. We say that $\mathbf{N}$ is the *outward normal* to S if for each $t < 0$ sufficiently small, $\mathbf{P} + t\mathbf{N}$ belongs to Ω and for each $t > 0$ sufficiently small, $\mathbf{P} + t\mathbf{N}$ lies outside Ω.

Theorem 3.7

Theorem 3.7. If S is a smooth surface in $\Re^3$ and if S is described parametrically in terms of one of the variables x, y, or z, then the surface integral of the continuous scalar field $F(x, y, z)$ is given by one of the following:

$$\iint_S F dA = \int\int_{U_{xy}} F(x, y, f(x, y)) \sqrt{f_x^2 + f_y^2 + 1}\ dxdy$$

$$= \int_{V_{xz}} F(x, g(x, z), z) \sqrt{g_x^2 + g_z^2 + 1}\ dxc$$

$$= \int_{W_{yz}} F(x(y, z), y, z) \sqrt{h_y^2 + h_z^2 + 1}\ dyc$$

The surface integral of the continuous vector field $\mathbf{V}$ is given by

$$\int_S V \cdot N\ dA = \iint_S v_1 dydz + v_2 dxdz + v_3 dxd$$

$$= \int_S (v_1 \text{ Cos } \alpha + v_2 \text{ Cos } \beta + v_3 \text{ Cos } \vartheta)\ d.$$

where $N = \text{Cos } \alpha i + \text{Cos } \beta j + \text{Cos } \vartheta k$ is the unit normal vector to S, and, thus,

$$\text{Sec } \alpha = \sqrt{h_y^2 + h_z^2 + 1};\quad \text{Sec } \beta = \sqrt{g_x^2 + g_z^2 + 1};\quad \text{Sec } \vartheta = \sqrt{f_x^2 + f_y^2 + 1}$$

IDENTITIES OF GAUSS, GREEN, AND STOKES

For Ω a closed bounded domain in $\Re^3$ with piecewise smooth boundary S, we have the following identities:

Theorem 3.8

Theorem 3.8. Let Ω be a closed bounded domain in $\Re^3$ with piecewise smooth boundary S. Then

1. If $V \in C^1(\Omega)$, $\iint_S V \cdot N \, dA = \iiint_\Omega \operatorname{div} V \, dxdydz$

(Gauss' Theorem)

2. If $F \in C^1(\Omega)$, $\iint_S FN dA = \iiint_\Omega \mathbf{grad} F \, dxdydz$

3. If $\vartheta, \psi \in C^2(\Omega)$: $\iiint_\Omega \nabla^2\vartheta \, dxdydz = \iint_S \nabla\vartheta \cdot N \, dA$

and (Green's identities)

$$\iiint_\Omega \psi\nabla^2\vartheta \, dxdydz = \iint_S \nabla\vartheta \cdot N \, dA - \iiint_\Omega \nabla\vartheta \cdot \nabla\psi \, dxdydz$$

$$\iiint_\Omega (\psi\nabla^2\vartheta - \vartheta\nabla^2\psi) \, dxdydz = \iint_S (\psi\nabla\vartheta \cdot N - \vartheta\nabla\psi) \, dA$$

ORIENTATION OF THE BOUNDARY OF A NONCLOSED SURFACE

We have also an identity which applies to a two sided surface S that is not the boundary of a solid region in $\Re^3$. If S is given parametrically by $S = \{(x, y, z)\colon x = x(u, v), y = y(u,v), z = z(u, v), (u, v) \in R\}$, where R denotes a closed bounded set in $\Re^2$, then the image in $\Re^3$ of the boundary of R is referred to as the boundary Γ of S. If the boundary γ of R is described parametrically in terms of parameter σ by $\gamma = \{(s(\sigma), t(\sigma)) \in \Re^2 : \sigma_1 \le \sigma \le \sigma_2$, then Γ, the boundary of S, is described parametrically in terms of σ by $\Gamma = \{(x, y, z) = (x(\sigma), y(\sigma), z(\sigma))\colon \sigma_1 \le \sigma \le \sigma_2\}$, where $x(\sigma) = x(s(\sigma), t\sigma))$, etc. If the simple closed plane curve γ is given the positive orientation, then this induces an orientation on Γ which we refer to as the positive orientation for Γ. It is usual to orient the normal N to S so that the "right hand rule" holds.

Theorem 3.9
Stokes' Theorem

Theorem 3.9. Suppose S is a smooth and bounded but not closed surface with oriented boundary Γ and unit normal N. If $V \in C^1(\Omega)$, then

$$\iint_S \operatorname{curl} V \bullet N \, dA = \int_\Gamma V \bullet T \, ds = \int_\Gamma V \bullet dr$$

IRROTATIONAL AND SOLENOIDAL FIELDS

Combining these integral identities with the equivalent properties of vector fields stated earlier, we can show that for a simply connected domain Ω in $\Re^3$, the following are equivalent on a vector field $V \in C^1(\Omega)$:

1. curl $V = 0$ in Ω.
2. $V = \textbf{grad}\, F$ in Ω for some $F \in C^2(\Omega)$.
3. $\int_C V \bullet T\, ds = 0$ for every simple closed curve C in Ω.

and the following are equivalent conditions on a vector field $V \in C^1(\Omega)$:

1. div $V = 0$ in Ω.
2. $V = $ curl G in Ω for some $G \in C^2(\Omega)$.
3. $\iint_S V \bullet N\, dA = 0$ for every closed surface S in Ω.

A vector field for which curl V is zero is called *irrotational* and when div V vanishes, then we say V is *solenoidal*.

SOLVED PROBLEMS

Systems Modeled by Laplace's Equation

PROBLEM 3.1

In Problem 1.1, we developed a discrete model for the equilibrium distribution of contaminant in a 2-dimensional diffusion cell. Use Taylor's theorem to derive a continuous version of the mass balance equation (1) in Problem 1.1.

SOLUTION 3.1

Let the diffusion cell, a bounded plane region, be denoted by Ω. We define a coordinate system on Ω with the x axis in the horizontal direction and the y axis in the vertical direction. Consider a small square subcell of sidelength $2h$ located inside the diffusion cell with its center at the point (x, y) relative to the origin of the coordinate system. Equation (1) in Problem 1.1 expresses the equilibrium condition that in any subcell in Ω, contaminant inflow equals contaminant outflow. This mass balance statement can be written in our notation as

$$2hF_h(x-h, y) - 2hF_h(x+h, y) + 2hF_v(x, y-h) - 2hF_v(x, y+h) = 0 \quad (1)$$

where $F_h(x, y)$ and $F_v(x, y)$ denote, respectively, the horizontal and vertical components of contaminant flux at the position (x, y) inside the diffusion cell. Then the four fluxes in equation (1) represent the horizontal fluxes through the left and right hand vertical sides and the vertical fluxes through the bottom and top horizontal sides of the square subcell. We continue to use the same sign convention for fluxes that we used in Problem 1.1. Note that each flux (expressed in units of mass per unit length per unit time) is multiplied by $2h$, the length of the side through which the flux passes.

Applying Taylor's theorem to each of the terms in equation (1) leads to

$$\begin{aligned} F_h(x-h, y) &= F_h(x, y) - \partial_x F_h(x, y)\, h + R_2(x, y)\, h^2 \\ F_h(x+h, y) &= F_h(x, y) + \partial_x F_h(x, y)\, h + R_2(x, y)\, h^2 \\ F_v(x, y-h) &= F_v(x, y) - \partial_y F_v(x, y)\, h + R_2(x, y)\, h^2 \\ F_v(x, y+h) &= F_v(x, y) + \partial_y F_v(x, y)\, h + R_2(x, y)\, h^2 \end{aligned} \tag{2}$$

Here $R_2(x, y)$ denotes a "generic" remainder term. The precise function in the remainder is not important, it only matters that the function is bounded on Ω and the remainder term is proportional to h^2 For convenience we denote all the remainder terms by $R_2(x, y)$. Using the expressions from (2) in (1), it follows that

$$F_h(x-h, y) - F_h(x+h, y) + F_v(x, y-h) - F_v(x, y+h)$$

$$= -2h\partial_x F_h(x, y) - 2h\partial_y F_v(x, y) + R_2(x, y)\, h^2 = 0$$

i.e.,

$$\partial_x F_h(x, y) + \partial_y F_v(x, y) = R_2(x, y)\, h$$

Note that the generic symbol $R_2(x, y)$ is not affected by algebraic operations. This last equation is satisfied at the center point (x, y) of the subcell for all $h > 0$. Letting h tend to zero, we obtain the continuous analogue of equation (1) in Problem 1.1,

$$\partial_x F_h(x, y) + \partial_y F_v(x, y) = 0 \tag{2}$$

In fact, if we define the *flux vector* by $F(x, y) = F_h(x, y)i + F_v(x, y)j$ then this equation can be written in the form

$$\operatorname{div} F(x, y) = 0 \tag{3}$$

This is the condition that must be satisfied at each point (x, y) in the diffusion cell if the distribution of contaminant in the cell is to remain independent of time.

PROBLEM 3.2

Use Taylor's theorem and the result of the previous problem to express the continuous equilibrium equation in terms of the contaminant concentration.

SOLUTION 3.2

Equation (2) from Problem 3.1 can be expressed in terms of $u = u(x, y)$, the contaminant concentration at the point (x, y) in Ω. We first use equation (2) from Problem 1.1, expressed in our notation, to write the fluxes in the balance equation

$$F_h(x-h, y) - F_h(x+h, y) + F_v(x, y-h) - F_v(x, y+h) = 0 \tag{1}$$

in terms of concentrations. That is,

$$\begin{aligned} F_h(x-h, y) &= -D\frac{u(x, y) - u(x-2h, y)}{2h} \\ F_h(x+h, y) &= -D\frac{u(x+2h, y) - u(x, y)}{2h} \\ F_v(x, y-h) &= -D\frac{u(x, y) - u(x-2h, y)}{2h} \\ F_v(x, y-h) &= -D\frac{u(x, y+2h) - u(x, y)}{2h} \end{aligned} \tag{2}$$

Substituting these into (1) leads to

$$u(x+2h, y) - 2u(x, y) + u(x-2h, y) + u(x, y+2h) - 2u(x, y) + u(x, y-2h) = 0 \tag{3}$$

Now Taylor's theorem implies

$$u(x+2h, y) = u(x, y) + \partial_x u(x, y)\,2h + \frac{1}{2}\partial_{xx}u(x, y)\,(2h)^2 + R_3(x, y)\,h^3$$

$$u(x-2h, y) = u(x, y) - \partial_x u(x, y)\,2h + \frac{1}{2}\partial_{xx}u(x, y)\,(2h)^2 + R_3(x, y)\,h^3$$

$$u(x, y+2h) = u(x, y) + \partial_y u(x, y)\,2h + \frac{1}{2}\partial_{yy}u(x, y)\,(2h)^2 + R_3(x, y)\,h^3$$

$$u(x, y-2h) = u(x, y) - \partial_y u(x, y)\,2h + \frac{1}{2}\partial_{yy}u(x, y)\,(2h)^2 + R_3(x, y)\,h^3$$

where, once again, $R_3(x, y)$ denotes a generic remainder term. It follows from these results that

$$u(x+2h,y) - 2u(x,y) + u(x-2h,y) = \partial_{xx}u(x,y)\,2h^2 + R_3(x,y)\,h^3$$

$$u(x,y+2h) - 2u(x,y) + u(x,y-2h) = \partial_{yy}u(x,y)\,2h^2 + R_3(x,y)\,h^3$$

Substituting these expressions into (3) then gives

$$2h^2(\partial_{xx}u(x,y) + \partial_{yy}u(x,y)) + R_3(x,y)\,h^3 = 0$$

i.e.,

$$\partial_{xx}u(x,y) + \partial_{yy}u(x,y) = R_3(x,y)\,h$$

Since this equation is satisfied at the center point (x, y) of the subcell for all $h > 0$, we may let h tend to zero to obtain the equation

$$\nabla^2 u(x,y) = \partial_{xx}u(x,y) + \partial_{yy}u(x,y) = 0. \tag{4}$$

This equation must be satisfied at each point (x, y) in Ω.

Note that we can also apply Taylor's theorem in connection with equations (2) to obtain

$$F(x,y) = -D \ \mathbf{grad}\ u(x,y) \tag{5}$$

Equation (3), known as Fick's law of diffusion, expresses the fact that the contaminant flows in the direction in which the concentration decreases most rapidly. Here D denotes a material dependent parameter which we assume is constant. Using (5) in equation (3) from Problem 3.1 leads to

$$\text{div}\ \mathbf{grad}\ u(x,y) = \partial_{xx}u + \partial_{yy}u = 0 \text{ in } \Omega$$

which is equation (4). This equation, Laplace's equation, is the continuous analogue of the discrete equilibrium $AX = B$ that was derived in Problem 1.1.

PROBLEM 3.3

Mimic the steps of Problem 3.1 for a diffusion cell in the shape of a circular disc in order to derive the polar coordinate form of equation (2) from Problem 3.1.

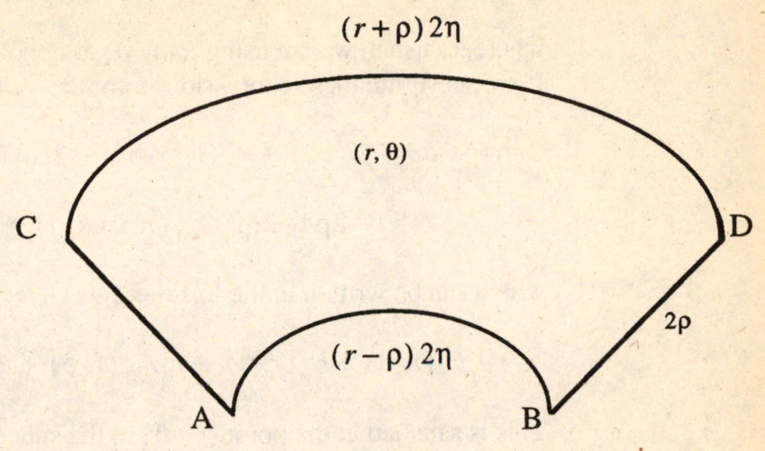

Figure 3.1

SOLUTION 3.3

Figure 3.1 shows a segment shaped subcell of the disc shaped diffusion cell. Note that the four sides of this subcell are of different lengths, unlike the square subcell of the previous problem where all sides had the same length. If the center of this subcell is at the position (r, ϑ), if the segment subtends an angular increment equal to 2η, and if the length of the radial sides of the segment equals 2ρ, then the lengths of the four sides of the subcell are given by:

Length of arc $AB = (r-\rho)2\eta$; Length of arc $CD = (r+\rho)2\eta$

Length of side $AC = 2\rho =$ Length of side BD.

Then the analogue of equation (1) from Problem 1.1 is

$$(r, \vartheta-\eta)2\rho - F_{\vartheta}(r, \vartheta+\eta)2\rho + F_r(r-\rho, \vartheta)(r-\rho)$$

$$\cdot F_r(r+\rho, \vartheta)(r+\rho)2\eta = (\qquad (1)$$

where $F_r(r, \vartheta)$ and $F_{\vartheta}(r, \vartheta)$ denote the radial and angular components of flux, respectively. Note that the inflows and outflows are equal to fluxes multiplied by the appropriate sidelengths. Then

$$(F_r(r-\rho, \vartheta) - F_r(r+\rho, \vartheta))2r\eta$$

$$-(F_r(r-\rho, \vartheta) + F_r(r+\rho, \vartheta))2\rho\eta$$

$$+ (F_{\vartheta}(r, \vartheta-\eta) - F_{\vartheta}(r, \vartheta+\eta))2\rho = 0 \qquad (2)$$

Now we use Taylor's theorem to write

$$F_r(r \pm \rho, \vartheta) = F_r(r, \vartheta) \pm \partial_r F_r(r, \vartheta)\rho + \frac{1}{2}\partial_{rr}F_r(r, \vartheta)\rho^2 + R_3(r, \vartheta)\rho^3$$

$$F_\vartheta(r, \vartheta \pm \eta) = F_\vartheta(r, \vartheta) \pm \partial_\eta F_\eta(r, \vartheta)\eta + \frac{1}{2}\partial_{\vartheta\vartheta}F_\vartheta(r, \vartheta)\eta^2 + R_3(r, \vartheta)\eta^3$$

where, as usual, we are using $R_3(r, \vartheta)$ to denote a generic remainder term. If we substitute these expressions into (2), we find

$$2r\eta(-2\rho\partial_r F_r(r, \vartheta) + R_3(r, \vartheta)\rho^3) - 2\rho\eta(2F_r(r, \vartheta) + R_2(r, \vartheta)\rho^2)$$
$$+ 2\rho(-2\eta\partial_\vartheta F_\vartheta(r, \vartheta) R_3(r, \vartheta)\eta^3) = 0$$

which can be written in the form

$$r\partial_r F_r(r, \vartheta) + F_r(r, \vartheta) + \partial_\vartheta F_\vartheta(r, \vartheta) + R(r, \vartheta)(\rho^2 + \eta^2) = 0.$$

This is satisfied at the point (r, ϑ) in the subcell for all positive ρ and η. Thus, letting ρ and η tend to zero, we obtain the polar coordinate version of the balance equation (2) from Problem 1.1;

$$r\partial_r F_r(r, \vartheta) + F_r(r, \vartheta) + \partial_\vartheta F_\vartheta(r, \vartheta) = 0 \tag{3}$$

PROBLEM 3.4

Express the balance equation (3) from the previous problem in terms of the contaminant concentration. This equation will then be the Laplace equation in polar coordinates.

SOLUTION 3.4

We begin by writing the polar coordinate version of equation (2) from Problem 1.1

$$F_r(r-\rho, \vartheta) = -D\frac{u(r, \vartheta) - u(r-2\rho, \vartheta)}{2\rho}$$

$$F_r(r+\rho, \vartheta) = -D\frac{u(r+2\rho, \vartheta) - u(r, \vartheta)}{2\rho}$$

$$F_\vartheta(r, \vartheta-\eta) = -D\frac{u(r, \vartheta) - u(r, \vartheta-2\eta)}{r2\eta}$$

$$F_\vartheta(r, \vartheta-\eta) = -D\frac{u(r, \vartheta+2\eta) - u(r, \vartheta)}{r2\eta}$$

Using these in equation (2) from Problem 3.3 leads to

$$\frac{2r\eta D}{2\rho}(u(r+2\rho, \vartheta) - 2u(r, \vartheta) + u(r-2\rho, \vartheta))$$

$$-\frac{2\rho\eta D}{2\rho}(u(r-2\rho,\vartheta)-u(r+2\rho,\vartheta))$$

$$+\frac{2\rho D}{2r\eta}(u(r,\vartheta+2\eta)-2u(r,\vartheta)+u(r,\vartheta-2\eta)) = 0$$

Now we can use Taylor's theorem as we have in previous problems to replace the expressions in parentheses. We obtain

$$\frac{r\eta}{\rho}(2\rho^2\partial_{rr}u(r,\vartheta)+R(r,\vartheta)\rho^3)+\eta(2\rho\partial_r u(r,\vartheta)+R(r,\vartheta)\rho^2)$$

$$+\frac{\rho}{r\eta}(2\eta^2\partial_{\vartheta\vartheta}u(r,\vartheta)+R(r,\vartheta)\eta^3) = 0$$

i.e.,

$$r\partial_{rr}u(r,\vartheta)+\partial_r u(r,\vartheta)+\frac{1}{r}\partial_{\vartheta\vartheta}u(r,\vartheta)+R(r,\vartheta)(\rho+\eta) = 0$$

Finally, letting ρ and η tend to zero, we get the polar form of Laplace's equation:

$$r\partial_{rr}u(r,\vartheta)+\partial_r u(r,\vartheta)+\frac{1}{r}\partial_{\vartheta\vartheta}u(r,\vartheta) = 0.$$

This equation must be satisfied at each point (r, ϑ) in the disc shaped cell.

Divergence and Curl

PROBLEM 3.5

Let $V(x, y, z)$ denote a smooth vector field in $\Re^3$; i.e., $V \in C^1(\Re^3)$. For P an arbitrary point in $\Re^3$, let $\Omega_r(P)$ denote the ball of radius r with center at P and let S denote the surface of this ball. Then show that at each P, div $V(P)$ satisfies

$$\text{div } V(P) = \lim_{r\to 0}\frac{\iint_S V\cdot N\, dA}{\text{vol }\Omega_r(P)} \tag{1}$$

where N denotes the unit outward normal to S. Interpret this result physically.

SOLUTION 3.5

Note first that for a smooth vector field $V(x, y, z)$, the function div $V(x, y, z)$ is continuous on $\Re^3$. Then the mean value theorem for integrals implies that

$$\iiint_{\Omega_r(P)} \operatorname{div} V = \operatorname{div} V(Q) \operatorname{vol} \Omega_r(P) \tag{2}$$

for some point Q inside $\Omega_r(P)$. But part 1 of Theorem 3.8 implies

$$\iiint_{\Omega_r(P)} \operatorname{div} V = \iint_S V \bullet N \, dA$$

and it follows that

$$\operatorname{div} V(Q) = \frac{\iint_S V \bullet N \, dA}{\operatorname{vol} \Omega_r(P)}$$

By continuity, div $V(Q)$ tends to div $V(P)$ as r tends to zero and (1) follows. Note that

$$\Phi[V;S] = \iint_S V \bullet N \, dA$$

can be interpreted as the outward *flux* of the vector field V through the surface S. For example, if V is the velocity field for a fluid, then $\Phi[V;S]$ may be interpreted as the fluid flow rate (outward) through the closed surface S. Dividing by vol $\Omega_r(P)$ and letting r tend to zero produces a result equal to the flow rate per unit volume away from P.

PROBLEM 3.6

Let $V(x, y, z)$ denote a smooth vector field in $\Re^3$. For P an arbitrary point in $\Re^3$, let $D_r(P)$ denote a disc of radius r with center at P and unit vector N normal to the plane of the disc. Finally, let C denote the circumference of this disc with the "right hand" orientation for C and N. Then show that at each P, curl $V(P)$ satisfies

$$\operatorname{curl} V(P) \cdot N = \lim_{r \to 0} \frac{\Gamma[V;C]}{\pi r^2} \tag{1}$$

where

$$\Gamma[V;C] = \int_C V \bullet T \, ds = \int_C V \bullet dr$$

denotes the circulation of V around C. Interpret this result physically.

SOLUTION 3.6

If V is a smooth vector field, then curl $V \cdot N$ is a continuous function and, thus, the mean value theorem for integrals implies

$$\iint_{D_r(P)} \operatorname{curl} V \bullet N \, dA = \operatorname{curl} V(Q) \bullet N \pi r^2$$

for some Q in $D_r(P)$. But Theorem 3.9, Stoke's theorem, implies

$$\iint_{D_r(P)} \text{curl}\, V \bullet N\, dA = \int_{C_r(P)} V \bullet T\, ds = \Gamma[V; C_r(P)]$$

and, thus,

$$\text{curl}\, V(Q) \cdot N = \frac{\Gamma[V; C_r(P)]}{\pi r^2}$$

Then as r tends to zero, curl $V(Q) \cdot N$ tends to curl $V(P) \cdot N$ by continuity and (1) follows. Note that $\Gamma[V; C_r(P)]$ represents a measure of the "rotational tendency" of the vector field V and dividing this quantity by the area enclosed by C reduces it to "rotational tendency per unit area." Finally, letting r tend to zero produces a quantity which may be interpreted as the rotational tendency per unit area of the vector field V at the point P.

Fluid Dynamics

PROBLEM 3.7 THE CONTINUITY EQUATION

Consider a fluid filled region D in $\Re^3$. The state of the fluid in D can be characterized in terms of the following fields: the velocity field $V = V(x, y, z, t)$, the density field $\rho = \rho(x, y, z, t)$, and the pressure field $p = p(x, y, z, t)$. The velocity field is a vector field whose value at each point P and time t is the velocity of the fluid particle in the position P at the time t. The scalar fields ρ and p reflect the fluid density and pressure at the point P at time t. Derive the so-called *continuity equation* which asserts the conservation of mass in the fluid flow.

SOLUTION 3.7

Let Ω denote an arbitrary subregion in D and let the boundary of Ω be denoted by S. Then fluid is flowing through the closed surface S at the rate

$$\phi[\rho V; S](t) = \iint_S \rho V \bullet N\, dA$$

where N denotes the outward unit normal to S. It follows that the quantity of fluid mass contained in Ω changes during the time interval (t_1, t_2) by the amount

$$\Delta m = \int_{t_1}^{t_2} \iint_S \rho V \bullet N\, dA dt.$$

According to Theorem 3.8 part 1, the divergence theorem, this integral is equal to

$$\Delta m = \int_{t_1}^{t_2}\iiint_{\Omega} \operatorname{div}(\rho V)\, dxdydzdt$$

But if mass is neither created nor destroyed in Ω during this time interval, then the mass difference is also equal to

$$\Delta m = \iiint_{\Omega} (\rho(x, y, z, t_1) - \rho(x, y, z, t_2))\, dxdydz$$

$$\Delta m = \int_{t_1}^{t_2}\iiint_{\Omega} -\partial_t \rho(x, y, z, t)\, dxdydzdt$$

Then, equating these last two expressions for Δm leads to

$$\int_{t_1}^{t_2}\iiint_{\Omega} (\partial_t \rho + \operatorname{div}(\rho V))\, dxdydzdt = 0 \tag{1}$$

Since the time interval (t_1, t_2) and the subregion Ω are arbitrary and the integrand in (1) is assumed to be a continuous function of (x, y, z, t), then it follows that

$$\partial_t \rho + \operatorname{div}(\rho V) = 0 \quad \text{at each time } t \text{ and every point } (x, y, z) \text{ in } \Omega \tag{2}$$

This is the *continuity equation* that must be satisfied by the velocity and density fields of the fluid flow if fluid mass is conserved. Note that (2) can be written in the form

$$\partial_t \rho + v_1 \partial_x \rho + v_2 \partial_y \rho + v_3 \partial_z \rho + \rho \operatorname{div} V = 0$$

or

$$\frac{d\rho}{dt} + \rho \operatorname{div} V = 0 \tag{3}$$

The derivative

$$\begin{aligned}\frac{d\rho}{dt} &= \partial_t \rho + v_1 \partial_x \rho + v_2 \partial_y \rho + v_3 \partial_z \rho \\ &= \partial_t \rho + \partial_x \rho \frac{dx}{dt} + \partial_y \rho \frac{dy}{dt} + \partial_z \rho \frac{dz}{dt}\end{aligned}$$

is called the *total derivative* of ρ. It accounts for the rate of change of ρ with respect to t through direct dependence on t (this is $\partial_t \rho$) and through indirect dependence on t through dependence on x, y and z.

A fluid for which the density field ρ is constant is referred to as an *incompressible fluid*. The continuity equation for an imcompressible fluid reduces to $\operatorname{div} V = 0$.

PROBLEM 3.8 THE NAVIER-STOKES EQUATIONS

The fields that characterize the flow of a fluid are assumed to conform to the conservation of momentum as well as the conservation of mass. Derive the equation that expresses the conservation of momentum.

SOLUTION 3.8

The conservation of momentum is expressed by Newton's second law which states that the time rate of change of momentum for a body is equal to the sum of the forces acting on the body. That is, for Ω an arbitrary subregion in the flow region D,

$$\frac{d}{dt}\iiint_{\Omega} \rho V \, dxdydz = \text{sum of the forces acting on } \Omega$$

We consider two types of forces acting on Ω:

$$\text{body forces} = \iiint_{\Omega} \rho F \, dxdydz$$

$$\text{surface tractions} = \iint_{S} [T] N \, dA$$

Here F denotes the body force (for example, gravity, an electric force or a magnetic force). The surface tractions on Ω are due to viscous fluid forces and $[T]$ denotes the stress tensor producing these tractions. Stress is a physical quantity that is specified by n^2 components in an n-dimensional space. In $\mathfrak{R}^3$, nine components are required as opposed to a vector quantity like velocity which is specified by three components in $\mathfrak{R}^3$ (or n components in $\mathfrak{R}^n$). Here we give the nine components of T in the form of a 3×3 matrix $[T]$. Then conservation of momentum requires

$$\frac{d}{dt}\iiint_{\Omega} \rho V \, dxdydz = \iiint_{\Omega} \rho F \, dxdydz + \iint_{S} [T] N \, dA$$

$$= \iiint_{\Omega} \rho F \, dxdydz + \iiint_{\Omega} \text{div } [T] \, dxdydz \qquad (1)$$

where we have used the divergence theorem to convert the surface integral to a volume integral. Since the subregion Ω remains fixed, we have that

$$\frac{d}{dt}\iiint_{\Omega} \rho V \, dxdydz = \iiint_{\Omega} \frac{d}{dt} \rho V \, dxdydz$$

$$= \iiint_{\Omega} \left(\partial_t(\rho V) + \partial_x(\rho V)\frac{dx}{dt} + \partial_y(\rho V)\frac{dy}{dt} + \partial_z(\rho V)\frac{d}{dt}z\right) dxdydz$$

But $V = \left[\frac{dx}{dt}, \frac{dy}{dt}, \frac{dz}{dt}\right]^T$ and, thus,

$$\partial_x(\rho V)\frac{dx}{dt} + \partial_y(\rho V)\frac{dy}{dt} + \partial_z(\rho V)\frac{dz}{dt} = (V \bullet \nabla)(\rho V) \tag{2}$$

Now the conservation of momentum statement can be written as

$$\iiint_{\Omega} (\partial_t + (V \bullet \nabla)((\rho V) - \rho F - \text{div}[T])) \; dxdydz = 0 \tag{3}$$

Using the usual argument that Ω is an arbitrary subregion in D and the integrand in (3) is continuous, we conclude

$$(\partial_t + (V \bullet \nabla))(\rho V) - \rho F - \text{div}[T] = 0 \tag{4}$$

Note also that we may use the continuity equation to obtain the following simplification of (4)

$$\begin{aligned}(\partial_t + (V \bullet \nabla))(\rho V) &= \rho(\partial_t + (V \bullet \nabla))V + V(\partial_t + (V \bullet \nabla))\rho \\ &= \rho\frac{d}{dt}V + V(\partial_t \rho + \text{div}(\rho V)) = \rho \frac{d}{dt} V\end{aligned}$$

Then the conservation of momentum equation takes the form

$$\rho\frac{d}{dt}V - \rho F - \text{div}\,[T] = \mathbf{0} \tag{5}$$

Equation (5) is a vector equation and this, together with the (scalar) continuity equation, provide four equations for the five unknowns, v_1, v_2, v_3, ρ, p. If we add the so-called *equation of state*

$$p = f(\rho) \tag{6}$$

relating p to ρ, then we have a complete system of equations. This system of equations (the continuity equation, the momentum equation, and the equation of state) together comprise the *Navier-Stokes equations* which are the governing equations of fluid dynamics.

PROBLEM 3.9

Consider the special case of irrotational flow of an incompressible perfect fluid. That is,

ρ = constant (Incompressible fluid assumption)
curl $V = \mathbf{0}$ (Irrotational flow assumption)
$[T] = -pI$ (Perfect fluid assumption; $I = 3 \times 3$ identity matrix)

Obtain the simplified version of the Navier-Stokes equations that govern this type of flow situation.

SOLUTION 3.9

The assumption that ρ is constant reduces the continuity equation to

$$\text{div } V = 0. \tag{1}$$

In addition, the assumption that curl $V = \mathbf{0}$ implies that $V = \mathbf{grad}\, F$ for some smooth scalar field $F = F(x, y, z)$. Substituting this into (1) shows that $F(x, y, z)$ must then satisfy

$$\text{div } \mathbf{grad}\, F = \partial_{xx} F + \partial_{yy} F + \partial_{zz} F = 0 \tag{2}$$

Now if $[T] = -pI$, then we compute

$$\text{div}[T] = [\partial_x, \partial_y, \partial_z]^T \begin{bmatrix} -p & 0 & 0 \\ 0 & -p & 0 \\ 0 & 0 & -p \end{bmatrix} = -\partial_x p\mathbf{i} - \partial_y p\mathbf{j} - \partial_z p\mathbf{k} = -\mathbf{grad}\, p.$$

Then the momentum equation reduces to

$$\rho\,(\partial_t + (V \cdot \nabla))\,V - \mathbf{grad}\ p = \mathbf{0} \tag{3}$$

It is not difficult to verify that

$$V \times (\nabla \times V) = \nabla (V \bullet V/2) - (V \bullet \nabla)\,V$$

and this, together with the assumption curl $V = \mathbf{0}$, implies that

$$(V \bullet \nabla)\,V = \mathbf{grad} \| V \|^2/2\,.$$

Then (2) becomes

$$\partial_t \mathbf{grad}\, F + \frac{1}{2}\ \mathbf{grad}\ \| V \|^2 - \frac{1}{\rho}\ \mathbf{grad}\, p = \mathbf{0}$$

i.e.,

$$\mathbf{grad}\left(\partial_t F + \frac{\| V \|^2}{2} - \frac{p}{\rho}\right) = \mathbf{0}.$$

But this implies that $\partial_t F + \| V \|^2/2 - p/\rho$ equals an arbitrary function of t.

If we suppose that $\partial_t F = 0$ and that the arbitrary function of t is the zero function, then the momentum equation reduces to

$$p = \rho \frac{\| V \|^2}{2} = \rho \frac{(F_x^2 + F_y^2 + F_z^2)}{2} \tag{4}$$

Then (2) together with (4) and ρ = constant are sufficient to completely determine the flow. A flow in which $V = \mathbf{grad}\, F$ for F satisfying (2) is referred to as a *potential flow* and we say the velocity field V is derivable from the *potential F.*

PROBLEM 3.10 ACOUSTIC WAVES

Acoustic waves (sound waves) are small amplitude variations in the density field. Thus, the assumption ρ = constant precludes the possibility of such waves. Consider then a model for flow in which

$$\rho = \rho_0(1 + s(x, y, z, t)) \quad \text{(acoustic wave assumption)} \tag{1}$$

where ρ_0 denotes a constant and $s = s(x, y, z, t)$ denotes a scalar field that is small compared to 1. Derive the equations governing the propagation of acoustic waves in an irrotational flow of a perfect fluid.

SOLUTION 3.10

The equation of state, when combined with the acoustic wave assumption, leads to

$$p = f(\rho) \approx f(\rho_0) + f'(\rho_0)(\rho - \rho_0) = p_0 + f'(\rho_0)\rho_0 s \tag{2}$$

In the previous problem, we showed that the momentum equation for a perfect fluid reduces to

$$\partial_t + (V \bullet \nabla))V - \frac{1}{\rho_0} \textbf{grad } p = \textbf{0} .$$

But (2) implies that **grad** $p = \rho_0 f'(\rho_0)$ **grad** s and if we now make the additional assumption that

$$(V \bullet \nabla) V << (\partial_t) V \tag{3}$$

then the momentum equation becomes

$$\partial_t V + f'(\rho_0) \textbf{grad } s = 0 \tag{4}$$

We will see later that assumption (3) is equivalent to the assumption that $\|V\|$, the fluid speed, is small compared to the speed of the acoustic waves. The continuity equation reads as follows:

$$\partial_t(\rho_0 + \rho_0 s) + \text{div}(\rho_0 V) = 0$$

where we have made use of the assumption that s is small compared to 1 in order to conclude that sV is negligible when compared to V. Then the continuity equation simplifies to

$$\partial_t s(x, y, z, t) + \text{div } V = 0 \tag{5}$$

Now, forming the divergence of equation (4) leads to

$$\text{div}(\partial_t V + f'(\rho_0) \textbf{grad } s) = \partial_t \text{ div } V + f'(\rho_0) \text{ div } \textbf{grad } s = 0$$

$$= \partial_t(-\partial_t s) + f'(\rho_0) \text{ div } \textbf{grad } s = 0$$

i.e.,

$$\partial_{tt} s(x, y, z, t) - c^2 \nabla^2 s(x, y, z, t) = 0 \tag{6}$$

where $c^2 = f'(\rho_0)$. In chapter 10, we will show that equation (6) governs the propagation of waves in 3-space and that the parameter c can be interpreted as the wave speed.

The density field is determined by solving equation (6). In order to solve for the velocity field, note that if we form the gradient of equation (5), we get

$$\mathbf{grad}\,(\partial_t s(x,y,z,t) + \text{div } V) = \partial_t\,\mathbf{grad}\, s + \mathbf{grad}\,\text{div } V$$

$$= \partial_t\left(-\frac{1}{c^2}\partial_t V\right) + \mathbf{grad}\,\text{div } V = \mathbf{0}$$

But we have from Theorem 3.5 that for an irrotational flow,

$$\mathbf{grad}\,\text{div } V = \text{curl curl } V + (\nabla \bullet \nabla)V = (\nabla \bullet \nabla)V$$

Then V satisfies

$$\partial_{tt}V - c^2(\nabla \bullet \nabla)V = (\partial_{tt} - c^2\nabla^2)V = \mathbf{0} \qquad (7)$$

i.e., each component of V satisfies the equation satisfied by $s(x, y, z, t)$. This is a partial differential equation known as the *wave equation*.

PROBLEM 3.11

Show that the assumption $(V \bullet \nabla))V \ll \partial_t V$ made in the previous problem is equivalent to the assumption that $\|V\|$ is small compared to c, the speed of the acoustic waves.

SOLUTION 3.11

The assumption that curl $V = \mathbf{0}$ implies that

$$(V \bullet \nabla)V = \mathbf{grad}\,\frac{\|V\|^2}{2} - V \times \text{curl } V = \mathbf{grad}\,\frac{\|V\|^2}{2}$$

and

$$V = \mathbf{grad}\, G \text{ for some smooth scalar field } G.$$

Then it follows that $(V \bullet \nabla)V \ll \partial_t V$ is equivalent to $\mathbf{grad}\,\|V\|^2 \ll 2\partial_t V$. In addition, the momentum equation implies

$$\partial_t(-\mathbf{grad}\, G) + c^2\,\mathbf{grad}\, s = \mathbf{grad}\,(-\partial_t G + c^2 s) = \mathbf{0}$$

This last result implies $\partial_t G - c^2 s = B(t)$ for $B(t)$ an arbitrary function of t. Thus,

$$2\partial_t V = 2\partial_t\,\mathbf{grad}\; G = \mathbf{grad}(2\partial_t G) = \mathbf{grad}(2c^2 s + 2B(t))$$

and $\mathbf{grad}\,\|V\|^2 \ll 2\partial_t V = \mathbf{grad}(2c^2 s + 2B(t)) = \mathbf{grad}(2c^2 s)$ is equivalent to

$$\|V\|^2 \ll 2c^2 s \ll 2c^2$$

which is to say $\|V\|/c$ is small compared to 1.

Maxwell's Equations

PROBLEM 3.12

Consider a region Ω in $\Re^3$ permeated by the following five vector fields:

E = electric intensity (volts/m)
D = electric displacement (coulombs/m^2)
H = magnetic intensity (amps/m)
B = magnetic induction (Webers/m)
J = current density (amps/m^2)

Derive the so-called Maxwell's equations governing these electromagnetic fields.

SOLUTION 3.12

We begin by listing four *conservation laws* that apply to these five fields. These are four empirical laws (i.e., based on experimental observation) that are analogous to the conservation of mass and momentum equations from fluid dynamics.

Gauss' Law for Electric Fields For an arbitrary subregion V in Ω with closed boundary S having unit outward normal vector N,

$$\iiint_V Q\, dsdydz = \iint_S \boldsymbol{D} \bullet \boldsymbol{N}\, dA \tag{1}$$

where Q denotes the (scalar) charge density in Ω in coulombs/m^3.

Gauss' Law for Magnetic Fields For an arbitrary subregion V in Ω

$$\iint_S \boldsymbol{B} \cdot \boldsymbol{N}\, dA = 0 \tag{2}$$

Faraday's Law For any nonclosed surface S having boundary curve C

$$\int_C \boldsymbol{E} \bullet \boldsymbol{T}\, ds = -\partial_t \iint_S \boldsymbol{B} \bullet \boldsymbol{N}\, dA \tag{3}$$

where N and T denote respectively the unit normal to S and the unit tangent to C.

Ampere's Law For any nonclosed surface S having boundary curve C,

$$\int_C \boldsymbol{H} \bullet \boldsymbol{T}\, ds = \iint_S \boldsymbol{J} \bullet \boldsymbol{N}\, dA + \partial_t \iint_S \boldsymbol{D} \bullet \boldsymbol{N}\, dA \tag{4}$$

In addition to these conservation laws, we have *constituitive relations* analogous to the equation of state from fluid dynamics. These constituitive relations describe how the materials or media involved react to the fields that are present. We have

$$\boldsymbol{D} = \chi \boldsymbol{E}$$

$$\boldsymbol{B} = \mu \boldsymbol{H}$$

$$J = \sigma E$$

where χ, μ, and σ denote material dependent scalars. In general, these quantities are second order tensors (i.e., 3×3 matrices), but we shall assume for simplicity that the media involved are homogeneous and isotropic which then amounts to supposing χ, μ, and σ are scalar constants.

First, (1) together with the divergence theorem implies

$$\iiint_V Q \; dxdydz = \iint_S D \bullet N \; dA = \iiint_V \operatorname{div}(\chi E) \, dxdydz$$

Then since V is arbitrary in Ω and the fields are assumed continuous,

$$\operatorname{div}(\chi E) = Q \quad \text{in } \Omega \tag{5}$$

Similarly, from (2) we obtain

$$\operatorname{div} B = 0 \quad \text{in } \Omega \tag{6}$$

Stokes' theorem applied to (3) leads to

$$\partial_t \iint_S B \bullet N \; dA = -\int_C E \bullet T \; ds = -\iint_S \operatorname{curl} \; E \bullet N \; dA$$

which implies in the usual way

$$\partial_t B + \operatorname{curl} \; E = 0 \tag{7}$$

The same reasoning produces the result

$$\operatorname{curl} B = \mu J + \partial_t (\mu \chi E) \tag{8}$$

Equations (5), (6), (7), and (8) are referred to collectively as *Maxwell's equations* for the unknown fields E and B in terms of the given fields J and Q.

PROBLEM 3.13

Show that Maxwell's equations are equivalent to the equations

$$\mu\chi\,(\partial_{tt} - c^2\nabla^2) E = -\mu\partial_t J - \operatorname{grad}\,(\frac{Q}{x}) \quad \text{where } c^2 = \frac{1}{(\mu\chi)} \tag{1}$$

$$\iota\chi\,(\partial_{tt} - c^2\nabla^2) B = \mu \; \operatorname{curl} \; J \tag{2}$$

SOLUTION 3.13

From equations (7) and (8) of Maxwell's equations, we have

$$\operatorname{curl}\operatorname{curl} \; E = -\partial_t \operatorname{curl} \; B = -\partial_t (\mu J + \partial_t (\mu \chi E))$$

Then the identity, curl curl E = **grad** div $E - \nabla^2 E$, together with (5) from Maxwell's equations leads to (1). Similarly, using Maxwell's equation (8)

$$\operatorname{curl}\operatorname{curl} B = \mathbf{grad} \operatorname{div} B - \nabla^2 B = \mu \; \operatorname{curl} J +)_t (\mu\chi \; \operatorname{curl} \; E)$$

and then equations (6) and (7) imply

$$-\nabla^2 \boldsymbol{B} = \mu \text{ curl } \boldsymbol{J} - \partial_{tt}(\mu\chi\boldsymbol{B}).$$

Equations (1) and (2) are a system of six partial differential equations for the two unknown fields $\boldsymbol{E}$ and $\boldsymbol{B}$ in terms of the prescribed fields $\boldsymbol{J}$ and Q.

PROBLEM 3.14

Find the simplified version of Maxwell's equations for static electromagnetic fields; i.e., suppose $\partial_t \boldsymbol{E} = \partial_t \boldsymbol{B} = \boldsymbol{0}$.

SOLUTION 3.14

If $\boldsymbol{E}$ and $\boldsymbol{B}$ are independent of time, then Maxwell's equations reduce to

$$\text{div } \boldsymbol{E} = \frac{Q}{\chi} \tag{1}$$

$$\text{div } \boldsymbol{B} = \boldsymbol{0} \tag{2}$$

$$\text{curl } \boldsymbol{E} = \boldsymbol{0} \tag{3}$$

$$\text{curl } \boldsymbol{B} = \mu \boldsymbol{J} \tag{4}$$

Using Theorem 3.7 together with (3) we conclude that $\boldsymbol{E} = \textbf{grad } \varphi$ for some smooth scalar field $\varphi = \varphi(x, y, z)$. Substituting this into (1), we see that φ must satisfy

$$\nabla^2 \varphi(x, y, z) = \frac{Q}{\chi} \tag{5}$$

Similarly (2) implies that $\boldsymbol{B} = \text{curl } \boldsymbol{A}$ for some smooth vector field $\boldsymbol{A}$. Note that we may assume without loss of generality that $\text{div } \boldsymbol{A} = 0$. For if $\text{div } \boldsymbol{A}$ is not zero, then note that for any smooth scalar field ψ, we may let $\boldsymbol{A}' = \boldsymbol{A} + \textbf{grad } \psi$. Then

$$\text{curl } \boldsymbol{A}' = \text{curl } (\boldsymbol{A} + \textbf{grad } \psi) = \text{curl } \boldsymbol{A} = \boldsymbol{B}$$

and

$$\text{div } \boldsymbol{A}' = \text{div } \boldsymbol{A} + \text{div } \textbf{grad } \psi = \text{div } \boldsymbol{A} + \nabla^2 \psi$$

Since ψ is arbitrary, we can choose ψ to satisfy $\nabla^2 \psi = -\text{div } \boldsymbol{A}$ to get $\text{div } \boldsymbol{A}' = 0$.

Thus, we are assured of the existence of a vector field $\boldsymbol{A}$ such that $\text{curl } \boldsymbol{A} = \boldsymbol{B}$ and $\text{div } \boldsymbol{A} = 0$. Now we use this in (4) to get

$$\text{curl curl } \boldsymbol{A} = \textbf{grad} \text{ div } \boldsymbol{A} - \nabla^2 \boldsymbol{A} = -\nabla^2 \boldsymbol{A} = \mu \boldsymbol{J} \tag{6}$$

Then $\boldsymbol{E}$ and $\boldsymbol{B}$ are known once we determine the scalar field ψ satisfying (5) and the vector field $\boldsymbol{A}$ which satisfies (6). These fields ψ and $\boldsymbol{A}$ are referred to as the scalar *electric potential* and the vector *magnetic potential*, respectively. Note that it is not necessary to find ψ since adding $\textbf{grad }\psi$ to $\boldsymbol{A}$ has no effect on $\boldsymbol{B}$.

To smooth scalar field f(x, y, z) and vector field V(x, y, z) we can apply the vector differential operator

$$\nabla = \frac{\partial}{\partial x}i + \frac{\partial}{\partial x}j + \frac{\partial}{\partial x}k$$

to form

$$\nabla f = \textbf{grad}\ f = f_x\textbf{i} + f_y\textbf{j} + f_z\textbf{k}$$
$$\nabla \bullet V = div\ \textbf{V} = \partial_x v_1 + \partial_y v_2 + \partial_z v_3$$
$$\nabla \times V = \text{curl } \textbf{V} = (\partial_y v_3 - \partial_z v_2)\textbf{i} - (\partial_x v_3 - \partial_z v_1)\textbf{j} + (\partial_x v_2 - \partial_y v_1)\textbf{k}$$

We have the following identities for these operations:

$$div\ \textbf{grad}\ f = \partial_{xx} f + \partial_{yy} f + \partial_{zz} f = \nabla^2 f$$

$$curl\ curl\ \textbf{V} = \textbf{grad}\ (div\ V) - (\nabla \bullet \nabla)\ V$$

*In addition, for smooth scalar fields f, g and smooth vector fields **U, V***

1. $\nabla\varphi(f) = \varphi'(f)\nabla f$ for φ a smooth function of one variable.

2. $\nabla(fg) = f\nabla g + g\nabla f.$

3. div (f**V**) = f div **V** + **V** • **grad** f.

4. curl (f**V**) = f curl **V** + **grad** $f \times V$.

5. div $(U \times V)$ = **V** • curl **U** – **U** • curl **V**.

6. curl $(U \times V) = (V \bullet \nabla)U - (U \bullet \nabla)V + (\text{div } V)U - (\text{div } U)V$.

For Ω a simply connected domain in $\Re^3$, then for **V** in $C^1(\Omega)$,

(a) div **V** = 0 if and only if **V** = curl **G** for some smooth **G**(x, y, z) then $\Phi[V;S] = 0$ for every closed surface S in Ω.

(b) curl **V** = 0 if and only if **V** = **grad** f for some smooth f(x, y, z) then $\Gamma[V;C] = 0$ for every simple closed curve C in Ω.

For Ω a closed bounded domain in $\Re^3$ with piecewise smooth boundary S, we have the integral identities:

1. If $\textbf{V} \in C^1(\Omega$ $\iint_S V \cdot N dA = \iiint_\Omega \text{div } V dx dy dz.$ (Guass' Theorem)

2. If $F \in C^1(\Omega)$ $\quad \iint_S FN\,dA = \iiint_\Omega \mathbf{grad}\, F\,dxdydz.$

3. If $\varphi, \psi \in C^2\Omega$ $\quad \iint_\Omega \nabla^2\varphi\; dxdydz = \iint_S \nabla\,\varphi \bullet N\; dA$

and (Green's identities)

$$\iiint_\Omega \psi\nabla^2\varphi\; dxdydz = \iint_S \nabla\varphi \bullet N\; dA - \iiint_\Omega \nabla\varphi \bullet \nabla\psi\; dxdydz$$

$$\iint (\psi\nabla^2\varphi - \varphi\nabla^2\psi)\; dxdydz = \iint (\psi\nabla\varphi \bullet N - \varphi\nabla\psi \bullet N)\, dA \quad .$$

For S a smooth and bounded but not closed surface with oriented boundary Γ and unit normal N, if $\mathbf{V} \in C^1(\Omega)$*, then*

$$\iint_S \text{curl}\, \mathbf{V} \bullet N\; dA = \int_\Gamma \mathbf{V} \bullet \mathbf{T}ds = \int_\Gamma \mathbf{V} \bullet d\mathbf{r} \qquad \textit{(Stokes' theorem)}$$

These identities play a significant role in the derivation of most of the equations of mathematical physics.

4

Ordinary Differential Equations: Linear Initial Value Problems

Now we will focus on three types of problems for ordinary differential equations: linear initial value problems, linear boundary value problems, and nonlinear problems. This chapter is devoted to linear initial value problems.

We begin by introducing a minimum amount of notation and terminology and then proceed to consider the initial value problem for a single homogeneous equation with particular attention to equations with constant coefficients.

The initial value problem for a single inhomogeneous equation is treated by means of the Laplace transform. This technique is particularly suited to analyzing the input-output relationship for physical systems modeled by ordinary differential equations.

Physical systems with more than a single degree of freedom are often modeled by systems of ordinary differential equations. If the number of degrees of freedom is small—say, two or three—then the Laplace transform is a feasible solution method. For systems involving larger numbers of equations, solution methods based on linear algebra concepts from Chapter 2 are liable to be more effective. Examples of both kinds are discussed.

INITIAL VALUE PROBLEMS FOR SINGLE EQUATIONS

Terminology and Notation

DIFFERENTIAL EQUATIONS

This chapter is devoted to the topic of ordinary differential equations. Differential equations involve an unknown function and one or more of its derivatives and if the unknown is a function of a single variable, then the differential equation is said to be an *ordinary differential equation.* We denote the derivatives of the single variable function $y(t)$ by

$$y'(t) = \frac{dy}{dt},\ y''(t) = \frac{d^2y}{dt^2}, \ldots, y^{(m)}(t) = \frac{d^my}{dt^m}$$

Differential equations in functions of several variables and their partial derivatives are called *partial differential equations.* These are considered in later chapters.

LINEAR HOMOGENEOUS EQUATIONS OF ORDER n

An ordinary differential equation in the unknown function $y(t)$ is said to be of *order n* if the equation contains the nth derivative of $y(t)$ but no derivatives of order higher than n. The most general equation of order n has the form

$$F(t, y(t), \ldots, y^{(n)}(t)) = 0$$

If the function F is linear in y and its derivatives, then we say the differential equation is *linear.* The most general linear ordinary differential equation of order n may be written as $L_n[y(t)] = f(t)$, where

$$L_n[y(t)] = a_n(t)y^{(n)}(t) + a_{n-1}(t)y^{(n-1)}(t) + \cdots + a_1(t)y'(t) + a_0(t)y(t)$$

We say the linear equation is in standard form if $a_n(t) = 1$. The equation is said to be *homogeneous* if $f(t) = 0$, otherwise it is *inhomogeneous.*

INITIAL VALUE PROBLEMS

In applications, we are often confronted with the problem of finding an unknown function $y = y(t)$ that satisfies a differential equation

$$L_n[y(t)] = f(t) \qquad \text{for all } t \text{ in } (t_0, t_1)$$

and also satisfies a set of *initial conditions*

$$y(t_0) = A_0,\ y'(t_0) = A_1, \ldots, y^{(n-1)}(t_0) = A_{n-1}$$

for given data $f(t), A_0, \ldots, A_{n-1}$. Such a problem is called an *initial value problem.*

LINEAR DEPENDENCE

A set of functions $y_1(t), y_2(t), \ldots, y_m(t)$ with common interval of definition I is said to be *linearly independent on I* if there exists constants c_1, .

. . , c_m, not all zero, such that

$$c_1y_1(t) + \cdots + c_my_m(t) = 0 \quad \text{for all } t \text{ in } I. \tag{4.1}$$

If the functions are not linearly dependent on I, they are said to be *linearly independent*. In this case, (4.1) holds if and only if $c_1 = c_2 = \cdots = c_m = 0$.

WRONSKIAN DETERMINANT

For an arbitrary set $y_1(t), y_2(t), \ldots, y_m(t)$ of functions it is not readily apparent whether they are independent. However, if the functions are sufficiently smooth, then there is a simple test for dependence. For this purpose we define the Wronskian determinant of the functions $y_1(t), y_2(t), \ldots, y_m(t)$ to be the following determinant:

$$W[y_1, \ldots, y_m](t) = \begin{bmatrix} y_1(t) & y_2(t) & \ldots & y_m(t) \\ y_1'(t) & y_2'(t) & \ldots & y_m'(t) \\ \vdots & \vdots & & \vdots \\ y_1^{(m-1)}(t) & & \ldots & y_m^{(m-1)}(t) \end{bmatrix}$$

Theorem 4.1

Theorem 4.1. Suppose the functions $y_1(t), y_2(t), \ldots, y_m(t)$ are continuous on the interval I together with all of their derivatives up to and including the order m. Then these functions are linearly dependent on I if and only if the Wronskian $W[y_1, \ldots, y_m](t) = 0$ for all t in I.

Homogeneous Initial Value Problem

Theorem 4.2. Let $L_n[y(t)]$ be given by

$$L_n[y(t)] = y^{(n)}(t) + a_{n-1}(t)y^{(n-1)}(t) + \cdots + a_1(t)y'(t) + a_0(t)y(t) \tag{4.2}$$

where each of the functions $a_m(t)$ $m = 0, 1, \ldots, n-1$ is continuous on $I = [t_0, t_1]$.

Then:

1. The homogeneous equation, $L_n[y(t)] = 0$ for t in I, has n solutions $y_1(t), y_2(t), \ldots, y_n(t)$ which are linearly independent on I

2. Every solution of $L_n[y(t)] = 0$ for t in I can be written in the form $y_g(t) = c_1y_1(t) + c_2y_2(t) + \cdots + c_ny_n(t)$ for some choice of the constants c_k. A solution $y_g(t)$ containing n arbitrary constants is called a *general solution* of the differential equation.

3. For any set of parameters $A_0, A_1, \ldots, A_{n-1}$, there exists a unique choice of $c_1, \ldots, c_n$ such that $c_1y_1(t) + c_2y_2(t) + \cdots + c_ny_n(t)$ satisfies

$$L_n[y(t)] = 0 \quad \text{for } t \text{ in } I$$
$$y(t_0) = A_0,\ y'(t_0) = A_1, \ldots, y^{(n-1)}(t_0) = A_{n-1}.$$

HOMOGENEOUS LINEAR EQUATIONS WITH CONSTANT COEFFICIENTS

There is no simple, general algorithm for solving the homogeneous equation $L_n[y(t)] = 0$ when the coefficients $a_m = a_m(t)$ are functions of t. When the coefficients are constant, the problem is much simpler. Note that for arbitrary constant r,

$$L_n[e^{rt}] = (r^n + a_{n-1}r^{n-1} + \cdots + a_1 r + a_0)e^{rt} = P_n(r)e^{rt}$$

The nth degree polynomial equation $P_n(r) = 0$ is called the *auxiliary equation* or *characteristic equation* associated with the differential equation $L_n[y(t)] = 0$.

Theorem 4.3

Theorem 4.3. The function $y(t) = e^{rt}$ satisfies the nth order linear constant coefficient equation $L_n[y(t)] = 0$ if and only if the constant r satisfies the auxiliary equation $P_n(r) = 0$. Suppose r is a root of multiplicity m for $P_n(r) = 0$.

(a) If r is real, then the m functions

$$y_1(t) = e^{rt},\ y_2(t) = te^{rt}, \ldots, y_m(t) = t^{m-1}e^{rt}$$

are linearly independent solutions of $L_n[y(t)] = 0$.

(b) If r is complex with $r = p + iq$, then $p - iq$ is also a root of $P_n(r) = 0$ and the $2m$ functions

$$u_1(t) = e^{pt}\text{Cos } qt,\ u_2(t) = te^{pt}\text{Cos } qt, \ldots, u_m(t) = t^{m-1}e^{pt}\text{Cos } qt$$
$$v_1(t) = e^{pt}\text{Sin } qt,\ v_2(t) = te^{pt}\text{Sin } qt, \ldots, v_m(t) = t^{m-1}e^{pt}\text{Sin } qt$$

are linearly independent solutions of $L_n[y(t)] = 0$.

Inhomogeneous Initial Value Problem

We will now consider the initial value problem in the case of an inhomogeneous differential equation. In connection with physical applications, this is sometimes referred to as the *forced problem*, and the function $f(t)$ on the right side of the inhomogeneous equation is called the *forcing term*.

Theorem 4.4

Theorem 4.4. Let $L_n[y(t)]$ be given by (4.2) and let $y_1(t), \ldots, y_n(t)$ denote n linearly independent solutions of the homogeneous equation $L_n[y(t)] = 0$. Let $f(t)$ be a given function, continuous on $I = [t_0, t_1]$.

1. Every solution of the inhomogeneous equation, $L_n[y(t)] = f(t)$ for t in I, can be written in the form $y_g(t) = Y(t) + c_1y_1(t) + c_2y_2(t) + \cdots +$

$c_n y_n(t)$ for some choice of the constants c_m and *any* choice of a function $Y(t)$ which solves $L_n[y(t)] = f(t)$. A solution $y_g(t)$ containing n arbitrary constants is called a *general solution* of the differential equation.

2. For any set of parameters $A_0, A_1, \ldots, A_{n-1}$, there exists a unique choice of $Y(t), c_1, \ldots, c_n$, such that $y(t) = Y(t) + c_1 y_1(t) + c_2 y_2(t) + \cdots + c_n y_n(t)$ satisfies

$$L_n[y(t)] = f(t) \quad \text{for } t \text{ in } I = [t_0, t_1]$$
$$y(t_0) = A_0,\ y'(t_0) = A_1, \ldots, y^{(n-1)}(t_0) = A_{n-1}$$

There are various procedures for solving the inhomogeneous equation including the methods of variation of parameters and undetermined coefficients. Here we will discuss only the Laplace transform approach to solving the inhomogeneous problem.

The Laplace Transform

Let $f(t)$ be defined on the semi-infinite interval, $t \geq 0$. Then the Laplace transform of $f(t)$ is defined to be

$$L[f(t)] = F(s) = \int_0^\infty f(t)\, e^{-st} dt \tag{4.3}$$

EXAMPLE 4.1 Laplace Transform

Consider the function $f(t) = 1$ for $t \geq 0$. Then, by (4.3)

$$L[f(t)] = \int_0^\infty 1 e^{-st} dt = \left.\frac{e^{-st}}{-s}\right|_{t=0}^{t=\infty} = \frac{1}{s} = F(s)$$

We say that the functions $f(t) = 1$ and $F(s) = 1/s$ are a *Laplace transform pair.* Additional Laplace transform pairs are found in the solved problems and in Table 4.1.

OPERATIONAL PROPERTIES OF THE LAPLACE TRANSFORM

Table 4.1 contains the Laplace transforms for some commonly occuring functions. We shall also have need of various *operational properties* of the transform. These are general facts that are true for the Laplace transform of any function.

	$f(t)$	$F(s)$
1.	1	$1/s$
2.	e^{kt}	$1/(s-k)$
3.	t^n	$\dfrac{n!}{s^{n+1}}$ $\quad n = \text{integer} > 0$
4.	$\sin at$	$\dfrac{a}{s^2+a^2}$
5.	$\cos at$	$\dfrac{s}{s^2+a^2}$
6.	$t \sin at$	$\dfrac{2as}{(s^2+a^2)^2}$
7.	$t \cos at$	$\dfrac{s^2-a^2}{(s^2+a^2)^2}$
8.	$e^{bt}t^n$	$\dfrac{n!}{(s-b)^{n+1}}$
9.	$e^{bt}\sin at$	$\dfrac{a}{(s-b)^2+a^2}$
10.	$e^{bt}\cos at$	$\dfrac{s-b}{(s-b)^2+a^2}$

Table 4.1
Laplace Transform Formulas

Theorem 4.5 Operational Properties of the Laplace Transform

Theorem 4.5. Let $f(t)$ and $g(t)$ denote arbitrary functions with $L[F(t)] = F(s)$ and $L[g(t)] = G(s)$, respectively.

1. $L[af(t) + bg(t)] = aF(s) + bG(s)$ for all constants a, b.
2. $L[f(at)] = (1/a)\,F(s/a)$ for a not equal to zero.
3. $L[f'(t)] = sF(s) - f(0)$.

   $L[f''(t)] = s^2F(s) - sf(0) - f'(0)$.

   $\vdots$

   $L[f^{(n)}(t)] = s^nF(s) - s^{n-1}f(0) - \cdots - f^{(n-1)}(0)$.
4. $L[tf(t)] = -F'(s)$.

5. $L[e^{bt}f(t)] = F(s-b)$ for all real constants b.
6. $L[H(t-b)f(t-b)] = e^{-bs}F(s)$ for all real constants b, where

$$H(t) = \textit{Heaviside step function} = \begin{cases} 0 \text{ if } t \le 0 \\ 1 \text{ if } t > 0 \end{cases}$$

7. $L[f*g(t)] = F(s)G(s)$, where

$$f*g(t) = \int_0^t f(t-T)\, g(T)dT.$$

THE CONVOLUTION PRODUCT

The product $f*g(t)$ defined in part 7 of Theorem 4.5 is called the *convolution product* of the functions f and g. The convolution product has the following properties:

$$f*g(t) = \int_0^t f(t-T)\, g(T)\, dT = \int_0^t f(T)\, g(t-T)\, dT = g*f(t) \qquad \text{(commutative)}$$

$$f*(g+h)(t) = f*g(t) + f*h(t) \qquad \text{(distributive)}$$

$$f*(g*h)(t) = (f*g)*h(t) \qquad \text{(associative)}$$

EXAMPLE 4.2

Consider the initial value problem for $y = y(t)$,

$$y'(t) - 3y(t) = t,\ y(0) = 10$$

If we let $Y(s) = L[y(t)]$, then $L[y'(t)] = sY(s) - y(0) = sY(s) - 10$. In addition, we find from Table 4.1 that $L[t] = 1/s^2$. Then

$$sY(s) - 10 - 3Y(s) = 1/s^2$$

and

$$Y(s) = \frac{10}{s-3} + \frac{1}{s^2}\frac{1}{s-3} = Y_h(s) + Y_p(s)\ .$$

Now it is evident from Table 4.1 that $L^{-1}[Y_h(s)] = 10e^{3t} = y_h(t)$, where we use the notation L^{-1} to denote the inverse Laplace transform operation. In this text, this operation is accomplished by looking in Table 4.1. To find the inverse transform of $Y_p(s)$, we first use the method of partial fractions to write

$$Y_p(s) = \frac{1}{s^2}\frac{1}{s-3} = \frac{1}{9}\frac{1}{s-3} - \frac{1}{3}\frac{1}{s^2} - \frac{1}{9}\frac{1}{s}$$

and next we use Table 4.2 to obtain

$$L^{-1}[Y_p(s)] = \frac{1}{9}e^{3t} - \frac{1}{3}t - \frac{1}{9} = y_p(t)$$

Alternatively, we could have used part 7 of Theorem 4.5 to write $y_p(t)$ as a convolution

$$y_p(t) = L^{-1}\left[\frac{1}{s^2}\right]*L^{-1}\left[\frac{1}{s-3}\right] = \int_0^t (t-T)\, e^{-sT}dt\ .$$

Performing this integration would lead to the result found previously by means of partial fractions. Then $y(t) = y_h(t) + y_p(t)$. Note that $y_h(t)$ satisfies $y_h'(t) - 3y_h(t) = 0$ and $y_h(0) = 10$, whereas $y_p(t)$ satisfies $y_p'(t) - 3y_p(t) = t$ and $y_p(0) = 0$. Thus, $y_h(t)$ represents the part of the solution that is caused by the initial conditions, $y_p(t)$ is the part of the solution that is due to the forcing term $f(t) = t$. We sometimes refer to $y_h(t)$ and $y_p(t)$, respectively, as the *response to initial state* and *response to forcing*. The Laplace transform approach to solving initial value problems is especially convenient for identifying the parts of the solution that are due to these respective inputs. Additional examples are found in the solved problems.

INITIAL VALUE PROBLEMS FOR SYSTEMS OF EQUATIONS

Mechanical systems with more than a single degree of freedom and electrical circuits consisting of more than a single loop lead to mathematical models involving systems of ordinary differential equations. In addition, any nth order differential equation can be reduced to a system of n equations of order one.

Theorem 4.6

Theorem 4.6. Consider the system of n linear equations in the n unknown functions, $x_1(t), \ldots, x_n(t)$:

$$\begin{aligned} x_1'(t) &= a_{11}(t)x_1(t) + \cdots + a_{1n}(t)x_n(t) + f_1(t) \\ &\;\;\vdots \\ x_n'(t) &= a_{n1}(t)x_1(t) + \cdots + a_{nn}(t)x_n(t) + f_n(t) \end{aligned} \tag{4.4}$$

If each of the functions $a_{jk}(t)$, $j, k = 1, \ldots, n$, and $f_k(t)$, $k = 1, \ldots, n$, is continuous on the interval $t_0 \le t \le t_1$, then for each set of n constants $C_1, \ldots, C_n$ there exists a unique set of functions $x = x_k(t)$ $k = 1, \ldots, n$, such that

(a) $x_1(t), \ldots, x_n(t)$ solves (4.4)

(4.5)

(b) $x_1(t_0) = C_1, \ldots, x_n(t_0) = C_n$.

Theorem 4.6 plays the same role for systems of linear differential equations that Theorem 4.2 plays for single equations. Note also that because second order systems can be transformed to equivalent first order systems, it is not necessary to have a separate existence theorem for systems of order two. Finally, the theorem implies that solutions for (4.4) can always be found and that when initial conditions are imposed, there is exactly one solution. However, the theorem does not suggest how solutions may be constructed. When the coefficients a_{jk} depend on t, there is no general algorithm for constructing solutions, but for the case of constant coefficients, there are various techniques available.

The Laplace transform is one approach to solving such systems. However, this approach is limited to small systems consisting of just two or three equations.

EXAMPLE 4.3

Consider the following initial value problem for a system of two equations in the two unknown functions $x_1(t)$ and $x_2(t)$,

$$x_1'(t) = 2x_1(t) - 5x_2(t), \qquad x_1(0) = 7$$

$$x_2'(t) = x_1(t) - 2x_2(t), \qquad x_2(0) = 3$$

If we let $X_1(s)$ and $X_2(s)$ denote the Laplace transforms of the two unknown functions, then it follows from results of the previous section that

$$sX_1 - 7 = 2X_1 - 5X_2$$

$$sX_2 - 3 = X_1 - 2X_2$$

Thus,

$$\begin{bmatrix} s-2 & 5 \\ -1 & s+2 \end{bmatrix} \begin{bmatrix} X_1 \\ X_2 \end{bmatrix} = \begin{bmatrix} 7 \\ 3 \end{bmatrix}$$

We can solve this system of two algebraic equations by Cramer's rule or any other method that is convenient. We obtain

$$X_1(s) = \frac{7s-1}{s^2+1} \quad \text{and} \quad X_2(s) = \frac{3s+1}{s^2+1}$$

It follows from entries 4 and 5 in Table 4.1 that

$$x_1(t) = 7\cos t - \sin t \quad \text{and} \quad x_2(t) = 3\cos t + \sin t.$$

Additional applications of the Laplace transform may be found in the solved problems.

For large systems, it is more efficient to apply techniques from linear algebra to construct solutions.

Theorem 4.7

Theorem 4.7. Suppose the $n \times n$ matrix A has n independent eigenvectors $\boldsymbol{X}_1$ to $\boldsymbol{X}_n$ corresponding to eigenvalues λ_1 to λ_n. Then for each given initial vector $\boldsymbol{C}$, the unique solution of the initial value problem,

$$\frac{d}{dt}U(t) = AU(t), \quad U(0) = \boldsymbol{C} \tag{4.6}$$

is given by

$$U(t) = c_1 e^{\lambda_1 t} X_1 + \ldots + c_n e^{\lambda_n t} X_n \tag{4.7}$$

where $c = [c_1, \ldots, c_n]^T$ is obtained by solving the system $Pc = C$, where P denotes the matrix whose columns are the eigenvectors of A. For each forcing function $F(t)$, the unique solution of the initial value problem,

$$\frac{d}{dt}U(t) = AU(t) + F(t), \quad U(0) = 0 \tag{4.8}$$

is given by,

$$U(t) = \int_0^t e^{\lambda_1 (t-\tau)} G_1(\tau)\, d\tau X_1 + \cdots + \int_0^t e^{\lambda_n (t-\tau)} G_n(\tau)\, d\tau X_n \tag{4.9}$$

where $G(t) = [G_1(t), \ldots, G_n(t)]^T$ is obtained by solving the system $PG = F$. In particular, if the eigenvectors form an orthonormal basis, then $G_k = F \cdot X_k$ for $k = 1, \ldots, n$.

Note that results discussed in chapters 1 and 2 allow us to write (4.7) in the form

$$U(t) = Pe^{tD}P^{-1}U(0)$$

and write (4.9) in the form

$$U(t) = \int_0^t Pe^{(t-\tau)D}P^{-1}F(\tau)\, d\tau$$

where D denotes the $n \times n$ diagonal matrix whose entries are the eigenvalues of A.

SOLVED PROBLEMS

Linear Differential Equations

PROBLEM 4.1

Determine whether each of the following sets of functions is linearly independent.

(a) $f(t) = \sin at$, $g(t) = \cos at$;
(b) $f(t) = t$, $g(t) = t^2$, $h(t) = t^3$;
(c) $f(t) = \sin 2t$, $g(t) = \sin t \cos t$;
(d) $f(t) = t^2$, $g(t) = t|t|$.

SOLUTION 4.1

The functions in part (a) are continuous together with their derivatives of all order for all values of t. Then Theorem 4.1 applies and we compute

$$W[f, g](t) = \det \begin{bmatrix} \sin at & \cos at \\ a\cos at & -a\sin at \end{bmatrix} = -a$$

Since the Wronskian is different from zero for all values of t, we conclude that the functions f and g are linearly independent on any interval I.

The three functions in part (b) are continuous together with their derivatives of all orders for all values of t and we can apply Theorem 4.1. We find

$$W[f, g, h](t) = \det \begin{bmatrix} t & t^2 & t^3 \\ 1 & 2t & 3t^2 \\ 0 & 2 & 6t \end{bmatrix} = 2t^3$$

Here the Wronskian is zero for $t = 0$ but is not identically zero on any interval. Thus, the functions f, g, and h are linearly independent on all intervals I.

The functions f and g in part (c) are similarly smooth and we compute

$$W[f, g](t) = \sin 2t(\cos^2 t - \sin^2 t) - 2\cos 2t \sin t \cos t.$$

But the identities, $\cos^2 t - \sin^2 t = \cos 2t$ and $2\sin t \cos t = \sin 2t$ imply that the Wronskian is zero for all values of t and thus, according to Theorem 4.1, these functions are linearly dependent on every interval. But this means $f(t)$ is some multiple of $g(t)$. In fact it is clear from the identities that $f(t) = 2g(t)$.

For the functions f and g in part (d) note that $W[f, g](t) = 0$ for all t. However,

$$g(t) = \begin{cases} t^2 & \text{if } t \geq 0 \\ -t^2 & \text{if } t < 0 \end{cases}, \quad g'(t) = \begin{cases} 2t & \text{if } t \geq 0 \\ -2t & \text{if } t < 0 \end{cases}, \quad g''(t) = \begin{cases} 2 & \text{if } t \geq 0 \\ -2 & \text{if } t < 0 \end{cases}$$

Thus, $g(t)$ and $g'(t)$ are continuous but $g''(t)$ is discontinuous at $t = 0$ so we may not apply Theorem 4.1 to determine whether these functions are dependent on any interval that includes $t = 0$. But on any such interval $f(t)$ is positive for all t, whereas $g(t)$ is positive for $t > 0$ and is negative where $t < 0$. Then there can be no single constant C such that $f(t) = Cg(t)$ for all t in the interval; hence, f and g are linearly independent on any interval that contains both positive and negative values of t.

PROBLEM 4.2

Find the general solution to each of the following differential equations:

(a) $y^{(3)}(t) - 6y''(t) + 11y'(t) - 6y(t) = 0;$

(b) $y^{(4)}(t) - y(t) = 0;$

(c) $y^{(4)}(t) + y(t) = 0;$

(d) $y^{(4)}(t) - y''(t) = 0.$

SOLUTION 4.2

The auxiliary equation associated with the equation of part (a) is

$$r^3 - 6r^2 + 11r - 6 = (r-1)(r-2)(r-3) = 0$$

Then Theorems 4.2 and 4.3 imply that the general solution of this equation can be written

$$y_g(t) = c_1 e^t + c_2 e^{2t} + c_3 e^{3t}$$

The equation in part (b) has as its auxiliary equation, $r^4 - 1 = 0$. The roots of this equation are $r = 1, -1, i, -i$ and since each has multiplicity one, we use parts (a) and (b) of Theorem 4.3 to conclude

$$y_g(t) = c_1 e^t + c_2 e^{-t} + c_3 \cos t + c_4 \sin t$$

For the differential equation in (c) we find the auxiliary equation $r^4 + 1 = 0$ has as its roots the four distinct fourth roots of -1. That is,

$$-1 = e^{i\pi + i2n\pi} \quad \text{for all integers } n$$

and

$$(-1)^{1/4} = e^{i\pi/4 + in\pi/2} \quad \text{for } n = 1, 2, 3, 4$$

Then the four roots of the auxiliary equation are as follows:

$$r_1 = \cos\frac{\pi}{4} + i\sin\frac{\pi}{4} = p + i, \quad p = \frac{\sqrt{2}}{2}$$

$$r_2 = \cos\frac{3\pi}{4} + i\sin\frac{3\pi}{4} = -p + ip$$

$$r_3 = \cos\frac{5\pi}{4} + i\sin\frac{5\pi}{4} = -p - ip$$

$$r_4 = \cos\frac{7\pi}{4} + i\sin\frac{7\pi}{4} = p - ip$$

According to part (b) of Theorem 4.3, these lead to four linearly independent solutions:

$$u_1(t) = e^{pt}\cos pt \quad \text{and} \quad v_1(t) = e^{pt}\sin pt$$

$$u_2(t) = e^{-pt} \cos pt \quad \text{and} \quad v_2(t) = e^{-pt} \sin pt$$

and the general solution has the form

$$y_g(t) = e^{pt}(c_1 \cos pt + c_2 \sin pt) + e^{-pt}(c_3 \cos pt + c_4 \sin pt)$$

Finally, for the differential equation of part (d), we have the auxiliary equation $r^2(r^2 - 1) = 0$ with roots $r = 0, 0, 1, -1$. The repeated root $r = 0$ generates two independent solutions $y_1(t) = 1$ and $y_2(t) = t$, while the roots $r = 1$ and $r = -1$ each contribute one solution each to the general solution which is

$$y_g(t) = c_1 + c_2 t + c_3 e^t + c_4 e^{-t}$$

Homogeneous Initial Value Problems

PROBLEM 4.3

A steel bearing that has been heated to a temperature of 300 °C is dropped into a liquid bath whose temperature is 20 °C. After 40 seconds, the temperature of the bearing has dropped to 292 °C. What will the temperature of the bearing equal after 10 minutes, assuming that the bath is large enough that the bearing does not cause the temperature of the bath to increase?

SOLUTION 4.3

Newton's law of cooling states that the rate at which the temperature of a body changes is proportional to the difference in temperature between the body and its surroundings. If we denote the temperature of the bearing at time t by $T(t)$ (where $t = 0$ is the instant the bearing is dropped into the bath) and if we let T^* denote the (constant) temperature of the bath, then

$$T'(t) = -k(T(t) - T^*) \tag{1}$$

Here k denotes a positive constant of proportionality. If we let $S(t) = T(t) - T^*$, then $S'(t) = T'(t)$ and

$$S'(t) = -kS(t) \tag{2}$$

The auxiliary equation for (2) is simply $P_1(r) = r + k = 0$ and then it follows from Theorem 4.3 that

$$S(t) = Ce^{-kt}$$

and

$$T(t) = T^* + Ce^{-kt} = 20 + Ce^{-kt}$$

Now $\quad T(0) = T^* + C = 300$

and, thus, $\quad T(t) = 20 + 280e^{-kt}$.

In order to find the constant k, recall that we are given $T(40) = 292$; i.e.,

$$T(40) = 20 + 280e^{-40k} = 292.$$

Then

$$k = \frac{1}{40} \ln\left(\frac{280}{272}\right) = 7.2468 \times 10^{-4}$$

After 10 minutes (600 seconds) we have

$$T(600) = 20 + 280e^{-.4347} = 201.268\ C.$$

PROBLEM 4.4

Consider a mass m suspended on a longitudinal spring with spring constant equal to K and suppose that the mass is submerged in a viscous fluid that retards the movement of the mass with a force that is proportional to its speed. Let the displacement of the mass from the equilibrium position be denoted by $x(t)$. Then Newton's second law, $F = ma$, implies that $x(t)$ satisfies equation

$$mx''(t) + Cx'(t) + Kx(t) = 0.(1)$$

If the mass is given an initial displacement and released from rest, compute the response in each of the following cases:

(a) Undamped case $C = 0$;
(b) Underdamped case $C > 0$ but $C^2 < 4Km$;
(c) Critically damped case $C^2 = 4Km$;
(d) Overdamped case $C^2 > 4Km$.

SOLUTION 4.4

If we denote the amount of the initial displacement by a, then the initial conditions that the mass is displaced by an amount a and released from rest are

$$x(0) = a, \quad x'(0) = 0 \tag{2}$$

The solution is now given in four parts:

(a) In the undamped case, $C = 0$, the auxiliary equation associated with (1) is

$$mr^2 + K = 0$$

with roots $r = i\omega, -i\omega$, where $\omega^2 = K/m$. Then, by part (b) of Theorem 4.3, the corresponding general solution for (1) is given by

$$x(t) = A \cos \omega t + B \sin \omega t \tag{3}$$

for arbitrary constants A and B. Then (3) implies that $x(0) = A$ and $x'(0) = \omega B$, and it follows that the initial conditions are satisfied if $A = a$, $B = 0$; hence,

$$x(t) = a \cos \omega t \tag{4}$$

is the undamped solution. This solution is periodic with frequency ω equal to $\sqrt{K/m}$; this frequency is referred to as the *natural frequency* of the undamped system. Note that ω is directly proportional to the square root of K and inversely proportional to the square root of m. So the system oscillates more rapidly if K is increased (i.e., the spring is made stiffer) and more slowly if the mass is increased.

(b) In the underdamped case $C > 0$ but $C^2 < 4Km$, the auxiliary equation associated with (1) is

$$mr^2 + Cr + K = 0$$

The roots of this equation are given by

$$r = \frac{-C \pm \sqrt{(C^2 - 4Km)}}{2m}$$

$$= -c \pm i\omega_0$$

where

$$c = \frac{C}{2m} \quad \text{and} \quad \omega_0^{\,2} = \frac{(4Km - C^2)}{4m^2} = \omega^2 - c^2$$

In this case, Theorem 4.3 suggests that the corresponding general solution of the differential equation (1) is

$$x(t) = e^{-ct}(A \cos \omega_0 t + B \sin \omega_0 t) \qquad (5)$$

According to (5), $x(0) = A$ and $x'(0) = -cA + \omega_0 B$, and it follows that the initial conditions will be satisfied if we choose the arbitrary constants A and B such that $A = a$ and $B = ac/\omega_0$. Then the solution of the initial value problem is

$$(t) = ae^{-ct}(\cos \omega_0 t + \frac{c}{\omega_0} \sin \omega_0 t) \qquad (6)$$

Note that the solution (6) reduces to (4) when C goes to zero. For C greater than zero but less than the critical value $C_c = \sqrt{4Km}$, the solution oscillates with steadily decreasing amplitude at the *damped natural frequency* $\omega_0 = \sqrt{\omega^2 - c^2}$. (See Figure 4.1)

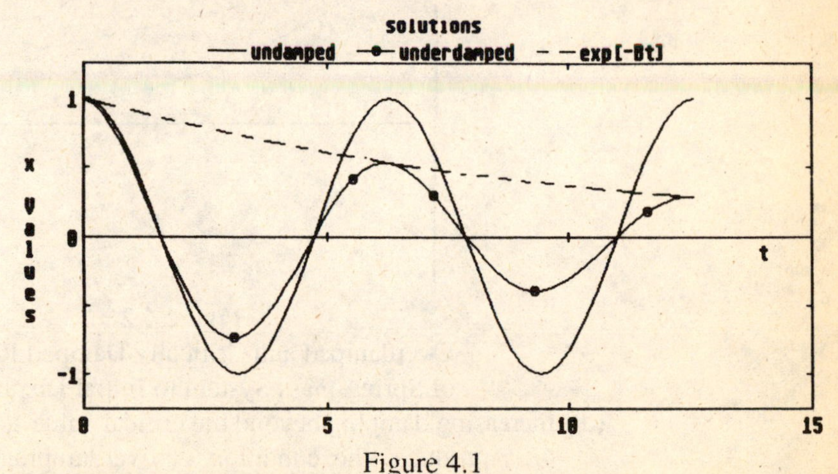

Figure 4.1
Undamped and Underdamped Response of a Spring-mass System

(c) Note that ω_0 is less than ω, the undamped natural frequency. As the damping increases from zero, the damped natural frequency continues to decrease until, finally, it reaches the value zero and no oscillation occurs. In the critically damped case $C^2 = 4Km$, the auxillary equation has a real root of multiplicity 2 at the value $r = \omega = \sqrt{K/m}$. According to part (a) of Theorem 4.3, the general solution to the differential equation (1) is given by

$$x(t) = Ae^{-\omega t} + Bte^{-\omega t} \tag{7}$$

Then $x(0) = A$ and $x'(0) = -\omega A + B$, which leads to $A = a$, $B = a\omega$, and

$$x(t) = a(e^{-\omega t} + \omega te^{-\omega t}) \tag{8}$$

is then the solution to the initial value problem. While the underdamped solution given by (5) oscillates with exponentially decreasing amplitude, the critically damped solution given by (8) decreases to zero without oscillating. Thus, when the damping reaches the critical value $\sqrt{4Km}$, the damping is just sufficient to suppress oscillation. Thus, the mass released from rest does not oscillate but merely returns slowly to the equilibrium position.

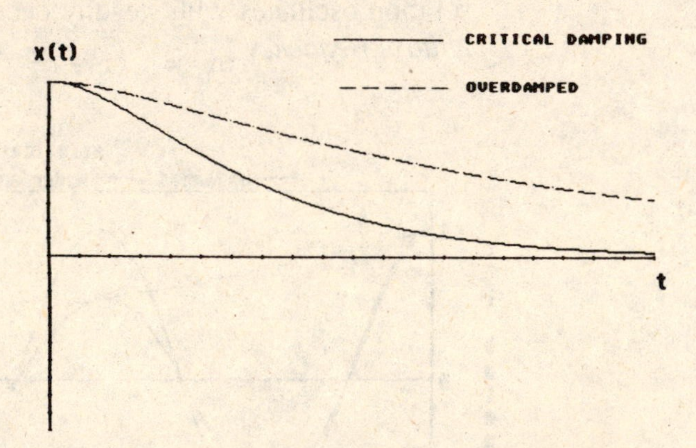

Figure 4.2
Overdamped and Critically Damped Response
of Spring-mass System to Initial Displacement

(d) Increasing damping beyond the critical value so that C^2 is greater than $4Km$ produces the condition of overdamping. Here the auxiliary equation associated with (1) has distinct real roots given by $r_{1,2} = -c \pm \lambda$, where $c = C/2m$ and $\lambda = \sqrt{c^2 - \omega^2}$. The corresponding solution for (1) is then

$$x(t) = Ae^{(-c-\lambda)t} + Be^{(-c+\lambda)t} \tag{9}$$

and since $x(0) = A + B$ and $x'(0) = (-c-\lambda)A + (-c+\lambda)B$, it follows that the solution of the initial value problem is

$$x(t) = \frac{a}{2\lambda}\left[(\lambda - c)e^{(-c-\lambda)t} + (\lambda + c)e^{(-c+\lambda)t}\right] \tag{10}$$

Note that by definition, c is greater than λ. Then $-c-\lambda < -c+\lambda < 0$ so both the exponents appearing in the solution $x(t)$ given by (10) are negative, and the solution decreases steadily to zero. Note further that the overdamped solution (10) decreases to zero less rapidly than the critically damped solution, (Figure 4.2). When the damping is increased beyond the critical value $\sqrt{4Km}$, the damping not only suppresses oscillation, it further impedes the motion of the mass released from rest so the return to the equilibrium position occurs more slowly.

Heavy doors are often equipped with spring devices that pull the door shut, but the motion is sufficiently damped that the door does not slam. For this to work propertly, the damping must equal or exceed the critical value.

PROBLEM 4.5

Consider the spring mass system of Problem 4.4 in the critically damped and overdamped cases if the initial conditions are changed to

$$x(0) = 0, \qquad x'(0) = b \tag{1}$$

These initial conditions correspond to the physical situation in which the mass is given a sharp blow so that it starts from the rest position with a nonzero velocity. This initial value problem for the spring-mass system is a reasonable approximation of the operation of a shock absorber in an automobile; i.e., the shock absorber is essentially a damped spring-mass system subject to the initial conditions (1) when the car hits a bump.

SOLUTION 4.5

In the critically damped case, we obtain the general solution

$$x(t) = Ae^{-\omega t} + Bte^{-\omega t}. \tag{2}$$

As in Problem 4.4, we compute $x(0) = A$, $x'(0) = -\omega A + B$. Then the initial conditions (1) lead to the result $A = 0$ and $B = b$ and the solution

$$x(t) = bte^{-\omega t} \tag{3}$$

In the overdamped case, the auxiliary equation has distinct, real roots and the general solution of the differential equation is

$$x(t) = Ae^{(-c-\lambda)t} + Be^{(-c+\lambda)t} \tag{4}$$

Then $x(0) = A + B$ and $x'(0) = (-c - \lambda)A + (-c + \lambda)B$ and the initial conditions (1) lead to the solution

$$x(t) = \frac{b}{2\lambda}\left(e^{(-c+\lambda)t} - e^{(-c-\lambda)t}\right) \tag{5}$$

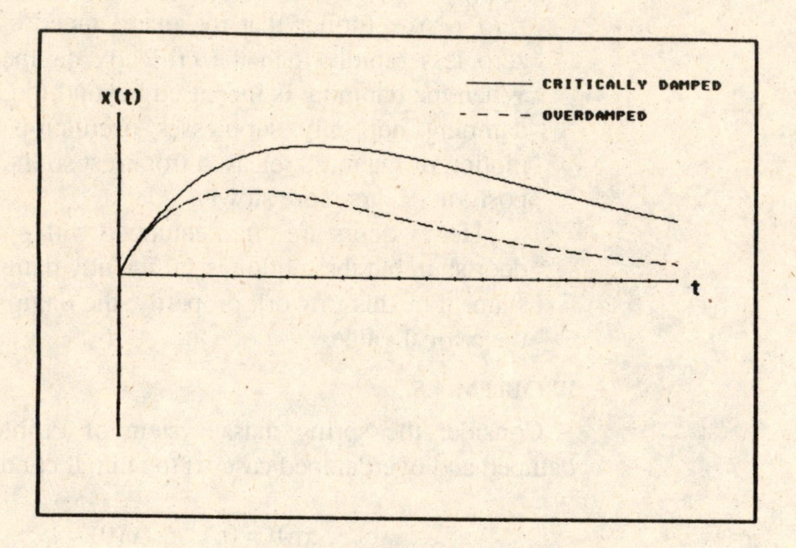

Figure 4.3
Overdamped and Critically Damped Response of Spring-mass System to Initial Velocity

Figure 4.3 shows the critically damped and overdamped solutions (3) and (5) for the initial conditions (1). Each increases to a maximum value and then decreases to zero. The critically damped solution rises to a greater maximum than the overdamped solution; thus, increasing the damping beyond the critical value needed to suppress oscillation further inhibits the motion of the mass. In both cases, note that the spring in the shock absorber is compressed by the bump but that the damping suppresses any oscillation. Then the car does not bob up and down after hitting a bump.

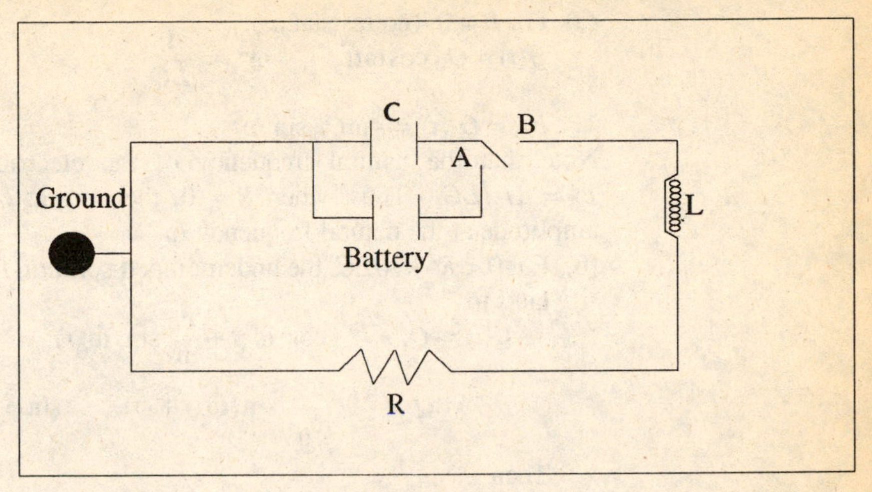

Figure 4.4
An L-C-R Driven by a Battery

PROBLEM 4.6

A circuit contains a capacitor, an inductor and a variable resistance in series (Figure 4.4). Such a circuit is called an *L-C-R* circuit. With the switch in position *A*, the battery places a charge of Q_0 on the capacitor. When the switch is moved to position *B*, the capacitor discharges into the circuit. Find the resulting current $I(t)$. Find the value of the resistance *R* in terms of the capacitance *C* and inductance *L* such that oscillation in $I(t)$ is just suppressed.

SOLUTION 4.6

When the switch is moved to position *B*, a charge of Q_0 is discharged into a circuit that is initially dead. We describe this situation with the initial conditions

$$Q(0) = Q_0 \quad \text{and} \quad Q'(0) = I(0) = 0 \tag{1}$$

Then $Q = Q(t)$, the charge in the circuit, varies with time $t > 0$ according to the equation

$$LQ''(t) + RQ'(t) + \frac{1}{C}Q(t) = 0 \tag{2}$$

But (1), (2) is the same initial value problem that arose in Problem 4.4 for the spring-mass system with the nominal differences that

- *L* takes the place of the mass *m*;
- *R* takes the place of the damping constant *C*;
- $1/C$ takes the place of the spring stiffness *K*.

Then the solution of (1), (2) can be obtained by using the solutions constructed in Problem 4.4 and replacing *m*, *C*, and *K* by *L*, *R*, and $1/C$, respectively.

(a) For $R = 0$ (no resistance):

$$Q(t) = Q_0 \cos \omega t, \qquad \omega^2 = \frac{1}{LC}$$

$$I(t) = Q'(t) = -\omega Q_0 \sin \omega t$$

Note that the natural frequency of the electrical circuit is equal to $\omega = 1/\sqrt{LC}$. Thus, when $R = 0$, the current oscillates with constant amplitude at the natural frequency ω.

(b) For $0 < R^2 < 4/LC$, the underdamped solution from Problem 4.4 translates to

$$Q(t) = Q_0 e^{-ct}(\cos \omega_0 t + \frac{c}{\omega} \sin \omega_0 t)$$

$$= Q_0 e^{-ct}\left(\frac{\omega}{\omega_0}\right)\sin(\omega_0 t + \varphi), \qquad \sin\varphi = \frac{\omega_0}{\omega}.$$

Then

$$I(t) = Q'(t) = Q_0 e^{-ct}\left(\frac{\omega}{\omega_0}\right)(\omega_0 \cos(\omega_0 t + \varphi) - c \sin(\omega_0 t + \varphi))$$

$$= Q_0 e^{-ct}\left(\frac{\omega^2}{\omega_0}\right)\sin \omega_0 t.$$

Thus, when R is positive but less than the critical value $2/\omega$, the current oscillates with decreasing amplitude at the damped natural frequency $\omega_0 = \sqrt{\omega^2 - c^2}$, where $\omega^2 = 1/LC$ and $c = R/2L$. $I(t)$ behaves like $x(t)$ in the underdamped spring-mass system, (Figure 4.1).

(c) When the resistance reaches the critical value $R^2 = 4/LC$, the oscillation in the current is suppressed and we obtain

$$Q(t) = Q_0(e^{-\omega t} + \omega t e^{-\omega t})$$

and $I(t) = Q'(t)$. Thus the value of resistance required to just suppress oscillation is

$$R = \frac{2}{\sqrt{LC}} = \frac{2}{\omega}$$

For R greater than this critical value, the current $I(t)$ decreases to zero without oscillating: i.e., it behaves like $x(t)$ in the overdamped spring mass system (Figure 4.2).

Laplace Transform

PROBLEM 4.7

Find the Laplace transform of each of the following functions:

(a) $f(t) = e^{kt}$, $\qquad k$ = real constant;

(b) $g(t) = \sin at$, a = real constant;
(c) $h(t) = \cos at$;
(d) $p(t) = e^{kt} \sin at$;
(e) $q(t) = t \cos at$.

SOLUTION 4.7

Note that $f(t) = e^{kt}$ satisfies $f'(t) = ke^{kt} = kf(t) = f(0) = 1$. Then

$$L[f'(t)] = L[kf(t)] = kF(s)$$

and

$$L[f'(t)] = sF(s) - f(0) = sF(s) - 1$$

Then $sF(s) - 1 = kF(s)$ and we find $F(s) = 1/(s - k)$.

Similarly, $g(t) = \sin at$ satisfies $g''(t) = -a^2 \sin at = -a^2 g(t)$ and $g(0) = 0$, $g'(0) = a$. Then

$$L[g''(t)] = L[-a^2 g(t)] = -a^2 G(s)$$

and

$$L[g''(t)] = s^2 G(s) - s \cdot 0 - a.$$

Then $s^2 G(s) - a = -a^2 G(s)$ and we obtain $G(s) = a / (s^2 + a^2)$. In addition, because $h(t) = \cos at$ satisfies $h'(t) = -a \sin at = -ag(t)$ and $h(0) = 1$, it follows that

$$L[h'(t)] = -aG(s) \quad \text{and} \quad L[h'(t)] = sH(s) - 1$$

Then $sH(s) = 1 - aG(s) = s^2 / (s^2 + a^2)$ and $H(s) = s / (s^2 + a^2)$. Finally,

$$L[t \cos at] = -\frac{d}{ds}([H(s)]) = \frac{(a^2 - s^2)}{(s^2 + a^2)} = Q(s)$$

PROBLEM 4.8

Find the functions whose Laplace transforms are equal to:

(a) $$F(s) = \frac{s}{(s^2 - 3s + 6)}$$

(b) $$G(s) = \frac{1}{(s^2 + a^2)(s^2 + b^2)}$$

(c) $$H(s) = \frac{s}{(s^2 + a^2)(s^2 + b^2)}$$

SOLUTION 4.8

Because $F(s) = s^2 - 3s + 6 = (s - 2)(s - 3)$, we can use partial fractions to write

$$\frac{s}{(s-2)(s-3)} = \frac{A}{s-2} + \frac{B}{s-3} = \frac{A(s-3) + B(s-2)}{(s-2)(s-3)}$$

This expression is an identity in s if $A = -2$ and $B = 3$. Then

$$L^{-1}\left[\frac{s}{(s-2)\,(s-3)}\right] = L^{-1}\left[\frac{3}{s-3}\right] - L^{-1}\left[\frac{2}{s-2}\right] = 3e^{3t} - 2e^{2t}.$$

Applying a similar approach to $G(s)$ leads to

$$G(s) = \frac{1}{s^2+a^2}\frac{1}{s^2+b^2} = \frac{As+B}{s^2+a^2} + \frac{Cs+D}{s^2+b^2}$$

$$= \frac{(A+C)\,s^3 + (B+D)\,s^2 + (a^2C+b^2A)\,s + b^2B + a^2D}{(s^2+a^2)\,(s^2+b^2)}$$

This is an identity in s if

$$A + C = 0, \quad B + D = 0, \quad a^2C + b^2A = 0, \quad b^2B + a^2D = 1$$

i.e.,

$$C = -A, \quad D = -B, \quad A(b^2 - a^2) = 0, \quad \text{and} \quad B(b^2 - a^2) = 1$$

Then

$$L^{-1}\,[G\,(s)\,] = \frac{1}{b^2-a^2}L^{-1}\left[\frac{1}{s^2+a^2}\right] - \frac{1}{b^2-a^2}L^{-1}\left[\frac{1}{s^2+b^2}\right]$$

$$= \frac{1}{b^2-a^2}\left(\frac{1}{a}\sin at - \frac{1}{b}\sin bt\right) = g\,(t)$$

Note that since $g(0) = 0$, we have $H(s) = sG(s) - g(0) = L[g'(t)]$. Then $h(t) = g'(t)$,

$$L^{-1}\,[H\,(s)\,] = \frac{1}{b^2-a^2}\,(\cos at - \cos bt)$$

PROBLEM 4.9

Let $\delta\,(t) = LT^{-1}\,[1]$. Then show that $\delta\,(t)$ must have the following properties:

(a) $LT\,[\delta\,(t-T)\,] = e^{-sT}$.

(b) $\delta\,(t-T) = H'\,(t-T)$, where $H(t)$ is the Heaviside step function.

(c) If $LT[L_n y(t)] = P(s)Y(s)$, then $LT^{-1}\,[1/\,(P\,(s)\,)\,]$ equals the *impulse response*; i.e., it is the solution of $L_n\,[y\,(t)\,] = \delta\,(t)$, $y^{(m)}(0) = 0$, for $m = 0, 1, \ldots, n-1$.

SOLUTION 4.9

In fact there is no function, in the classical meaning of the term, whose Laplace transform equals 1. However, there is a mathematically consistent theory of generalized functions in which $\delta\,(t)$ is perfectly well defined. We will only use the "δ-function" in a formal way without becoming involved in this theory.

By part 6 of Theorem 4.5, we have

$$L[H(t-T)\delta(t-T)] = L[\delta(t-T)] = e^{-sT}L[\delta(t)] = e^{-sT}$$

i.e., $\int_0^\infty \delta(t-T)e^{-st}dt = e^{-sT}$ for all $T > 0$ (1)

Parts 3 and 6 of Theorem 4.5 also imply

$$L[H'(t-T)] = sL[H(t-T)] - 0 = s\frac{e^{-sT}}{s} = e^{-sT} = L[\delta(t-T)]$$

Then

$$\delta(t-T) = H'(t-T) \tag{2}$$

although the Heaviside function is not differentiable in the classical sense. Finally, let L_n denote a linear constant coefficient differential operator of order n. Then the initial value problem

$$L_n[y(t)] = \delta(t), \quad y(0) = y'(0) = \cdots = y^{(n-1)}(0) = 0 \tag{3}$$

transforms to $P(s)Y(s) = 1$ where $P(s)$ is nth degree characteristic polynomial for L_n; i.e., $P(r) = 0$ is the auxiliary equation associated with $L_n[y(t)] = 0$. Then the solution $y(t)$ for the initial value problem (3) is equal to

$$y(t) = L^{-1}\left[\frac{1}{P(s)}\right].$$

We refer to $y(t)$ as the impulse response associated with the operator L_n.

Inhomogeneous Initial Value Problems

PROBLEM 4.10

A damped spring-mass system as described in Problem 4.4 is driven by a periodic forcing term $F(t) = F\cos\Omega t$. If the initial displacement of the mass is equal to d and the initial velocity is v, then find the subsequent motion of the mass. Identify the part of the response due to the initial state and the response to the forcing. Also identify the transient and steady-state components of the solution and explain the meaning of these terms.

SOLUTION 4.10

The displacement $x(t)$ of the mass from the equilibrium position satisfies the initial value problem,

$$x''(t) + 2cx'(t) + \omega^2 x(t) = \frac{F}{m}\cos(\Omega t) \tag{1}$$

$$x(0) = d, \quad x'(0) = v \tag{2}$$

Here $2c = C/m$ and $\omega^2 = K/m$, where C, K, and m have the same meaning as in Problem 4.4. We are supposing Ω does not equal ω. Applying the Laplace transform to this problem, we find $X(s)$, the transform of the solution to b,

$$X(s) = \frac{sd + v + 2cd}{s^2 + 2cs + \omega^2} + \frac{F}{m}\frac{s}{s^2 + \Omega^2}\frac{1}{s^2 + 2cs + \omega^2} \tag{3}$$

$$= X_i(s) + X_f(s)$$

It is apparent that $X_i(s)$ and $X_f(s)$ denote the transforms of the response to the initial state and the forced response, respectively. That is, $x_i(t) = L^{-1}[X_i(s)]$ solves the homogeneous version of the differential equation (1) and the inhomogeneous initial conditions (2), whereas $x_f(t) = L^{-1}[X_f(s)]$ satisfies the inhomogeneous differential equation (1) and homogeneous initial conditions of the form (2). Note that $X_i(s)$ can be written as

$$X_i(s) = d\frac{s + c}{(s + c)^2 + \omega_0^2} + \frac{v + cd}{\omega_0}\frac{\omega_0}{(s + c)^2 + \omega_0^2} \tag{4}$$

and partial fractions can be used to write $X_f(s)$ in the form

$$X_f(s) = \frac{F}{m}\frac{(\Omega^2 - \omega^2)s - 2c\omega^2}{(\omega^2 - \Omega^2)^2 + (2c\Omega)^2}\frac{1}{s^2 + 2cs + \omega^2}$$

$$- \frac{F}{m}\frac{(\Omega^2 - \omega^2)s - 2c\Omega^2}{(\omega^2 - \Omega^2)^2 + (2c\Omega)^2}\frac{1}{s^2 + \Omega^2}$$

$$= \frac{F}{m}\frac{1}{A^2 + B^2}\left(\frac{A(s + c) - B'\omega_0}{(s + c)^2 + \omega_0^2} - \frac{As - B\Omega}{s^2 + \Omega^2}\right) \tag{5}$$

where $A = (\Omega^2 - \omega^2)$, $B = 2c\omega$, and $B' = B(\Omega^2 + \omega^2)/(2\Omega\omega_0)$. Then using the Laplace transform formulas in Table 4.1 to find the inverses of (4) and (5), we obtain

$$x(t) = e^{-ct}\left(d \cos \omega_0 t + \frac{v + cd}{\omega_0} \sin \omega_0 t\right)$$

$$+ \frac{F}{m}e^{-ct}\frac{(A \cos \omega_0 t - B' \sin \omega_0 t)}{(A^2 + B^2)} \tag{6}$$

$$- \frac{F}{m}\frac{(A \cos \Omega t - B \sin \Omega t)}{(A^2 + B^2)}$$

If we denote the three parts in the sum on the right in equation (6) by $x_1(t)$, $x_2(t)$, and $x_3(t)$, respectively, then it is clear that $x_1(t)$ is just $x_i(t)$, the response to the initial state. Then the sum of $x_2(t)$ and $x_3(t)$ represents the response of the spring-mass system to the forcing function; i.e., these terms are proportional to F, the amplitude of the forcing function.

The transient portion of the solution "dies out" with increasing time (its amplitude tends to zero). Then the steady-state component of the solution is the part of the solution that remains after all transient behavior has died out. It is evident that the sum of $x_1(t)$ and $x_2(t)$ must be the transient part of the solution $x(t)$ because these are the terms that tend to zero as t tends to infinity. The transient part of the solution is seen to be composed of $x_i(t)$, the response to the initial state, and $x_2(t)$ which reflects the transient behavior of the spring-mass system as it becomes synchronized with the forcing term. Finally, $x_3(t)$ is the steady-state part of the solution which is the response that remains after the influence of the initial conditions has died out and the system has become synchronized with the forcing function. Figure 4.5 shows the solution $x(t)$ and its transient part plotted against time. At a sufficiently large time T, the transient is seen to be essentially zero and the solution becomes periodic with period equal to that of the forcing term.

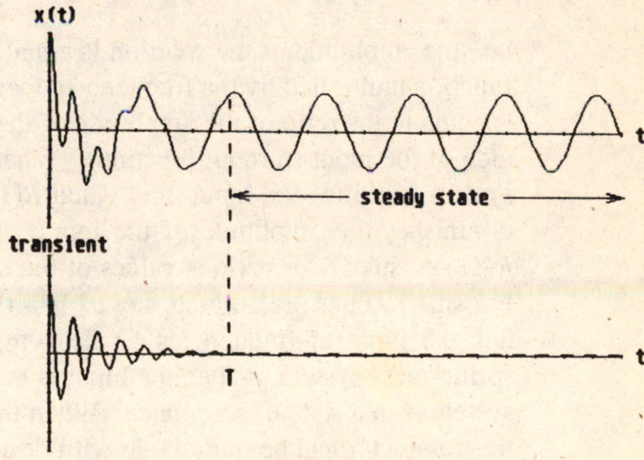

Figure 4.5
Transient and Steady State Response

PROBLEM 4.11

Show how to write the steady-state component of the solution to the previous problem in the form $x(t) = FR(\Omega)\cos(\Omega t + \varphi)$, where $R(\Omega)$ and φ denote the *amplification factor* and *phase angle*, respectively. Plot $R(\Omega)$ versus Ω for various values of the damping factor c.

SOLUTION 4.11

Recall that the steady-state component of the solution to the previous problem could be written as

$$x_3(t) = \frac{F}{m}\frac{(A\cos\Omega t - B\sin\Omega t}{(A^2+B^2)}$$

Then for φ such that $\tan\varphi = B/A$ we can express this as

$$x_3(t) = \frac{F}{m}\frac{1}{\sqrt{A^2+B^2}}(\cos\Omega t\cos\varphi - \sin\Omega t\sin\varphi)$$

$$= \frac{F}{m}\frac{1}{\sqrt{A^2+B^2}}\cos(\Omega t+\varphi)$$

Thus, the steady-state response is periodic with the same period as the forcing term, but there is a "phase shift" equal to φ. In addition, the amplitude of this solution is equal to

$$\frac{F/m}{\sqrt{A^2+B^2}} = \frac{F/m}{\sqrt{(\omega^2-\Omega^2)^2+(2c\Omega)^2}} = FR(\Omega)$$

i.e., the amplitude of the solution is equal to the amplitude of the forcing function multiplied by the frequency dependent factor $R(\Omega)$. Thus, $R(\Omega)$ is equal to the ratio of the amplitude of the output (solution) to the amplitude of the input (forcing function). When $R(\Omega)$ is greater than one, the system amplifies the input, and when $R(\Omega)$ is less than one, the system diminishes the amplitude of the input. In Figure 4.6 we have plotted $R(\Omega)$ versus Ω for various values of the damping parameter $c = C/2m$. For small values of damping, say c^2 less than $Km/4$, the system amplifies the input at frequencies Ω near ω, the natural frequency of the spring-mass system. When the input is magnified in this way, we say the system is in a state of *resonance*. When the damping is small, the magnification factor can be quite large with destructive results. For this reason it is important to know the resonant frequencies of a physical system and the magnification factor that applies at each frequency. The system in this example has just a single degree of freedom and, thus, just one resonant frequency. Systems with more degrees of freedom have more resonant frequencies. At the value $c^2 = 4Km$, the system is critically damped, and for larger values of c, it is overdamped. At such levels of damping there is no amplification of input for any value of input frequency Ω.

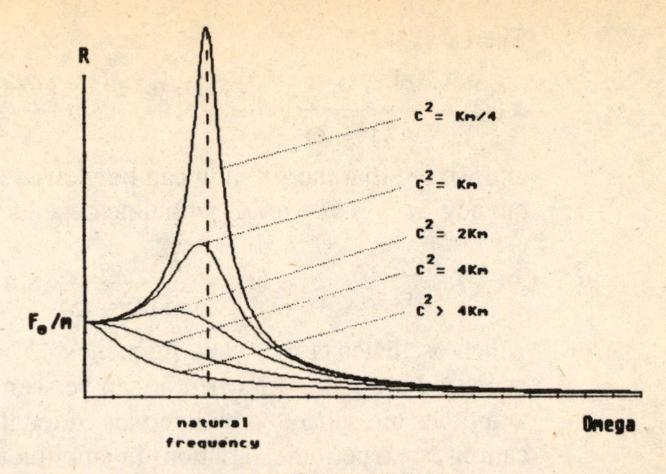

Figure 4.6
Amplitude magnification factor, R versus forcing frequency,
Omega for various levels of damping, c

PROBLEM 4.12

Solve the inhomogeneous initial value problem for the undamped spring-mass system

$$mx''(t) + Kx(t) = F\cos\Omega t, \qquad x(0) = x'(0) = 0 \tag{1}$$

for Ω^2 not equal to $\omega^2 = K/m$.

SOLUTION 4.12

Applying the Laplace transform to (1) leads to

$$X(s) = \frac{F}{m}\,\frac{1}{s^2+\omega^2}\,\frac{s}{s^2+\Omega^2}$$

and then, using result (c) from Problem 4.8, we find

$$x(t) = \frac{F}{m}\,\frac{1}{\Omega^2-\omega^2}(\cos\omega t - \cos\Omega t)$$

Note that if we let $2a = \Omega + \omega$ and $2b = \Omega - \omega$, then $\Omega = a + b$ and $\omega = a - b$ and it follows that

$$\cos\omega t = \cos(a-b)t = \cos at\cos bt + \sin at\sin bt$$
$$\cos\Omega t = \cos(a+b)t = \cos at\cos bt - \sin at\sin bt$$

i.e.,

$$\omega t - \cos\Omega t = 2\sin at\sin bt = 2\sin\left(\frac{(\Omega-\omega)t}{2}\right)\sin\left(\frac{(\Omega+\omega)t}{2}\right)$$

Then

$$x(t) = \frac{F}{m}\frac{2}{\Omega^2-\omega^2}\sin\frac{(\Omega-\omega)t}{2}\sin\frac{(\Omega+\omega)t}{2} = A(t)\sin\frac{(\Omega+\omega)t}{2}$$

and we see that the solution can be viewed as oscillating at the "fast frequency" $a = (\Omega+\omega)/2$ with time dependent amplitude $A(t)$,

$$A(t) = \frac{F}{m}\frac{2}{\Omega^2-\omega^2}\sin\frac{(\Omega-\omega)t}{2}$$

which oscillates at the "slow frequency" $b = (\Omega-\omega)/2$. This solution is plotted in Figure 4.7 where it can be seen that the maximum values of amplitude are separated by intervals of length $2\pi/b$. In acoustical applications, such periodic variations in amplitude are called *beats*.

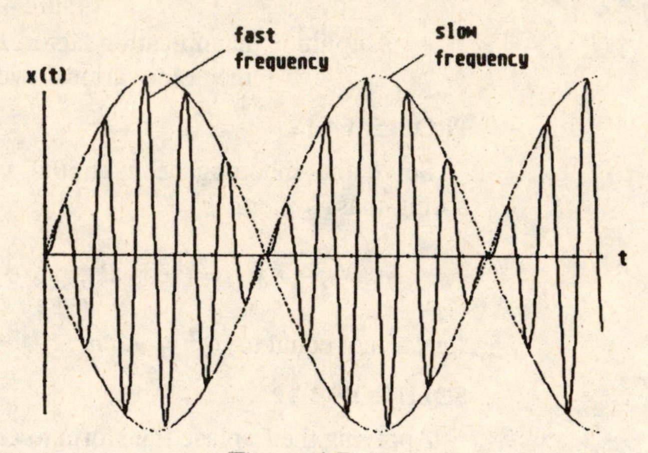

Figure 4.7
Response to Periodic Forcing: Nonresonant Response (Beats)

Initial Value Problems For Systems

PROBLEM 4.13

A transformer consists of a pair of electrical circuits that influence one another by means of "magnetic induction." Figure 4.8 shows a pair of such circuits. The circuit on the left is driven by an electromotive force $E(t) = E_0 \cos \Omega t$, which then induces a current in the circuit on the right. Find the current induced in the right half of the circuit if

$L_1 = 2$ henrys, $L_2 = 1$ henry,
$R_1 = 10$ ohms, $R_2 = 1$ ohm,
$C = 10^{-1}$ farads, $\Omega = 10$ cps.

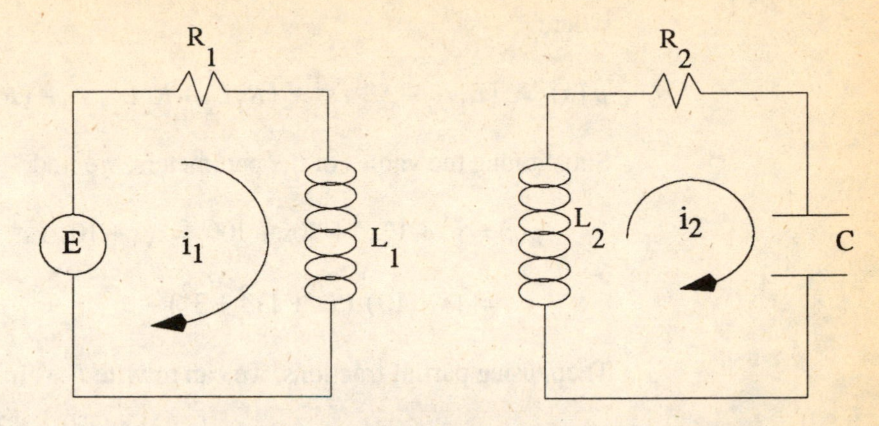

Figure 4.8
Transformer Circuit for Problem

SOLUTION 4.13

If we let $i_1(t)$ and $i_2(t)$ denote the current in the left and right parts of the circuit, respectively, then it can be shown that these functions satisfy

$$L_1 i_1'(t) + M i_2'(t) + R_1 i_1(t) = E(t) \qquad (1)$$

$$L_2 i_2'(t) + M i_1'(t) + R_2 i_2(t) + \int^t i_2 = 0$$

where M denotes the so-called "mutual inductance" between the two parts of the circuit. We suppose that $L_1 L_2 > M^2$; in particular, let us assume that $M^2 = L_1 L_2 - 1$. We can express (1) equivalently in the form

$$L_1 i_1''(t) + M i_2''(t) + R_1 i_1'(t) = E'(t) = -E_0 \Omega \sin \Omega t \qquad (2)$$

$$L_2 i_2''(t) + M i_1''(t) + R_2 i_2'(t) + \frac{1}{C} i_2(t) = 0$$

If the circuit is initially dead, then

$$i_1(0) = i_2(0) = 0 \quad \text{and} \quad i_1'(0) = i_2'(0) = 0 \qquad (3)$$

and $I_1(s)$, $I_2(s)$, the Laplace transforms of $i_1(t)$, $i_2(t)$, respectively, satisfy

$$(L_1 s^2 + R_1 s) I_1(s) + M s^2 I_2(s) = -\frac{E_0 \Omega^2}{(s^2 + \Omega^2)} \qquad (4)$$

$$M s^2 I_1(s) + (L_2 s^2 + R_2 s + \frac{1}{C}) I_2(s) = 0$$

We can solve these equations for $I_2(s)$ to obtain

$$I_2(s) = \frac{E_0 \Omega^2}{s^2 + \Omega^2} \frac{Ms}{p(s)} \qquad (5)$$

where

$$p(s) = (L_1L_2 - M^2)s^3 + (R_1L_2 + R_2L_1)s^2 + (R_1R_2 + \frac{L_1}{C})s + \frac{R_1}{C}.$$

Substituting the values of the parameters, we find

$$p(s) = s^3 + 12s^2 + 30s + 100 = (s+10)(s^2+2s+10)$$

$$= (s+10)((s+1)^2 + 3^2)$$

Then, using partial fractions, we can rewrite $I_2(s)$ in the form

$$I_2(s) = E_0\left[-\frac{1}{18}\frac{1}{s+10} - \frac{1}{170}\frac{11s+70}{s^2+100} + \frac{2}{765}\frac{46s+37}{s^2+2s+10}\right].$$

Finally, we invert the Laplace transform

$$i_2(t) = E_0\left[2e^{-t}\frac{46\cos 3t + \frac{37}{3}\sin 3t}{765} - \frac{e^{-10t}}{18} - \frac{11\cos 10t + 7\sin 10t}{170}\right]$$

Note that the steady-state current $i_2(t)$ is equal to

$$i_2(t) = -\left(\frac{E_0}{\sqrt{170}}\right)\sin(10t + \varphi),$$

where $\varphi = \arctan 11/7$. Then the steady-state voltage drop across the capacitor is given by

$$\frac{1}{C}\int^t i_2 = \left(\frac{E_0}{\sqrt{170}}\right)\cos(10t + \varphi)$$

PROBLEM 4.14

Suppose that the $n \times n$ matrix A has n independent eigenvectors X_1 to X_n corresponding to eigenvalues λ_1 to λ_n. Show that for a given initial vector C, the solution of the initial value problem

$$\frac{d}{dt}U(t) = AU(t), \quad U(0) = C \tag{1}$$

is given by

$$U(t) = c_1 e^{\lambda_1 t} X_1 + \cdots + c_n e^{\lambda_n t} X_n, \tag{2}$$

where $\boldsymbol{c} = [c_1, \ldots, c_n]^{\mathrm{T}}$ satisfies $P\boldsymbol{c} = \boldsymbol{C}$ for P, the matrix whose columns are the eigenvectors of A.

SOLUTION 4.14

The assumption that A has n linearly independent eigenvectors implies that P^{-1} exists. Then it follows that

$$A = PDP^{-1} \tag{3}$$

and

$$\frac{d}{dt}U(t) = PDP^{-1}U(t)$$

Because $P^{-1}P = I$ and P^{-1} is independent of t, multiplying both sides of this equation by P^{-1} leads to

$$\frac{d}{dt}P^{-1}U(t) = DP^{-1}U(t)$$

Now we let

$$V(t) = P^{-1}U(t) \tag{4}$$

Then $V(t)$ solves the initial value problem

$$\frac{d}{dt}V(t) = DV(t) \quad V(0) = P^{-1}\boldsymbol{C} = \boldsymbol{c} \tag{5}$$

Since the matrix D is a diagonal matrix, each differential equation in the system (5) contains only one unknown function; thus the equations are said to be *uncoupled*. For $k = 1, \ldots, n$,

$$\frac{d}{dt}v_k(t) = \lambda_k v_k(t) \quad v_k(0) = c_k \tag{6}$$

Note: The system (1) is a system of *coupled* differential equations; each equation may contain as many as n of the unknown functions. The change of variable (4) reduces the system to the *uncoupled* system (5), which is equivalent to the n single equations (6). Each of the equations (6) is readily solved; the solutions are

$$v_k(t) = C_k e^{\lambda_k t} \tag{7}$$

and

$$V(t) = \left[c_1 e^{\lambda_1 t}, \ldots, c_n e^{\lambda_n t}\right]^{\mathrm{T}} \tag{8}$$

It follows from (4) that $U(t) = PV(t)$ and, since the columns of P are the eigenvectors of A, it follows from (1.6) that

$$PV(t) = U(t) = c_1 e^{\lambda_1 t}\boldsymbol{X}_1, \ldots, c_n e^{\lambda_n t}\boldsymbol{X}_n$$

This establishes (2) and, incidentally, proves the first part of Theorem 4.7.

PROBLEM 4.15

Solve the initial value problem

$$\frac{d}{dt}U(t) = \begin{bmatrix} 2 & 3 \\ 1 & 4 \end{bmatrix} U(t), \quad U(0) = [5, 1]^T$$

SOLUTION 4.15

Using the techniques of chapter 2 we find the coefficient matrix for this system has distinct eigenvalues $\lambda_1 = 1$ and $\lambda_2 = 5$, with corresponding independent (but not orthogonal) eigenvectors

$$X_1 = \begin{bmatrix} -3 \\ 1 \end{bmatrix} \quad \text{and} \quad X_2 = \begin{bmatrix} 1 \\ 1 \end{bmatrix}$$

According to Theorem 4.7, the solution of the initial value problem is

$$U(t) = c_1 e^t X_1 + c_2 e^{5t} X_2$$

where the constants c_1 and c_2 are obtained by solving $Pc = C$; i.e.,

$$\begin{bmatrix} -3 & 1 \\ 1 & 1 \end{bmatrix} \begin{bmatrix} c_1 \\ c_2 \end{bmatrix} = \begin{bmatrix} 5 \\ 1 \end{bmatrix}$$

This system may be solved (using Cramer's rule, for example) to obtain $c_1 = -1$ and $c_2 = 2$. Then

$$U(t) = 2X_2 e^{5t} - X_1 e^t$$

i.e.,

$$\begin{bmatrix} u_1(t) \\ u_2(t) \end{bmatrix} = \begin{bmatrix} 2e^{5t} + 3e^t \\ 2e^{5t} - e^t \end{bmatrix}$$

PROBLEM 4.16

In Problem 1.3 we derived the discrete model for 1-dimensional time dependent diffusion. In a similar fashion, we can show that if we let $u_m(t)$ denote the concentration of contaminant in cell number m at time t then $U(t) = [u_1(t), \ldots, u_n(t)]^T$ satisfies

$$\frac{d}{dt}U(t) = rAU(t), \quad U(0) = C \tag{1}$$

where $r = D/w^2$, $C = [C_1, \ldots, C_n]^T$ is a vector containing the initial concentrations in cells 1 through n, and if we suppose the ends of the tube are in contact with reservoirs in which the contaminant level is zero for all t, then A denotes the $n \times n$ symmetric matrix from Problem 2.5. Solve the initial value problem (1) and find the steady-state solution.

SOLUTION 4.16

Time dependent 1-dimensional diffusion may be modeled by an initial value problem of the form (1). We are assuming the diffusion tube has been divided into n cells of width w and $r = D/w^2$, where D denotes a material dependent parameter called the diffusivity. In Problem 2.5, we found the symmetric matrix A to have the following real eigenvalues:

$$\lambda_m = 2\left(1 - \cos\frac{m}{n+1}\right), \quad m = 1, \ldots, n \tag{2}$$

and corresponding mutually orthogonal eigenvectors

$$\boldsymbol{X}_m = \boldsymbol{K}_m\left[\sin\frac{m\pi}{n+1}, \ldots, \sin\frac{nm\pi}{n+1}\right]^{\mathrm{T}}, \quad m = 1, \ldots, n \tag{3}$$

The arbitrary constants K_m can be chosen so that $\boldsymbol{X}_m \cdot \boldsymbol{X}_m = 1$; i.e., so that the eigenvectors are normalized eigenvectors. Then, according to Theorem 4.7, the solution of (1) is given by

$$U(t) = C \cdot X_1 e^{r\lambda_1 t} X_1 + \cdots + C \cdot X_n e^{r\lambda_n t} X_n \tag{4}$$

Note that for each fixed n, the eigenvalues λ_m given by (2) satisfy

$$0 > \lambda_1 > \lambda_2 > \cdots > \lambda_n > -4. \tag{5}$$

Then, since the parameter r is positive, (5) implies that each of the exponential terms in (4) tends to zero as t tends to infinity. Thus, in the case that A is the matrix from Problem 2.5, the steady-state solution of (1) is $\boldsymbol{U} = \boldsymbol{0}$.

PROBLEM 4.17

Consider a system of n identical masses connected to one another and to a pair of rigid supports by $n + 1$ springs as shown in Figure 4.9. Let $x_k(t)$ denote the displacement of the kth mass from its equilibrium position. Then it is not hard to show that these displacements satisfy

$$\frac{d^2}{dt^2}X(t) = rAX(t) \tag{1}$$

where $r = K/m$ and A denotes the $n \times n$ matrix from Problem 2.5.

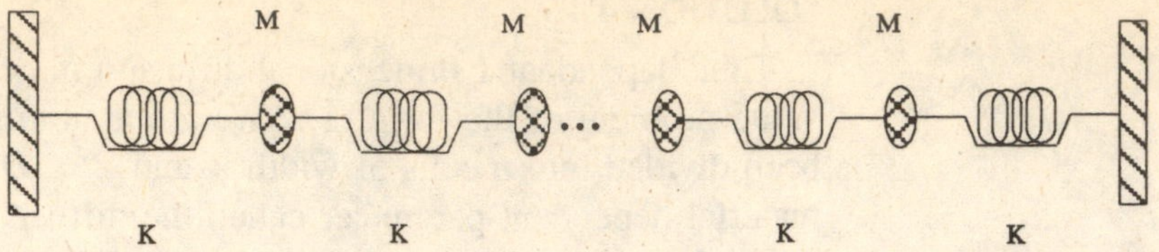

Figure 4.9
Spring - mass System With N Identical Springs and Masses
For Problem 4.17

Here m and K denote the common values of mass and spring stiffness for the system. If the displacement and velocity of each mass are given initially by

$$X(0) = d \quad \text{and} \quad X'(0) = v \tag{2}$$

then solve the second order initial value problem for $X(t)$.

SOLUTION 4.17

The eigenvectors of the symmetric matrix A are given by (3) in the previous problem. When the constants K_m are chosen such that $X_m \cdot X_m = 1$, these eigenvectors form an orthonormal basis for $\Re^n$. If we let P denote the matrix whose columns are these orthonormal eigenvectors, then $A = PDP^{-1}$, where D denotes the diagonal matrix whose diagonal entries are the eigenvalues λ_m of the matrix A. Then the system (1) can be written as

$$X''(t) = rPDP^{-1}X(t)$$

or

$$\frac{d^2}{dt^2}(P^{-1}X(t)) = rDP^{-1}X(t)$$

If we let $Y(t) = P^{-1}X(t)$ (3)

then the system has been reduced to

$$Y''(t) = rDY(t) \tag{4}$$

Since the matrix D in the system (4) is a diagonal matrix, each equation in the system contains just a single unknown function; i.e., the system is uncoupled. So, for $k = 1, \ldots, n$,

$$\frac{d^2}{dt^2}y_k(t) = r\lambda_k y_k(t)$$

Recalling from Problem 4.16 that the eigenvalues λ_k of the matrix A are all negative, let

$$\omega_k^2 = -r\lambda_k \quad \text{for } k = 1, \ldots, n \tag{5}$$

Then

$$y_k''(t) = -\omega_k^2 y_k(t) \tag{6}$$

and it follows at once that $y_k(t) = a_k \cos \omega_k t + b_k \sin \omega_k t$ for arbitrary constants a_k, b_k and $k = 1, \ldots, n$.

Then

$$Y(t) = [a_1 \cos \omega_1 t, \ldots, a_n \cos \omega_n t]^T + [b_1 \sin \omega_1 t, \ldots, b_n \sin \omega_n t]^T$$

and $X(t) = PY(t)$. Since the columns of P are the orthonormal eigenvectors of A it follows from (1.6) that

$$\begin{aligned} PY(t) = X(t) = a_1 \cos \omega_1 t X_1 + \cdots + a_n \cos \omega_n t X_n + \\ b_1 \sin \omega_1 t X_1 + \cdots + b_n \sin \omega_n t X_n. \end{aligned} \tag{7}$$

Since the eigenvectors X_k are an orthonormal set, it follows from (2) that

$$\begin{aligned} X(0) \cdot X_k = a_k = d \cdot X_k \quad &\text{for } k = 1, \ldots, n \\ X'(0) \cdot X_k = \omega_k \beta_k = b \cdot X_k \quad &\text{for } k = 1, \ldots, n \end{aligned} \tag{8}$$

Using these values for a_k and b_k in (7) produces a solution $X(t)$ that satisfies the system of equations (1) and the initial conditions (2).

Note that the system (1) of second order equations considered in this problem has the same coefficient matrix as the first order system considered in the previous problem. While the solution to the first order systems of the previous problem decayed steadily with increasing time, the solution to the second order system (1) oscillates in time with constant amplitude. Note also that the eigenvalues λ_k of the coefficient matrix are related to the natural frequencies ω_k of the spring-mass system by (5) and that there are n different natural frequencies for this n degree of freedom system.

The linear homogeneous equation $L_n[y(t)] = 0$ where

$$L_n[y(t)] = y^{(n)}(t) + a_{n-1}(t)y^{(n-1)}(t) + \cdots + a_1(t)y'(t) + a_0(t)y(t)$$

has n linearly independent solutions (provided the coefficients are continuous functions of t). A set of functions is said to be linearly independent if none of the functions can be written as a linear combination of the others in the set. If the functions are sufficiently smooth then they are linearly independent on an interval I if and only if their Wronskian determinant is zero at all points of I.

If the coefficients in L_n are constant, then $y(t) = e^{rt}$ is a solution of the homogeneous equation $L_n[y(t)] = 0$ if and only if r is a root of the auxiliary equation $P_n(r) = 0$. The solutions of the homogeneous equation are *completely determined by finding all the zeroes of the characteristic polynomial $P_n(r)$.*

The Laplace transform is an effective method for solving the initial value problem for the inhomogeneous equation. Variation of parameters and undetermined coefficients are two other ways of dealing with inhomogeneous equations, but these methods are not discussed here. It is often helpful to simplify a transform expression by partial fractions before attempting the inversion. Alternatively, the convolution formula for the inverse of the product of two transforms may be used.

Systems of differential equations often arise in connection with modeling physical systems with more than one degree of freedom. When the number of equations is small, the Laplace transform is effective, but this method becomes unwieldy for large systems. In such cases, the solution to the system can be expressed in terms of the eigenvectors and eigenvalues of the coefficient matrix.

5

Nonlinear Ordinary Differential Equations

In the previous chapter we saw that the general solution to a linear differential equation of order n (or an $n \times n$ system of first order equations) contains n arbitrary constants of integration. Each choice of values for these constants produces a particular solution to the equation and every solution corresponds to some choice of values for the constants. The situation for nonlinear equations is less simple. Even for those special nonlinear equations where it is possible to construct solutions containing arbitrary constants, it may happen that there exist other solutions to the equation that correspond to no value of the constants. In general it is not even possible to find closed form solutions for nonlinear differential equations in terms of elementary functions. For that reason we will often have to be content with deducing qualitative properties of the solution from observations about the equation or system of equations.

It is often possible, for example, to determine the ultimate behavior of a nonlinear system without knowing precisely how the system evolves. This is the question of asymptotic stability. Similarly we may be able to determine if periodic solutions to a system exist without being able to construct such a solution exactly. Some simple tools for considering such questions are presented in this chapter.

For nonlinear equations where the exact solution cannot be found and a qualitative description of the solution is insufficient, we may construct an approximate solution by numerical methods. Numerical solutions are discussed in Chapter 12 and have been used here to construct the figures in the solved problems which relate to nonlinear systems.

NONLINEAR EQUATIONS

The most general differential equation of order n in the unknown function $y(t)$ and independent variable t is of the form $F(t, y, y', \ldots, y^{(n)}) = 0$. In general this equation has no solution expressible in terms of elementary functions, even in the case $n = 1$. However, there are a few special classes of nonlinear equations for which a solution may be found.

Separable Equations

A differential equation that can be written in the form $y'(t) = f(y)g(t)$ is said to be *separable*. Since $y'(t)dt = dy$, it follows that

$$\int \frac{dy}{f(y)} = \int g(t)\, dt$$

i.e.,

$$F(y(t)) = G(t)$$

where $G' = g$ and $F' = 1/f$.

Exact Equations

Suppose that for smooth functions $F = F(t, y)$ and $y = y(t)$, we have $F(t, y(t)) = 0$ for t in some interval I. Then, by the chain rule,

$$\frac{dF}{dt} = \partial_t F(t, y) + \partial_y F(t, y)\, y'(t) = 0 \quad \text{for } t \text{ in } I$$

and $y'(t) = -\dfrac{\partial_t F}{\partial_y F}$

Now consider a differential equation of the form $y'(t) = -P(t, y)/Q(t, y)$ for given functions $P = P(t, y)$ and $Q = Q(t, y)$. Then this equation is said to be *exact* if there exists a smooth function $F = F(t, y)$ such that $P = \partial_t F$ and $Q = \partial_y F$. In this case, the differential equation has a solution that can be written in the implicit form, $F(t, y(t)) = 0$. Note that in order to have $P = \partial_t F$ and $Q = \partial_y F$ for a smooth F, it is necessary that P and Q satisfy $\partial_y P = \partial_t Q$. If this condition is satisfied, then the equation is exact and has a solution of the form $F(t, y(t)) = 0$.

Autonomous Equations

An equation of the form $y'(t) = f(y)$ is said to be *autonomous*. Clearly, first order autonomous equations are a subset of the separable equations and are, therefore, solvable by the means described previously. A second order autonomous equation has the form $y''(t) = f(y, y')$. If we let $u(t) =$

$y'(t)$, then

$$y'(t) = u'(t) = \frac{du}{dy}\frac{dy}{dt} = u\frac{du}{dy} = f(y, u)$$

This first order equation in the unknown function $u = u(y)$ may now be solvable by one of the previously discussed methods. If it is, then $y'(t) = u(y)$ is an autonomous equation for $y = y(t)$.

Bernoulli Equations

A nonlinear equation of the form $y'(t) + p(t)y(t) = q(t)y(t)^r$ is called a *Bernoulli equation*. Equations of this special form occur in connection with various applications. The equation is linear in the special cases $r = 0$ and $r = 1$; for all other values of r, it is nonlinear but the change of variable $u(t) = (y(t))^{1-r}$ reduces the Bernoulli equation to a linear equation for $u = u(t)$:

$$u'(t) + p(t)u(t) = (1 - r)q(t)$$

RICCATI EQUATIONS

Another important special class of nonlinear equations is the class of *Riccati equations* which have the form $y'(t) = f(t) - p(t)y(t) + q(t)y(t)^2$. When $q = 0$, this equation is linear, and when $f = 0$, it is a Bernoulli equation. Apart from these two special cases, there is no general solution method for Riccati equations. However, if one solution $y = Y(t)$ to the Riccati equation can be found, then the change of variable $u(t) = y(t) - Y(t)$ reduces the Riccati equation to a Bernoulli equation for $u = u(t)$,

$$u'(t) = (2q(t)Y(t) - p(t))u(t) + q(t)u(t)^2$$

NONLINEAR SYSTEMS

We consider nonlinear systems of the form

$$X'(t) = F(X(t), t) \tag{5.1}$$

for $X = [x_1(t), \ldots, x_n(t)]^T$ and $F = [F(X,t), \ldots, F_n(X, t)]^T$. Here for $m = 1, \ldots, n$, F_m denotes a given smooth function of $n + 1$ variables and the functions $x_m(t)$ are the unknowns in the system of equations (5.1). If $n = 1$, then (5.1) reduces to the scalar case considered in the previous section.

Autonomous Systems

The system (5.1) is said to be a *nonautonomous system*. By contrast, a system in which F does not depend explicitly on the independent variable t is said to be *autonomous*. An autonomous system has the form

$$X'(t) = F(X(t)) \tag{5.2}$$

i.e.,
$$x_1'(t) = F_1(x_1(t), \dots, x_n(t))$$
$$\vdots$$
$$x_n'(t) = F_n(x_1(t), \dots, x_n(t))$$

Note that the introduction of the new dependent variable $x_{n+1}(t)$ such that $x_{n+1}'(t) = 1$ reduces a nonautonomous system of order n to the following autonomous system of order $n + 1$:

$$x_1'(t) = F_1(x_1(t), \dots, x_n(t), x_{n+1}(t))$$
$$\vdots$$
$$x_n'(t) = F_n(x_1(t), \dots, x_n(t), x_{n+1}(t))$$
$$x_{n+1}'(t) = 1$$

Because nonautonomous systems can always be thus replaced by an autonomous system of one order higher, we will focus our analysis on autonomous systems.

Existence of a Solution

For each initial vector C, the linear initial value problem $X'(t) = AX(t)$, $X(0) = C$ has a unique solution $X(t)$ existing for all t. On the other hand, a solution for a nonlinear initial value problem may develop spontaneous singularities and, thus, fail to exist for all t.

Theorem 5.1

Theorem 5.1. Consider the nonlinear initial value problem $X'(t) = F(X)$, $X(0) = C$, where $X = [x_1(t), \dots, x_n(t)]^{\mathrm{T}}$ and $F = [F_1(x_1, \dots, x_n), \dots, F_n(x_1, \dots, x_n)]^{\mathrm{T}}$.

(a) If $F(X)$ is sufficiently smooth in a neighborhood of $X = C$, then there exists a positive number T such that a solution $X(t)$ exists for $-T < y < T$.

(b) For each initial vector C, there exists a maximum interval of existence, $T_0 < t < T_1$, on which the initial value problem has a unique solution. If T_1 is finite, then for every positive number M, there is a t in (T_0, T_1) such that $\|X(t)\| > M$.

EXAMPLE 1

Theorem 5.1 states that if $F(X)$ is smooth, then a solution exists in some interval of positive length, and that if the maximum length of the interval is finite, then it must happen that the solution becomes unbounded. We have not been precise about the smoothness requirements on F that are needed to assure existence of a solution to the initial value problem. While it is necessary that F have some degree of smoothness, this example shows that smoothness is not the controlling factor in the development of spontaneous singularities. We consider the initial value problem

$$x'(t) = x(t)^2 \qquad x(0) = C > 0$$

The solution is easily found to be $x(t) = C/(1 - Ct)$ which exists only on the interval $(-\infty, 1/C)$. Thus, in spite of the fact that $F(x) = x^2$ is infinitely smooth, the solution develops a spontaneous singularity at a point which depends on the initial value C.

ORBITS

For functions $x_1(t), \ldots, x_n(t)$ defined and smooth for t in the interval I, the set of points $\Gamma = \{X(t) = [x_1(t), \ldots, x_n(t)]^T : t \text{ in } I\}$ describes a curve in $\Re^n$. Then Γ is said to be an *orbit* or *trajectory* or *solution curve* of the system (5.2) if the functions $x_m(t)$ solve the system of differential equations. We sometimes say that Γ is a curve parameterized by t in the n-dimensional *phase space*.

We will devote much of our attention to system (5.2) in the special case $n = 2$ where it will be possible to visualize the solution behavior by plotting solution curves in the x_1x_2 plane. Because the number of equations and unknowns is just two, it will be convenient to dispense with the subscript notation in this case and consider systems of the form

$$x'(t) = P(x, y) \quad \text{and} \quad y'(t) = Q(x, y) \tag{5.3}$$

Then $\Gamma = \{(x,y): x = x(t), y = y(t), t \text{ in } I\}$ is a curve in the xy plane and Γ is an orbit of the system (5.3) if $x(t)$, $y(t)$ are solutions for the equations (5.3) for all t in I. We refer to the xy plane as the *phase plane*.

We suppose the interval I contains $t = 0$ and if $x(0) = x_0$, $y(0) = y_0$, then we refer to $\Gamma = \{(x,y): x = x(t), y = y(t), t \text{ in } I\}$ as the orbit through the point (x_0, y_0).

EXAMPLE 2

We consider the autonomous system

$$x'(t) = y$$
$$y'(t) = -x$$

and note that we have then $xx'(t) + yy'(t) = 0$. But

$$xx'(t) + yy'(t) = \frac{1}{2}\frac{d}{dt}(x^2(t) + y^2(t))$$

and it follows that $x^2(t) + y^2(t)$ = constant on solution curves of this system. Then the orbits of this system are a family of concentric circles with center at the origin. Note that the orbits are traversed in the clockwise direction since $x'(t)$ is positive (and thus x is increasing) when y is positive. Similarly, x is decreasing when y is negative.
Note also that distinct orbits of this system do not intersect. This is true for autonomous systems in general since to each point $\boldsymbol{X}(t)$ on a solution curve, there corresponds a unique tangent vector $\boldsymbol{X}'(t)$ equal to $\boldsymbol{F}(\boldsymbol{X}(t))$. Similarly, an orbit of an autonomous system does not cross itself unless that orbit is a simple closed curve. A closed orbit corresponds to a periodic solution to the system of equations.

Linearization

CRITICAL POINTS

Although it is generally not possible to find a solution for a nonlinear system, we can often learn a great deal about the solution by examining the orbits near special points in the phase space called *critical points*. The point $\boldsymbol{X}$ in $\Re^n$ is a critical point for the system (5.2) if $\boldsymbol{F}(\boldsymbol{X}) = \boldsymbol{0}$ and (a, b) is a critical point for the system (5.3) if $P(a, b) = Q(a, b) = 0$. Note that $\boldsymbol{X}'(t) = \boldsymbol{0}$ at a critical point, so that the system is at rest at a critical point. We can show further that a point moving along an orbit passing through a critical point cannot reach the critical point in a finite time. For a linear system, the one and only critical point is the origin. Critical points are also called *equilibrium points*.

APPROXIMATION BY A LINEAR SYSTEM

Suppose (a, b) is a critical point for the system (5.3). We can expand $P(x, y)$ and $Q(x,y)$ in a Taylor series about the point (a, b) to obtain

$$\begin{aligned} P(x,y) &= P(a,b) + \partial_x P(a,b)\,(x-a) + \partial_y P(a,b)\,(y-b) + \cdots \\ &= 0 + \partial_x P(a,b)\,(x-a) + \partial_y P(a,b)\,(y-b) + \cdots \\ Q(x,y) &= Q(a,b) + \partial_x Q(a,b)\,(x-a) + \partial_y Q(a,b)\,(y-b) + \cdots \\ &= 0 + \partial_x Q(a,b)\,(x-a) + \partial_y Q(a,b)\,(y-b) + \cdots \end{aligned}$$

Then since a and b are constant, we can approximate the system (5.3) in a neighborhood of the critical point by the linear system

$$\frac{d}{dt}(x(t)-a) = \partial_x P(a,b)\,(x-a) + \partial_y P(a,b)\,(y-b)$$
$$\frac{d}{dt}(y(t)-b) = \partial_x Q(a,b)\,(x-a) + \partial_y Q(a,b)\,(y-b)$$

JACOBIAN MATRIX

This linear system can be written in matrix notation as

$$\mathbf{Z}'(t) = J(a,b)\mathbf{Z}(t) \tag{5.4}$$

where $\mathbf{Z}(t) = [x(t)-a, y(t)-b]^T$ and $J(a,b)$ denotes the *Jacobian matrix*

$$J(a,b) = \begin{bmatrix} \partial_x P(a,b) & \partial_y P(a,b) \\ \partial_x Q(a,b) & \partial_y Q(a,b) \end{bmatrix} \tag{5.5}$$

This linear system (5.4) is called the *linearization* of the system (5.3) about the critical point (a, b). Since $\mathbf{Z} = \mathbf{0}$ corresponds to $x = a$ and $y = b$, it follows that the orbits of the linear system (5.4) near the origin may approximate the orbits of the nonlinear system (5.3) near the critical point (a, b). There are a limited number of possible configurations for the orbits of (5.4) near $\mathbf{Z} = \mathbf{0}$.

ORBITS OF LINEAR SYSTEMS

Consider the linear system $X'(t) = AX(t)$ for A a 2×2 real matrix with distinct eigenvalues λ_1 and λ_2 and corresponding independent eigenvectors X_1 and X_2. Then the general solution of this system is given by

$$X(t) = C_1 e^{\lambda_1 t} X_1 + C_2 e^{\lambda_2 t} X_2$$

or, if $\lambda_{1,2} = p \pm iq$, by

$$X(t) = e^{pt}(C_1 \cos qtX_1 + C_2 \sin qtX_2)$$

The origin is the only critical point for this linear system and the behavior of the orbits is determined by the character of the eigenvalues λ_1 and λ_2. Note that since A is real, complex eigenvalues can only occur in conjugate pairs.

Real Eigenvalues

1. Unstable Node: $0 < \lambda_1 < \lambda_2$. In this case the positive eigenvalues cause the length of $X(t)$ to increase without bound as t increases; all orbits escape to infinity. The critical point at the origin is classified as

an *unstable node* when A has only real, positive eigenvalues.

2. Saddle Point: $\lambda_1 < 0 < \lambda_2$. Since λ_1 is negative, orbits approach the origin in the direction of the eigenvector X_1 and orbits escape to infinity along the direction X_2 corresponding to the positive eigenvalue λ_2. The critical point at the origin is said to be a *saddle point* if A has both positive and negative real eigenvalues.
3. Stable Node: $\lambda_1 < \lambda_2 < 0$. When the eigenvalues are real and both negative, the length of $X(t)$ decreases to zero with increasing t; all orbits approach the origin. We say the origin is a *stable node*.
4. Degenerate Node: $\lambda_1 = \lambda_2$. repeated eigenvalue with geometric multiplicity equal to 1. When there is only a single eigenvector X, all orbits approach the origin or escape along this direction according to whether the eigenvalue is negative or positive.

Complex Eigenvalues

5. Center: Re $\lambda = 0$ (i.e., $\lambda_{1,2} = \pm iq$). When the eigenvalues have real part equal to zero, the solution $X(t)$ is periodic and the orbits are closed curves. The critical point at the origin is said to be a *center* when the eigenvalues of A have zero real part.
6. Stable Focus: Re $\lambda < 0$. If the complex eigenvalues of A have negative real part then the orbits spiral in toward the origin with increasing t. We say that the origin is a *stable focus*.
7. Unstable Focus: Re $\lambda > 0$. When A has complex eigenvalues with positive real part, the orbits spiral outward, away from the origin, and we say that the critical point is an *unstable focus*.

Typical orbit configurations for these seven cases are pictured in the solved problems.

HYPERBOLIC POINTS

We can similarly classify critical points for the system (5.2) in terms of the eigenvalues of the Jacobian matrix

$$J(X) = \begin{bmatrix} \partial_1 F_1 & \partial_2 F_1 \ldots \partial_n F_1 \\ \partial_1 F_2 & \ldots \\ \partial_1 F_n & \ldots \partial_n F_n \end{bmatrix}$$

where $\partial_j F_m$ denotes the partial derivative of F_m with respect to the jth variable x_j and the entries in the Jacobian $J(X)$ are all evaluated at the point X. We find that it is convenient to distinguish between critical points where purely imaginary eigenvalues occur and those where this does not happen. The critical point X_0, where $F(X_0) = \mathbf{0}$, is said to be a *hyperbolic critical point* if none of the eigenvalues of $J(X_0)$ has real part equal to zero. Then the hyperbolic critical point X_0 is further classified as a:

1. *Sink* if Re $\lambda < 0$ for all eigenvalues λ of the Jacobian matrix $J(X_0)$;
2. *Source* if Re $\lambda > 0$ for all eigenvalues λ of the Jacobian matrix $J(X_0)$;
3. *Saddle* if $J(X_0)$ has at least one eigenvalue with positive real part and at least one with negative real part.

Stability

EQUILIBRIUM STATES

Many physical systems are modeled by autonomous systems of differential equations. For such systems, a critical point represents an *equilibrium state* of the physical system. Once the system has been placed in a state corresponding to a critical point, it will remain in that state. Some equilibrium states are such that if the system is disturbed slightly when it is in the equilibrium state, the system returns to the equilibrium state. Such states are said to be *stable*. If a slight disturbance of the system in an equilibrium state causes the system to assume a different state, then the state is said to be *unstable*. A motionless pendulum hanging straight down is in a stable equilibrium state. If the pendulum is pulled slightly to one side it swings back and forth until friction eventually brings it to rest again hanging motionless straight down. A long rod, balanced on end is in an unstable equilibrium state. Pushing the rod to one side will cause it to topple and come to rest in a new equilibrium state. For mathematical purposes, we need more precise definitions of these terms.

ASYMPTOTIC STABILITY

Let $\Gamma = \{(x, y)\colon\ x = x(t), y = y(t)\}$ denote an orbit for the system (5.3) and suppose that (a, b) is a critical point for this system. For $(x(t), y(t))$ on Γ, let

$$d(t) = \sqrt{(x(t) - a)^2 + (y(t) - b)^2}$$

i.e., $d(t)$ is the distance between the critical point and the point $(x(t), y(t))$ on the orbit Γ. Then the critical point (a, b) is said to be *stable* if for every positive number ε there exists another positive number δ such that $d(0) < \delta$ implies $d(t) < \varepsilon$ for all $t > 0$. We can also define a more restrictive type of stability. We say that the critical point is *asymptotically stable* if it is stable and, in addition, there is a positive number η such that $d(0) < \eta$ implies that $d(t)$ tends to zero as t tends to infinity.

If a critical point is stable, then any orbit that begins close to the critical point will remain near the critical point. The critical point is asymptotically stable if it is stable and, in addition, any orbit that starts out close to the critical point is eventually "attracted to" the critical point. An asymptotically stable critical point is said to be an *attractor* for all orbits sufficiently near it. Any critical point that is not stable is said to be *unstable*.

A stable node and stable focus are asymptotically stable critical points for a linear autonomous system. A center is stable but not asymptotically stable. The unstable node and focus and the saddle are examples of unstable critical points for linear autonomous systems. As to the stability of critical points for nonlinear autonomous systems, we have the following theorem.

Theorem 5.2

Theorem 5.2. Let $X = (a, b)$ be a critical point for the autonomous system (5.3) and let (5.4) denote the linearization of (5.3) about this critical point. If $\mathbf{Z} = \mathbf{0}$ is an asymptotically stable critical point for (5.4), then (a, b) is an asymptotically stable critical point for (5.3). If $\mathbf{Z} = \mathbf{0}$ is an unstable critical point for (5.4), then (a, b) is an unstable critical point for (5.3).

In short, Theorem 5.2 says that if X is a hyperbolic critical point for (5.3) then X is either asymptotically stable or it is unstable and the stability of X is then the same as the stability of origin for the linearized system (5.4); i.e., sinks are asymptotically stable, sources and saddles are unstable. Only nonhyperbolic critical points can be stable without being asymptotically stable. Determining the stability of a nonhyperbolic critical point is often a delicate matter.

Limit Cycles

There are a limited number of possibilities for the behavior of orbits of linear autonomous systems in the plane:

1. The orbit is attracted to the critical point at the origin.
2. The orbit is repelled by the critical point at the origin.
3. The orbit is a closed curve around the critical point at the origin.

A nonlinear autonomous planar system may have more than a single critical point and an orbit may exhibit any of these behaviors near each critical point. For a nonlinear system there is also an additional possibility:

4. The orbit is attracted to a closed curve around the critical point.

   The closed curve attracting the orbit is called a *limit cycle*. Limit cycles do not occur for linear systems. The following facts about limit cycles are based on topological considerations for the plane and, thus, do not extend to higher dimensions. We recall briefly that a region D in the plane is *connected* if any two points of D can be connected by a curve lying inside D and we say D is *simply connected* if any simple closed curve in D encloses only points of D.

Theorem 5.3

Theorem 5.3. Consider the nonlinear autonomous planar system (5.3) where P and Q are assumed to be smooth in a connected and simply connected set D in the plane. Then

(a) If $\partial_x P + \partial_y Q$ does not change sign in D, then there can be no closed orbit for (5.3) lying entirely inside the domain D.
(b) Every closed orbit for (5.3) must contain at least one critical point. If the closed orbit contains just one critical point, then the critical point cannot be a saddle. If the critical points inside the closed orbit are all hyperbolic, then there must be an odd number of them.
(c) If D_1 is a closed bounded subset of D containing no critical points and Γ is an orbit which originates in D_1 and remains in D_1 for all t, then either Γ is a closed orbit or else Γ spirals toward a closed curve in D_1. In this case the closed curve attracting Γ is a limit cycle.

Part (a) of the theorem is a simple consequence of Green's theorem in the plane; part (b) is based on a topological approach to differential equations called degree theory. Finally, part (c) is a version of the so-called Poincare-Bendixson theorem. The set D_1 in part (c) is connected because it contains an orbit; but if it contains a closed orbit and no critical points, then it cannot be simply connected because of part (b) of the theorem. Thus, D_1 must be annular.

EXAMPLE 3

Consider the second order nonlinear equation $x''(t) + p(x)x'(t) + Kx = 0$ where the function $p(x)$ is assumed to be positive for all x. This equation describes the motion of a mass on a spring with a nonlinear damping force and it is equivalent to a system of the form (5.3) with $P(x, y) = y$ and $Q(x, y) = -Kx - p(x)y$. Then $\partial_x P + \partial_y Q = -p(x)$ is negative for all x, and, by part (a) of Theorem 5.3, this system has no periodic orbits.

In the case of the so-called VanderMonde oscillator, $p(x)$ is given by $p(x) = C(1 - x^2)$, and because this function changes sign, periodic orbits for this system are possible. The only critical point for this system is an unstable focus at the origin and, thus, any periodic orbit must contain the origin and cross into at least one of the two regions $R_1 = \{(x, y)\colon\ x < -1\}$ and $R_2 = \{(x,y)\colon\ x > 1\}$. See also Problem 5.11.

SOLVED PROBLEMS

Nonlinear Equations

PROBLEM 5.1

Find a family of solutions containing one arbitrary constant for the nonlinear differential equation $y'(t) = 3(y(t))^{2/3}$. Show that this equation has another solution which corresponds to no value of the constant.

SOLUTION 5.1

This differential equation is separable. Thus

$$\int \frac{dy}{3y^{2/3}} = \int dt$$

and integration leads to $y(t)^{1/3} = t + C$, or $y(t) = (t+C)^3$. This is a family of solutions producing one solution for each choice of the constant C. Note that $y(t) = 0$ is also a solution for the differential equation, and this solution is not obtained from $y(t) = (t+C)^3$ for any choice of the constant C.

PROBLEM 5.2

Solve the following nonlinear equation:

$$y(t)\,y'(t) = t\left(y(t)^{-2} - y(t)^2\right)$$

SOLUTION 5.2

This equation can be rewritten in the form of a Bernoulli equation with $n = -3$,

$$y'(t) + ty(t) = ty(t)^{-3}$$

i.e.,

$$4y(t)^3 y'(t) + 4ty(t)^4 = 4t$$

Then the change of variable $u(t) = y(t)^4$ reduces this to the linear equation

$$u'(t) + 4tu(t) = 4t$$

with the general solution $u(t) = y(t)^4 = Ce^{-2t^2} + 1$.

PROBLEM 5.3

Solve the following nonlinear equation: $y'(t) = 1 - \frac{y}{t} + \left(\frac{y}{t}\right)^2$

SOLUTION 5.3

This equation has the form of a Riccati equation and $Y(t) = t$ is seen to be one solution. Then substituting $y(t) = u(t) + t$ into the differential equation leads to

$$u'(t) = t^{-1}u(t) + t^{-2}u(t)^2$$

which is a Bernoulli equation with $n = 2$ for $u(t)$. Then $v(t) = u(t)^{-1}$ is seen to satisfy

$$v'(t) - t^{-1}v(t) = t^{-2}$$

This linear equation is found to have the general solution $v(t) = Ct - 1/(2t)$ and this leads to $u(t) = 2t/(2Ct^2 - 1)$. Then for each choice of the constant C,

$$y(t) = t + \frac{2t}{2Ct^2 - 1}$$

solves the original nonlinear equation.

Orbits of Linear Systems

PROBLEM 5.4

We saw in Problem 4.4 that $x(t)$, the unforced displacement of a mass on a damped spring, satisfies

$$mx''(t) = -Kx(t) - Cx'(t) \tag{1}$$

Letting $y(t) = x'(t)$ reduces this second order equation to a first order system:

$$\begin{aligned} x'(t) &= y(t) \\ y'(t) &= -(K/m)\,x(t) - (C/m)\,y(t). \end{aligned} \tag{2}$$

For $C = 0$, the undamped case, plot the orbits through $(a_k, 0)$, $k = 1, 2, 3$, where $0 < a_1 < a_2 < a_3$.

SOLUTION 5.4

The system (2) is a linear system with coefficient matrix

$$A = \begin{bmatrix} 0 & 1 \\ -\frac{K}{m} & -\frac{C}{m} \end{bmatrix}$$

The origin is the single critical point for a linear system and the character of the critical point is determined by the eigenvalues of A. If we let $C/m = 2b$ and $K/m = \omega^2$, then the eigenvalues of A are equal to

$$\lambda_{1,2} = -b \pm i\sqrt{\omega^2 - b^2} = -b \pm i\omega_0 \tag{3}$$

If $C = 0$, then $b = 0$, and the eigenvalues in this case are $\lambda_{1,2} = \pm i\omega$.

Because the eigenvalues of A are complex with real part equal to zero, the critical point at the origin is a *center*, and the orbits are going to be closed curves. To see what the orbits are, note that for $b = 0$, the system (2) has the form

$$\begin{aligned} x'(t) &= y \\ y'(t) &= -\omega^2 x \end{aligned}$$

Multiplying the first equation by $2\omega^2 x$ and the second by $2y$, we can add them to obtain

$$2\omega^2 xx' + 2yy' = \frac{d}{dt}(\omega^2 x^2 + y^2) = 0$$

It follows from this that

$$\omega^2 x^2 + y^2 = \text{constant} \tag{4}$$

The expression (4) is called a *first integral* for equation (1) and the system (2). For each choice of a value for the constant, the locus of points (x, y) in the plane for which (4) is satisfied is an ellipse. Each such ellipse is an orbit for the system (2). Choosing the constant in (4) such that constant $= \omega^2 a_k^{\ 2}$ for $k = 1, 2, 3$ produces the elliptic orbit through the points $(a_k, 0)$, $k = 1, 2, 3$. These initial states correspond to the situation in which the mass is given an initial displacement equal to a_k and is then released from rest. The orbits describe the subsequent motion of the spring-mass system. These orbits are pictured in Figure 5.1.

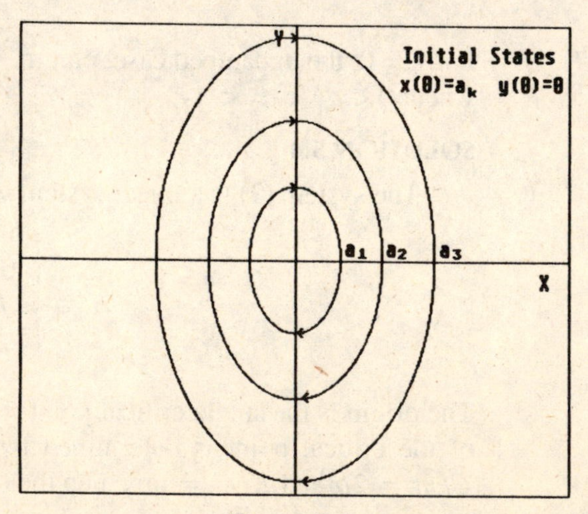

Figure 5.1
Orbits near a Center

Note that there are two points on each of the three orbits where $x = 0$. According to the equation for the orbit,

$$\omega^2 x^2 + y^2 = \omega^2 a_k^{\ 2}, \quad k = 1, 2, 3 \tag{5}$$

these occur when $y = \pm\omega a_k$, $k = 1, 2, 3$. These points on the orbit correspond to the instant when the mass passes through the equilibrium position, moving first in one direction and then in the other. Note that the derivative x' has the same sign as y. Thus, x is increasing when y is positive and decreasing when y is negative. It follows that the elliptical orbits must be traversed in the *clockwise* direction.

Note also that there are two points on each orbit where $y = 0$. It follows from (4) that these occur when $x = \pm a_k$; i.e., the velocity of the oscillating mass equals zero when the mass is at the points of extreme amplitude on either side of the equilibrium position.

It is evident from Figure 5.1 that orbits that start near the origin remain in a neighborhood of thc origin. However, an orbit that starts at a point outside the origin remains a positive distance from the origin for all time. Thus, a center is stable, but it is not asymptotically stable.

PROBLEM 5.5

We continue to consider the linear system (2) of the previous problem. For the underdamped case, where C satisfies $0 < C^2 < 4Km$, plot the orbits through the points $(0, a_k)$, $k = 1, 2, 3$.

SOLUTION 5.5

We continue to use the notation associated with the linear system (2) of Problem 5.4. If $0 < C^2 < 4Km$, then $\omega^2 > b^2$ and, according to (3) of the previous problem, the eigenvalues of A are complex with negative real part. Then the critical point at the origin is a *stable focus*. We can solve this linear system with the initial conditions $x(0) = 0$, $y(0) = a_k$, to obtain

$$x(t) = \left(\frac{a_k}{\omega_0}\right) e^{-bt} \sin \omega_0 t$$

$$y(t) = a_k e^{-bt} \left(\cos \omega_0 t - \frac{b}{\omega_0} \sin \omega_0 t\right)$$

These are parametric equations for the orbit through the point $(0, a_k)$. While these are not the equations of any readily recognizable curve, it is evident that $x(t)$ and $y(t)$ oscillate with decreasing amplitude so that the curve is some sort of spiral, converging toward the origin. Figure 5.2 pictures the orbits for $k = 1, 2, 3$.

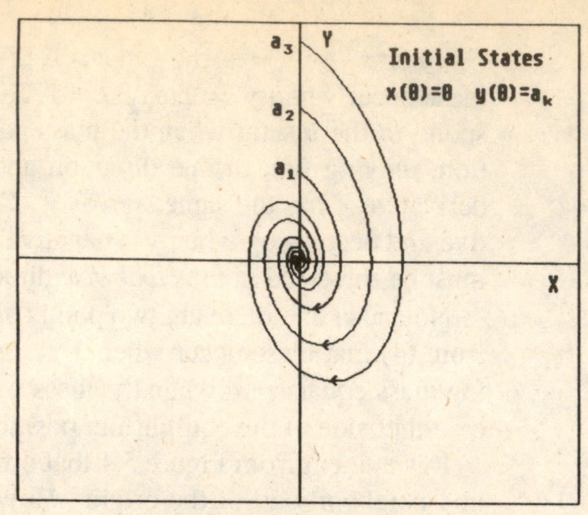

Figure 5.2
Orbits near a Stable Focus

Since the first equation in the system (2) implies that x' and y are of the same sign, the spirals, like the ellipses in part (a), go around the origin in the clockwise direction. Alternatively, it follows that since the eigenvalues have *negative* real part, points move along the orbits *toward* the origin; i.e., in the clockwise direction.

It appears from the figure that an orbit that starts near the origin remains in a neighborhood of the origin. Moreover, all orbits appear to eventually spiral into the origin. Thus, a stable focus is both stable and asymptotically stable.

Note that a point where an orbit crosses the x axis corresponds to a point where the mass has maximum amplitude and zero velocity, while a crossing of the y axis is a point where the mass is passing through the equilibrium position and its velocity at that point is locally maximal.

Each time one of the spiral orbits in Figure 5.2 crosses the x axis, the crossing is closer to the origin than the previous crossing. This corresponds to the fact that the damped spring-mass system oscillates with decreasing amplitude. Similarly, each time one of the spiral orbits crosses the y axis, the crossing is closer to the origin than the previous crossing of the y axis, i.e., the oscillating mass is slowing down due to damping.

PROBLEM 5.6

For the linear system (2) of Problem 5.4, show that the critically damped case, $C^2 = 4Km$, corresponds to a stable degenerate node at the origin. Plot several orbits.

SOLUTION 5.6

If $C^2 = 4Km$, then $\omega^2 = b^2$, and it follows from (3) of Problem 5.4

that the only eigenvalue for A is the negative number $\lambda = -b$. It is not difficult to show that dim $N[A + bI]$ is equal to one; hence, the origin is a *degenerate, stable node* for the linear system (2). We can easily solve the linear system in this case to obtain

$$x(t) = C_1 e^{-\omega t} + C_2 t e^{-\omega t}$$

$$y(t) = -\omega C_1 e^{-\omega t} + C_2 e^{-\omega t}(1 - \omega t)$$

Once again, for each choice of C_1 and C_2, these are the parametric equations of an orbit for the system (2), although the equations are not readily recognizable as the equations of any simple curve. Several orbits are plotted in Figure 5.3, and since λ is negative, the direction of motion along any orbit is always toward the origin. It may help to refer to solved Problems 4.4 and 4.5 at this point in order to reconcile the phase plane portrait in Figure 5.3 with the motion of a critically damped system.

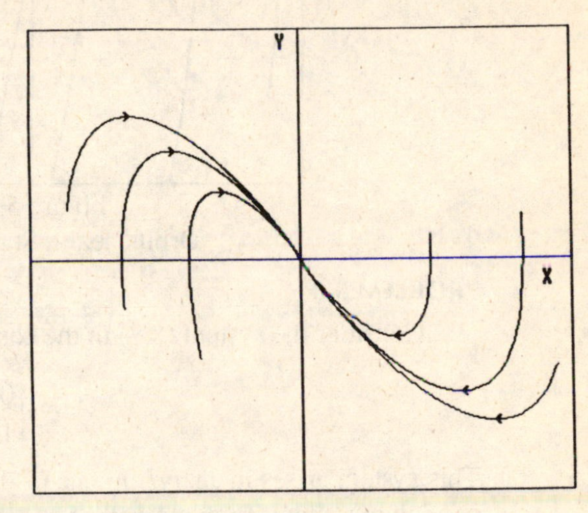

Figure 5.3
Orbits near a Degenerate Node

PROBLEM 5.7

For the linear system (2) of Problem 5.4, show that the overdamped case where $C^2 > 4Km$ corresponds to a stable node at the origin. Plot several orbits.

SOLUTION 5.7

In the overdamped case, $C^2 > 4Km$, we have $b^2 > \omega^2$, and then it follows from (3) of Problem 5.4 that A has two real, negative eigenvalues

$$\lambda_2 = -b - \sqrt{b^2 - \omega^2} < \lambda_1 = -b + \sqrt{b^2 - \omega^2} < 0$$

The solution in this case is given by

$$X(t) = C_1 e^{\lambda_1 t} X_1 + C_2 e^{\lambda_2 t} X_2 \tag{1}$$

and, since both eigenvalues are negative, it follows that $X(t)$ tends to **0** as t tends to infinity; i.e., all orbits approach the origin, which is a *stable node*. Several orbits are shown in Figure 5.4 where points move along the orbits toward the origin. The directions of the eigenvectors X_1 and X_2 are indicated in the figure.

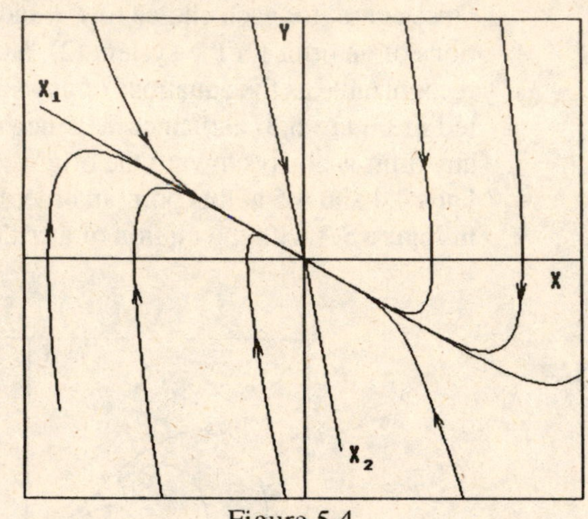

Figure 5.4
Orbits near a Stable Node

PROBLEM 5.8

Consider the system (8.7) in the case that A is given by

$$A = \begin{bmatrix} 0 & 1 \\ 1 & 0 \end{bmatrix}$$

This system arises in analyzing one of the singular points of the nonlinear pendulum system (Problem 5.9). Show that the origin is a saddle point for this linear system and sketch some of the orbits.

SOLUTION 5.8

The eigenvalues of the matrix A are easily found to be $\lambda_1 = -1$ and $\lambda_2 = 1$ with corresponding eigenvectors

$$X_1 = [1, -1]^T \quad \text{and} \quad X_2 = [1, 1]^T$$

Since the eigenvalues are real and of opposite sign, the origin is a *saddle point* for this system. The solution to the system is given by equation (1) from the previous problem from which it is clear that points will move along orbits *toward* the origin along the direction X_1, the eigenvector associated with the negative eigenvalue -1. Points will move *away* from the origin along the direction X_2, which is associated with the positive eigenvalue λ_2. Several orbits are pictured in Figure 5.5 on which the directions X_1 and X_2 are indicated. Since every orbit appears to escape eventually to infinity, a saddle point is an example of an *unstable* critical point.

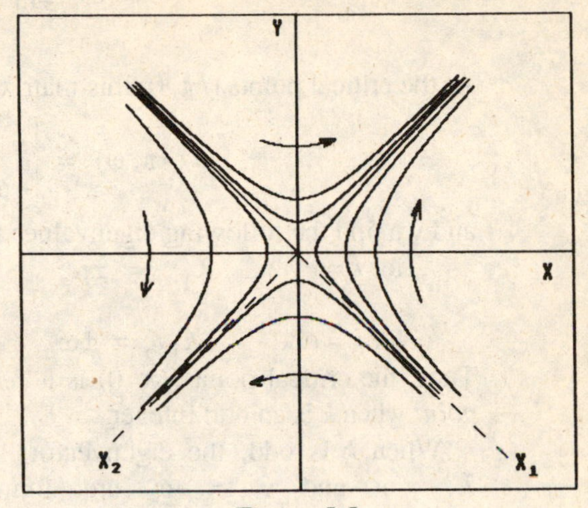

Figure 5.5
Orbits near a Saddle Point

Phase Plane Analysis of Nonlinear Systems

PROBLEM 5.9

We can show that the angular displacement ϑ of a mass m on a frictionless pendulum of length L is governed by the equation

$$mL\vartheta''(t) = -mg \sin \vartheta(t) \tag{1}$$

If we let $\varphi(t) = \vartheta'(t)$, then we can replace this equation by the first order autonomous system

$$\begin{aligned} \vartheta'(t) &= \varphi(t) \\ \varphi'(t) &= -\omega^2 \sin \vartheta(t) \end{aligned} \tag{2}$$

where $\omega^2 = g/L$. Find and classify all the critical points of this system. Sketch the orbits around the critical points and "fill in the blanks" between critical points to get an approximate global picture of the phase plane portrait for the nonlinear system.

SOLUTION 5.9

For this system we have $P(\vartheta, \varphi) = \varphi$ and $Q(\vartheta, \varphi) = -\omega^2 \sin \vartheta$, from which it follows that $P = Q = 0$ if and only if $\varphi = 0$ and $\vartheta = k\pi$, k = integer. Thus, (2) has infinitely many critical points located at integer multiples of π along the ϑ axis (horizontal axis). While linear systems have only one critical point, nonlinear systems may have any number of them (from zero to infinity). Computing the partial derivatives of P and Q

with respect to ϑ and φ, we find

$$J(\vartheta, \varphi) = \begin{bmatrix} 0 & 1 \\ -\omega^2 \cos \vartheta & 0 \end{bmatrix} \tag{1}$$

At the critical points $(k\pi, 0)$ this matrix reduces to

$$J(k\pi, \varphi) = \begin{bmatrix} 0 & 1 \\ -\omega^2 (-1)^k & 0 \end{bmatrix}$$

and we find the following eigenvalues for the matrix $J(k\pi, 0)$:

for k = even, $\lambda_{1,2} = \pm i\omega$

for k = odd, $\lambda_{1,2} = \pm\omega$

Thus, the critical point $(k\pi, 0)$ is a *center* when k is even, and a *saddle point* when k is an odd integer.

When k is odd, the eigenvectors corresponding to the eigenvalues $\lambda_1 = \omega$ and $\lambda_2 = -\omega$ are found to be $X_1 = [1, \omega]^T$ and $X_2 = [1, -\omega]^T$. Then orbits approach the saddle points from the direction X_2 and depart in the direction X_1. The saddle points are unstable critical points for the linearized system and, by Theorem 5.2, they are unstable for the nonlinear system as well.

The stability of the nonhyperbolic critical points corresponding to even values of k is not determined by the theorem. Instead, we multiply both sides of the equation (1) by ϑ' and note that

$$\frac{d}{dt}\left(\frac{\vartheta'(t)^2}{2}\right) = \vartheta'(t)\vartheta''(t) \quad \text{and} \quad \frac{d}{dt}(1 - \cos\vartheta(t)) = \vartheta'(t)\sin\vartheta(t)$$

Then (1) implies that for (ϑ, φ) a solution for (2), we have

$$E(\vartheta, \varphi) = (1/2)\varphi(t)^2 + \omega^2(1 - \cos \vartheta(t)) = \text{constant} \tag{8}$$

The quantity E is related to the total energy of the frictionless pendulum, and (8) states that the energy is constant on orbits. More precisely, the kinetic energy of the pendulum is proportional to $1/2\ \varphi(t)^2$, and the potential energy due to position is proportional to $\omega^2(1 - \cos \vartheta)$. So E is a first integral for the system (2) and the level curves of the function $E(\vartheta, \varphi)$ are the orbits for (2) near the critical points. Since these orbits are closed curves, the critical points are centers.

We can examine the orbits about the center at the origin for various values of the constant E. First, $E = 0$ implies that $\vartheta(t) = \varphi(t) = 0$ for all t;

i.e., the orbit corresponding to zero energy consists of the pendulum remaining motionless in the straight-down equilibrium position.
For small values of E, ϑ is small and it follows that $1 - \cos\vartheta \approx \vartheta^2/2$. Thus, for small values of E, we have E approximately equal to $(\varphi^2 + \omega^2\vartheta^2)/2$ and the level curves of this function are ellipses; i.e., the small amplitude orbits are elliptical closed curves corresponding to periodic oscillations of the pendulum. As the magnitude of E increases, the maximum values of ϑ and φ on the curves $E(\vartheta, \varphi) = E$ also increase, and the approximation of $(1 - \cos\vartheta)$ by $\vartheta^2/2$ becomes less and less accurate. The orbits becomes less and less like ellipses until finally, when E reaches the value $2\omega^2$, the corresponding orbit connects the saddle point at $(-\pi, 0)$ to the saddle point $(\pi, 0)$. A second orbit with $E = 2\omega^2$ connects the saddle point at $(\pi, 0)$ to the saddle point at $(-\pi, 0)$. Since the saddle points are critical points, it cannot happen that a point moving along one of these orbits will pass *through* the critical point onto the other orbit.

For values of E exceeding $2\omega^2$, the orbits are no longer closed curves, and the motion of the pendulum is then not periodic. For each value of E greater than $2\omega^2$, there are two possible orbits, one on which $\varphi = \vartheta'$ is always positive and one on which ϑ' is steadily negative. These correspond to orbits on which the pendulum revolves about the pivot point in the clockwise or counterclockwise direction, respectively. Physically, we interpret this to mean that if a frictionless pendulum is set in motion with sufficiently high energy, it will revolve about the pivot point in one direction for all time.

The phase plane portrait of the region $-2\pi < \vartheta < 2\pi, -4\omega^2 < \varphi < 4\omega^2\}$ is sketched in Figure 5.6. The orbits corresponding to $E = 2\omega^2$ are called *separatrices*. The orbits with $E < 2\omega^2$ are closed curves of more or less elliptic shape. These correspond to periodic solutions, while the orbits with $E > 2\omega^2$ are the wavelike curves running parallel to the horizontal axis; i.e., ϑ tends to plus or minus infinity according to whether φ is positive or negative. These two types of orbits are separated from one another by the separatrices. Since the separatrices pass through the saddle points at $(\pm\pi, 0)$, any motion that begins at a point on one of these curves will end up at one of the saddle points after an infinite amount of time. For example, the orbit that begins at the point $(0, 2\omega^2)$ moves in the clockwise direction toward $(\pi, 0)$. Since the velocity decreases toward zero as the critical point is approached, it can be shown that the critical point cannot be reached in a finite time. Similarly, an orbit beginning at $(0, -2\omega^2)$ would move toward $(-\pi, 0)$ but could not reach this point in finite time.

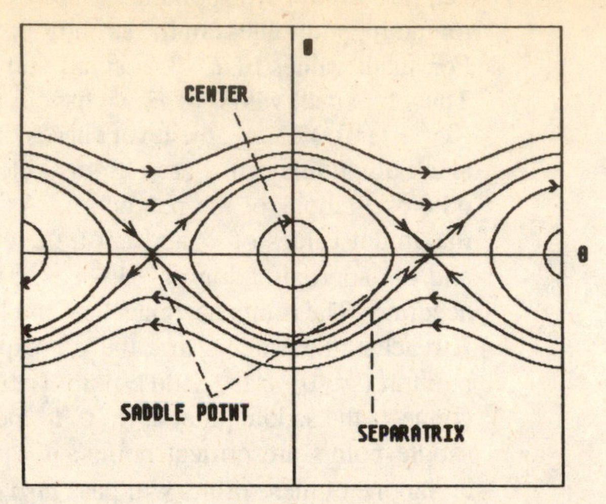

Figure 5.6
Phase plane for undamped nonlinear pendulum

PROBLEM 5.10

A two species population model in which the population denoted by $Y(t)$ preys on the population denoted by $X(t)$ is described by the system of equations

$$X'(t) = (b - \delta Y(t))X(t) \tag{1}$$
$$Y'(t) = (\beta X(t) - d)Y(t)$$

The equations in (1) express the fact that population X, the prey, has a constant birth rate b, but the death rate is directly proportional to the size of the predator population. Similarly, the Y population, the predators, has a constant death rate d, but the birth rate is directly proportional to the size of the prey population. Roughly speaking, this amounts to saying that there is an unlimited food supply for the prey population and that there is no population that is preying on the predators. Find all the equilibrium solutions to this system and determine whether any of them is stable.

SOLUTION 5.10

Note that if the predator population Y equals zero, then the prey population grows exponentially without bound. Similarly, if the prey population X equals zero, then the predator population decreases exponentially to zero. We are interested in determining whether there are any stable solutions other than these two extremes. In particular, we want to know if it is possible for the two populations to exist indefinitely without either of them dying out.

The equations

$$(b - \delta Y(t))X(t) = 0$$
$$(\beta X(t) - d)Y(t) = 0$$

are satisfied for $X = Y = 0$ and $X = d/\beta$, $Y = b/\delta$. We compute

$$J(X, Y) = \begin{bmatrix} b - \delta Y & -\delta Y \\ \beta Y & \beta X - d \end{bmatrix}$$

Then

$$J(0, 0) = \begin{bmatrix} b & 0 \\ 0 & -d \end{bmatrix}$$

and since the eigenvalues of this matrix are equal to $\lambda_{1,2} = b, -d$, it follows that the origin is a saddle point for this system. Note that the eigenvector corresponding to $\lambda_1 = b$ is the vector $\boldsymbol{E}_1 = [1, 0]^T$, and $\boldsymbol{E}_2 = [0, 1]^T$ is the eigenvector associated with the negative eigenvalue $\lambda_2 = -d$. Thus, if the predator population Y is zero, then the prey population grows to infinity along $\boldsymbol{E}_1$, whereas if the prey population X is zero, then the predator population decreases to zero along $\boldsymbol{E}_2$. So (0, 0) is an unstable, saddle-type equilibrium point.

Since negative values for either population have no physical significance, we restrict our attention to the first quadrant of the phase plane. The X and Y axes are orbits for this system, and since distinct orbits never cross, no orbit that begins in the first quadrant can cross into one of the other quadrants.

Now

$$J\left(\frac{d}{\beta}, \frac{b}{\delta}\right) = \begin{bmatrix} 0 & -\frac{d\delta}{\beta} \\ \frac{b\beta}{\delta} & 0 \end{bmatrix}$$

has eigenvalues

$$\lambda_{1,2} = \pm i\sqrt{bd}$$

hence, $(d/\beta,\ b/\delta)$ is a center for the linearization of (1) about this critical point. We must employ a trick to determine the stability of this equilibrium point.

After some trial and error, we find that if we multiply the first equation in (1) by $(\beta X - d)/X$ and multiply both sides of the second equation by $(b - \delta Y)/Y$, then subtraction leads to

$$X'(t)\left(\beta - \frac{d}{X(t)}\right) + Y'(t)\left(\delta - \frac{b}{Y(t)}\right) = 0 \tag{2}$$

This can be written as

$$\frac{d}{dt}(\beta X(t) - d \ln X(t) + \delta Y(t) - b \ln Y(t)) = 0 \tag{3}$$

Then along each orbit of the system (1), we have that

$$F(X, Y) = \beta X(t) + \delta Y(t) - \ln (X^d Y^b) = \text{constant}; \tag{4}$$

i.e., (4) is a first integral for (1). While it is not easy to see what type of curves are the level curves for the function $F(X, Y)$ in (4), we can sketch the phase plane plot in the first quadrant using numerical solution methods. Using a more delicate argument, we can even prove that the level curves of F are necessarily closed curves.

The phase plane portrait for this system is shown in Figure 5.7. The existence of closed orbits in the first quadrant implies that for predator-prey population interactions described by (1), the populations can vary periodically with each population surviving indefinitely. Note that the orbits in the first quadrant are not attracted to the critical point there. The equilibrium point in the first quadrant is stable but not asymptotically stable. The orbits in quadrants two, three, and four near the unstable critical point at the origin have no significance in the context of population models but have been sketched anyway to emphasize the saddle point nature of this critical point.

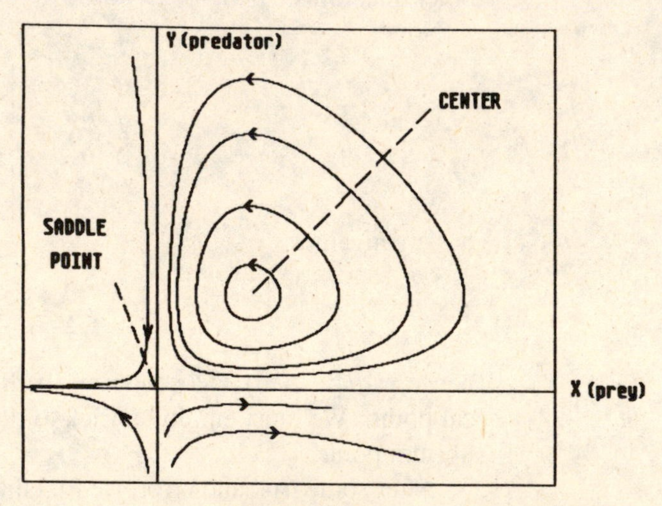

Figure 5.7
Phase Plane for Predator Prey System

PROBLEM 5.11

A spring-mass system with an unusual friction force is governed by

$$mx''(t) = -Kx(t) - m(px'(t)^2 - 2d)x'(t) \tag{1}$$

where p and d denote positive constants. This friction force is unusual because it changes sign depending on the size of $x'(t)$. Equation (1) is equivalent to the autonomous system

$$\begin{aligned} x'(t) &= y(t) \\ y'(t) &= -\omega^2 x(t) - (py^2 - 2d)\,y \end{aligned} \tag{2}$$

Show that the origin is the only critical point for this system and that the system possesses a limit cycle enclosing the origin.

SOLUTION 5.11

It is easy to see that $x = y = 0$ is the only point where $y = -\omega^2 x - y(py^2 - 2d) = 0$. We compute

$$J(0,0) = \begin{bmatrix} 0 & 1 \\ -\omega^2 & 2d \end{bmatrix}$$

for which the eigenvalues are

$$\lambda_{1,2} = d \pm \sqrt{d^2 - \omega^2} \tag{3}$$

For d such that $0 < d < \omega$, the origin is an unstable focus; i.e., the eigenvalues are a complex conjugate pair with positive real part. Then the orbits spiral outward from the origin. Note, however, that if we multiply the first equation in (2) by $2kx$ and the second equation by $2y$, then adding the result leads to

$$\frac{d}{dt}(kx^2 + y^2) = -2(py^4 - dy^2) \tag{4}$$

Here we have used the fact that

$$\frac{d}{dt}(kx^2 + y^2) = 2kxx'(t) + 2yy'(t)$$

It is apparent from (4) that for y^2 sufficiently large (i.e., for $y^2 > d/p$), the quantity $kx^2 + y^2$ is a decreasing function of time. Since $kx^2 + y^2$ is related to the distance of the point (x, y) from the origin, it follows that for y^2 large, the orbits behave like inward spirals. On the other hand, for small values of y^2 (i.e., for $y^2 < d/p$), it is also evident from (4) that $kx^2 + y^2$ is increasing with time.

This last observation is consistent with the fact that the origin is an unstable focus. However, the fact that orbits on which y^2 assumes large values are drawn back toward the origin makes it difficult to imagine how the phase plane portrait for this system must look.

Consider then the roughly annular region D_1 defined by

$$\frac{d}{2p} < kx^2 + y^2 < \frac{2d}{p}$$

The region D_1 excludes the "hole" where $kx^2 + y^2 \leq \frac{d}{2p}$. In particular, D_1 excludes the origin so that D_1 contains no critical points of the system (2). Equation (4) implies that orbits originating in the "hole" are forced to spiral outward away from the origin. Similarly, (4) implies that orbits that originate outside of D_1, where $kx^2 + y^2 \geq d/2p$, are attracted back toward the origin and toward D_1. This does not *prove* that orbits that originate in D_1 remain inside D_1, but it is suggestive that Theorem 5.3 part (c), the Poincare-Bendixson theorem, may be applied in order to conclude that the region D_1 contains a limit cycle. The phase plane portrait for this system is sketched in Figure 5.8. Since the limit cycle to which the orbits are attracted is a closed curve, the solutions of this system tend with time to periodic solutions.

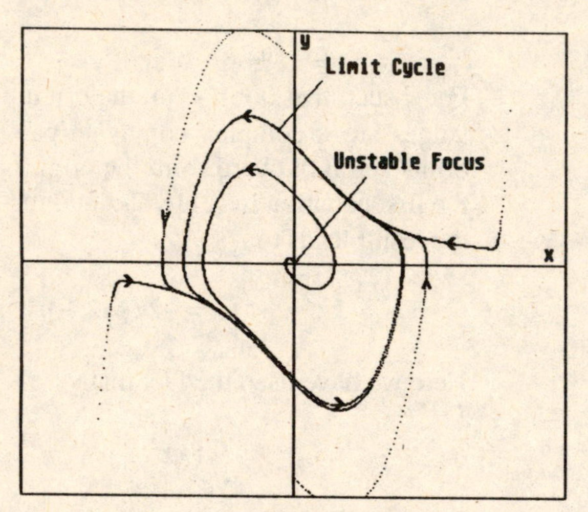

Figure 5.8
Phase plane portrait of limit cycle behavior

Note that when the parameter d is negative, then the critical point at (0, 0) is an asymptotically stable focus; all orbits are attracted to the origin. For d positive, we have just shown that the origin is an unstable focus and the system has a limit cycle attractor. That is, for $d < 0$ all solutions tend to zero, whereas for $d > 0$, all solutions tend to a periodic solution. In the context of the physical system, this may be interpreted to mean that when $d < 0$ every initial state produces a motion that is oscillatory with amplitude decaying steadily to zero and when $d > 0$, there are initial states that produce a motion that tends to a self-sustaining periodic motion.

There is no general method for constructing solutions to nonlinear differential equations, even in the simplest case of equations of order one. Solutions can be found in certain special cases but even then it is not appropriate to speak of the general solution of the equation as we do for linear equations. This is because there may be singular solutions that correspond to no values of the arbitrary constants appearing in the constructed solution.

In cases where we are unable to construct a solution it may still be possible to obtain significant qualitative information about the solution. Since an equation of order n can always be reduced to a first order system of n equations, this qualitative analysis is focused on nonlinear systems. We can further limit our attention to systems that do not depend explicitly on the independent variable, so-called autonomous systems, since it is always possible to eliminate explicit dependence by introducing an additional dependent variable.

Information about the behavior of the nonlinear system $\mathbf{X}'(t) = \mathbf{F}(\mathbf{X})$ *can be obtained by finding and classifying the critical points where* $\mathbf{F}(\mathbf{X}) = \mathbf{0}$. *The classification is based on the eigenvalues of the associated Jacobian matrix. A critical point is said to be hyperbolic if none of the eigenvalues of the Jacobian at the point have real part equal to zero. Hyperbolic critical points are classified further as sources, sinks, or saddles according to whether the real parts of the eigenvalues are all positive, negative, or a mixture of both positive and negative, respectively. A sink is an asymptotically stable fixed point for the nonlinear system, whereas sources and saddles are unstable fixed points. The stability of a nonhyperbolic fixed point is usually more difficult to determine. Problems 5.10 and 5.11 provide examples of how this can be done. Problem 5.10 is an example of a nonlinear system with a nonhyperbolic fixed point that is stable but not asymptotically stable, whereas Problem 5.11 illustrates the purely nonlinear phenomenon of limit cycle behavior.*

6

Linear Boundary Value Problems

When solving differential equations in connection with applications, we are rarely interested in finding the general solution. Usually we seek a particular solution of the differential equation that satisfies some auxiliary conditions. In chapter 4, we considered one type of problem common in applications, so-called initial value problems. In this chapter, we will consider another type of problem that arises in applied mathematics, the boundary value problem.

Boundary value problems occur both as inhomogeneous problems that are analogous to the systems of linear algebraic equations considered in chapter 1, and as eigenvalue problems analogous to the problems considered in chapter 2. Of particular interest among the eigenvalue problems are the special class of eigenvalue problems called Sturm-Liouville problems.

Sturm-Liouville problems are the source of a deep and rich mathematical theory which can only be hinted at here, but even a cursory discussion of this topic can best be carried out in the setting of a function space. We define then a vector space in which functions play the role of the vectors and we define an inner product on this vector space. The inner product carries with it the notion of orthogonality and orthogonal families of vectors (functions). Perhaps the most significant property of Sturm-Liouville problems is that they generate orthogonal families of eigenfunctions which can then be used as the basis of a series expansion called a generalized Fourier series. We will study these series in some detail in the next chapter.

Sturm-Liouville problems can be separated into two classes: (1) regular problems and (2) those that are said to be singular. Some of the most important Sturm-Liouville problems are singular and we conclude the chapter with two singular examples, Legendre's equation and Bessel's equation. The families of eigenfunctions generated by these two singular Sturm-Liouville problems are referred to as Legendre polynomials and Bessel functions, respectively, two examples of so-called special functions of applied mathematics.

LINEAR BOUNDARY VALUE PROBLEMS

BOUNDARY CONDITIONS

Theorem 4.2 states that the general solution of a linear ordinary differential equation of order n contains n arbitrary constants. The theorem states further that there exists a unique choice of these n parameters for which the solution satisfies n *initial conditions* imposed on the unknown function $y(t)$ and its derivatives of order less than n. Problems in which the n conditions are all imposed at a single point are called *initial value problems*. It is also possible to impose conditions at more than a single point. In applications, we frequently encounter problems in which the differential equation is required to hold at each point of a bounded interval and additional conditions are imposed on the solution and its derivatives at the endpoints of the interval. Such problems are called *boundary value problems* and the conditions imposed at the endpoints are called *boundary conditions*. The independent variable in boundary value problems often is interpreted as a spatial variable and will be denoted by x.

Example 6.1 Boundary Value Problems

(a) Let I denote the bounded interval (a, b). For $p(x)$ and $q(x)$ given functions, continuous on I, consider the linear second order differential operator

$$L_2[y(x)] = y''(x) + p(x)y'(x) + q(x)y(x), \quad a < x < b \qquad (6.1)$$

Then the problem of finding an unknown function $y = y(x)$ that satisfies $L_2[y] = 0$ on I and also satisfies the boundary conditions $y(a) = A$, $y(b) = B$ is an example of a second order boundary value problem. More generally, we consider boundary conditions

$$B[y; a] = C_1 y(a) + C_2 y'(a) = A \qquad (6.2)$$
$$B[y; b] = C_3 y(b) + C_4 y'(b) = B$$

where C_1, C_2 are not both zero nor are the constant coefficients in the other boundary condition both zero. If A and B are both zero, we say the boundary conditions are *homogeneous*; otherwise they are said to be *inhomogeneous boundary conditions*.

(b) The problem of finding an unknown function $y = y(x)$ such that

$$y^{(4)}(x) + Ky(x) = f(x) \quad \text{for} \quad a < x < b$$
$$y(a) = 0, y''(a) = 0 \quad \text{and} \quad y'(b) = 0, y^{(3)}(b) = 0$$

is an example of a fourth order boundary value problem. Here the equation is inhomogeneous and the boundary conditions are homogeneous. Note that the total number of boundary conditions is equal to the order of the equation. To uniquely determine the n arbitrary constants in the general solution to an equation of order n it is necessary to have n boundary conditions. As the next theorem shows, however, having n conditions is not necessarily sufficient to uniquely determine the solution.

Theorem 6.1

Theorem 6.1.Suppose $y_1(x)$ and $y_2(x)$ are linearly independent solutions of the equation $L_2[y(x)] = 0$ on $I = (a, b)$ for L_2 given by (6.1). Then the boundary value problem

$$L_2[y(x)] = F(x), \quad B[y;a] = A, \quad B[y;b] = B \tag{6.3}$$

has a unique solution if and only if

$$D = B[y_1;a]B[y_2;b] - B[y_1;b]B[y_2;a] \tag{6.4}$$

does not equal zero. If $D = 0$, then either the boundary value problem has no solution or else it has infinitely many solutions.

Corollary 6.2

Corollary 6.2.The homogeneous problem

$$L_2[y(x)] = 0, \quad B[y: a] = 0, \quad B[y; b] = 0 \tag{6.5}$$

has nontrivial solutions if and only if $D = 0$.

Eigenvalue Problems

Consider the problem of finding a function $y = y(x)$ and a value λ such that

$$L_2[y(x)] = \lambda y(x), \quad B[y: a] = 0, \quad B[y; b] = 0 \tag{6.6}$$

Clearly, this boundary value problem has the trivial solution $y(x) = 0$ for all choices of the parameter λ. By Corollary 6.2, it follows that for certain special values λ, the problem may have nontrivial solutions. Any λ for which (6.6) has a nontrivial solution is called an *eigenvalue* of the problem and the corresponding solution $y(x)$ is called an *eigenfunction* corresponding to the eigenvalue λ. Note that if $y = y(x)$ is an eigenfunction for (6.6) corresponding to the eigenvalue λ, then the same is true of the function $Cy(x)$ for every nonzero constant C. Thus, eigenfunctions are determined only up to a multiplicative constant.

STURM-LIOUVILLE PROBLEMS

We shall be interested here in eigenvalue problems of a special form. The eigenvalue problem for unknown function $y = y(x)$ and parameter λ such that

$$\begin{aligned} &-(p(x)y'(x))' + q(x)y(x) = \lambda r(x)y(x), \quad a < x < b \\ &B[y; a] = C_1 y(a) + C_2 y'(a) = 0 \\ &B[y; b] = C_3 y(b) + C_4 y'(b) = 0 \end{aligned} \tag{6.7}$$

is called a *Sturm-Liouville problem*. We suppose here that:

$$\begin{aligned} &p(x),\ p'(x),\ q(x) \text{ and } r(x) \text{ are continuous on } (a, b) \\ &p(x) \text{ and } r(x) \text{ are strictly positive on } (a, b) \\ &\text{neither } C_1, C_2 \text{ nor } C_3, C_4 \text{ both vanish} \end{aligned} \tag{6.8}$$

Theorem 6.3

Theorem 6.3. Under the stated assumptions (6.8), the Sturm-Liouville problem (6.7) has infinitely many eigenvalues λ and corresponding eigenfunctions $y(x; \lambda)$. These eigenpairs $\{\lambda, y(x; \lambda)\}$ have the following special properties:

(a) the eigenvalues are all real;

(b) to each eigenvalue there corresponds a single independent eigenfunction;

(c) the eigenvalues form an increasing infinite sequence $\{\lambda_n\}$, with λ_n tending to plus infinity with increasing n.

The importance of Sturm-Liouville problems lies in some additional special properties of the family of eigenfunctions. Discussion of these properties can be carried out most efficiently in the setting of a vector space of functions.

A FUNCTION SPACE

SQUARE INTEGRABLE FUNCTIONS

A function $f = f(x)$ is said to be *square integrable* on the interval $I = (a, b)$ if

$$\int_I f(x)^2 dx \text{ is finite}$$

Every function that is continuous on the closed bounded interval $[a, b]$ is square integrable on I, but a function that is square integrable on I need not be continuous on I.

Vector Space of Square Integrable Functions

In chapter 1 we considered the vector space of n-tuples of real numbers. Here we consider the vector space of square integrable functions. The square integrable functions play the role of the vectors with the real numbers as the scalars. For arbitrary square integrable functions f and g and arbitrary scalars C_1 and C_2, the linear combination $C_1f + C_2g$ is a square integrable function whose value at x in I is defined to be $(C_1f + C_2g)(x) = C_1f(x) + C_2g(x)$. With vector addition and multiplication by a scalar so defined, the set of square integrable functions is a vector space which we denote by $R^2(I)$. A vector space in which functions play the role of the vectors is also referred to as a *function space*.

INNER PRODUCTS

We can define an *inner product* of two functions in the vector space $R^2(I)$ in the following way: For f and g square integrable on I, let

$$(f, g) = \int_I f(x)\, g(x)\, dx \tag{6.9}$$

Then (f, g) is a scalar valued product that is analogous to the dot product of two vectors in $\Re^n$. It is easy to show that

$$(f, g) = (g, f) \quad \text{and} \quad (C_1f + C_2g, h) = C_1(f, h) + C_2(g, h)$$

for all square integrable functions f,g, and h and all scalars C_1 and C_2. We say that the inner product is *commutative* and *associative*.

We can also define a so-called *weighted inner product* by

$$\langle f, g\rangle = \int_I f(x)\, g(x)\, r(x)\, dx \tag{6.10}$$

where $r(x)$ denotes a function which we will suppose is continuous and positive on I. We refer to $r(x)$ as the *weight function* in the weighted inner product. The weighted inner product is commutative and associative.

NORMS

Associated with the inner products we have the notion of the *norm* of a square integrable function

$$\|f\| = (f, f)^{1/2}$$

We may also define the *weighted norm*

$$|f| = \langle f, f\rangle^{1/2}$$

These are analogous to the length of a vector in $\Re^n$. The norms $|f|$ and $\|f\|$ are zero if and only if

$$\int_I f(x)^2 dx = 0$$

This does not necessarily imply $f(x) = 0$ at each x in I. For example, $f(x)$ may differ from zero at finitely many points x in I and zero everywhere else in I and yet $|f| = \|f\| = 0$. When $|f|$ and $\|f\|$ are zero, we say f is zero *almost everywhere* and we say f equals g almost everywhere if $|f-g| = \|f-g\| = 0$.

Orthogonality

Just as the dot product in $\Re^n$ allows us to define the notion of orthogonality for vectors, we can use the inner product to define the notion of functions that are orthogonal on I. We say f and g are *orthogonal* on I if $(f, g) = 0$ and we say they are orthogonal on I with respect to the weighted inner product if $\langle f, g\rangle = 0$.

ORTHOGONAL FAMILIES

A family of functions $f_1(x), f_2(x), \ldots$ in $R^2(I)$ is said to be an *orthogonal family* if $(f_i, f_j) = 0$ for i not equal to j. The orthogonal family is said to be an *orthonormal family* if, in addition, we have $\|f_j\| = 1$ for each j. Of course, we can also define families that are orthogonal and orthonormal with respect to the weighted inner product.

Example 6.2 Orthogonal Families

(a) For I = $[0, \pi]$, consider the family $g_n(x) = \sin nx$, n = 1, 2, ... These functions are continuous; hence, they are square integrable on I and elementary integration shows that

$$(g_m, g_n) = \int_0^\pi \sin mx \sin nx \, dx = 0 \quad \text{if } m \neq n$$

so this is an orthogonal family. Since $\|g_n\| = \sqrt{\pi/2}$ for all n, $\{g_n\}$ is not an orthonormal family. The family

$$G_n(x) = \sqrt{\frac{2}{\pi}} \sin nx, \quad n = 1, 2, \ldots$$

is an orthonormal family.

(b) The family $f_n(x) = \cos nx, n = 0, 1, \ldots$, is an orthogonal family on $I = [0, \pi]$ since elementary integration shows

$$(f_m, f_n) = \int_0^\pi \cos mx \cos nx \, dx = 0 \quad \text{for } m \neq n$$

The family is not orthonormal since $\|f_0\| = \sqrt{\pi}$ and $\|f_n\| = \sqrt{\pi/2}$ for $n > 0$. The family

$$F_n(x) = \frac{f_n(x)}{\|f_n\|}, \quad n = 0, 1, \ldots$$

is an orthonormal family in I.

Using this new terminology we can now state an additional important property of the family of eigenfunctions generated by a Sturm-Liouville problem.

Theorem 6.4

Theorem 6.4. Under the assumptions (6.8) the Sturm-Liouville problem (6.7) has an infinite sequence of real eigenvalues $\lambda_1 < \lambda_2 < \cdots$. The corresponding family of eigenfunctions $y_1(x), y_2(x), \ldots$ is an orthogonal family with respect to the weighted inner product (6.10) with the weight function $r(x)$ from (6.7).

Generalized Fourier Series

Let $\{g_1, g_2, \ldots\}$ be a given orthonormal family of square integrable functions. Then for arbitrary square integrable function $f = f(x)$, we define the *generalized Fourier series* for f with respect to the family $\{g_n\}$ to be the series

$$\sum_{n=1}^{\infty} (f, g_n) g_n(x) \tag{6.11}$$

Generalized Fourier series based on orthogonal families of eigenfunctions that are generated by Sturm-Liouville problems will be used extensively in chapter 9 for solving problems in partial differential equations. Convergence properties of these series will be discussed in chapter 7 where we will show that if an orthogonal family $\{g_n\}$ is generated by a Sturm-Liouville problem, then the series (6.11) converges to $f(x)$ in a sense to be defined.

SINGULAR STURM-LIOUVILLE PROBLEMS

A point x in I where $p(x) = 0$ is said to be a *singular point* for the differential equation in (6.7). A point where $p(x)$ is not zero is called an *ordinary point*. The assumptions (6.8) imposed on the Sturm-Liouville problem (6.7) ensure that the problem has no singular points on the closed interval $[a, b]$. However, there are Sturm-Liouville problems we wish to consider where the conditions (6.8) do not all hold. We call these problems *singular* Sturm-Liouville problems. In case an endpoint of the interval I is a singular point for the differential equation then the usual boundary condition at the point is replaced by the condition that the solution tends to a finite limit as x approaches the singular point.

Legendre's Equation

Consider the following eigenvalue problem for λ and $y = y(x)$

$$(1-x^2)\,y''(x) - 2xy'(x) + \lambda y(x) = 0, \quad -1 < x < 1 \tag{6.12}$$

$y(x)$ approaches finite limits as x tends to 1, –1

The points $x = 1, -1$ are singular points for this equation which is known as Legendre's equation. Since the differential equation has variable coefficients, we look for a general solution in the form of a power series expansion about the ordinary point at $x = 0$.

LEGENDRE POLYNOMIALS

We can show that the power series solutions for (6.12) become unbounded at $x = 1, -1$ unless λ equals $n(n+1)$ for non-negative integer values n. Then (6.12) has eigenvalues $\lambda_n = n(n+1)$, $n = 0, 1, \ldots$, and eigenfunctions $P_n(x)$ where

$$P_n(x) = \sum_{m=0}^{N(n)} \frac{(-1)^m (2n-2m)!\,x^{n-2m}}{2^n m!\,(n-m)!\,(n-2m)!}, \quad n = 0, 1, \ldots \tag{6.13}$$

with $N(n) = n/2$ if n is even and $N(n) = (n-1)/2$ if n is odd. For each n, $P_n(x)$ is a polynomial of degree n in x referred to as the *Legendre polynomial* of degree n. We have

$$P_0(x) = 1, \quad P_1(x) = x, \quad P_2(x) = \frac{3x^2-1}{2}, \quad P_3(x) = \frac{5x^3-3x}{2}$$

$$P_4(x) = \frac{35x^4 - 30x^2 + 3}{8}, \quad P_5(x) = \frac{63x^5 - 70x^3 + 15x}{8}, \quad \text{etc.}$$

We have also *Rodrigues' formula* for generating Legendre polynomials

$$P_n(x) = \frac{1}{2^n n!}\frac{d^n}{dx^n}(((x^2-1)^n)), \quad n = 0, 1, \ldots$$

It is not hard to show that for $n = 0, 1, \ldots$

$$P_n(1) = 1 \quad \text{and} \quad P_n(-1) = (-1)^n$$

$$P_n'(1) = \frac{n(n+1)}{2}, \quad P_n'(-1) = \frac{(-1)^{n+1} n(n+1)}{2}$$

$$-1 \le P_n(x) \le 1 \quad \text{for } |x| \le 1$$

Note that Legendre's equation can be written in the Sturm-Liouville form

$-((1-x^2)y'(x))' = \lambda y(x)$ from which it is easy to see that $p(x) = 1-x^2$, $q(x) = 0$, $r(x) = 1$. The problem (6.12) is a singular Sturm-Liouville problem for which all the conclusions of Theorems 6.3 and 6.4 hold. In particular, the eigenfunctions form an orthogonal family on (–1, 1) with respect to the weight functions $r(x) = 1$; i.e.,

$$\int_{-1}^{1} P_n(x) P_m(x)\, dx = 0 \quad \text{for } m \text{ not equal to } n$$

Bessel's Equation

Consider the differential equation

$$x^2 y''(x) + xy'(x) + (x^2 - a^2)y(x) = 0 \tag{6.14}$$

This equation, known as Bessel's equation, has a singular point at $x = 0$. If we seek a solution in the form of a power series expansion about the singular point, then it must have the so-called *Frobenius form*

$$y(x) = x^s \sum_{n=0}^{\infty} c_n x^n$$

BESSEL FUNCTIONS OF THE FIRST KIND

We find that one solution of (6.14) is given by

$$y(x) = \sum_{n=0}^{\infty} \frac{(-1)^n \Gamma(a+1) x^{2n+a}}{2^{2n} n! \Gamma(a+n+1)} = J_a(x) \tag{6.15}$$

where

$$\Gamma(x) = \int_0^{\infty} e^{-t} t^{x-1} dt, \quad x > 0$$

Here $\Gamma(x)$ denotes the so-called Gamma function which we use here as an extension of the factorial function to noninteger values. That is, $\Gamma(x)$ has the property that $\Gamma(x+1) = x\Gamma(x)$ for all $x > 0$ and $\Gamma(1) = 1$ which implies that for all positive integers n, $\Gamma(n+1) = n!$.

The function $J_a(x)$ defined in (6.15) is called the *Bessel function of the first kind of order a.* For all real a, $J_a(x)$ is an oscillating function whose amplitude decreases with increasing x. In particular, $J_a(x)$ has infinitely many positive zeroes whose values depend on a.

BESSEL FUNCTIONS OF THE SECOND KIND

For noninteger values of a, the functions $J_a(x)$ and $J_{-a}(x)$ are linearly independent and, thus, the general solution of (6.14) may be written as

$$y(x) = C_1 J_a(x) + C_2 J_{-a}(x) \tag{6.16}$$

Alternatively, the *Bessel function of the second kind of order a* defined by

$$Y_a(x) = \frac{J_a(x)\cos a\pi - J_{-a}(x)}{\sin a\pi}$$

is independent of $J_a(x)$ for all a and can, therefore, be used in place of $J_{-a}(x)$ for the general solution (6.16) for any a. Clearly, $J_a(0) = 0$ and it can be shown that $J_{-a}(x)$ and $Y_a(x)$ both become unbounded as x tends to zero.

PROPERTIES OF BESSEL FUNCTIONS

It follows from the definition (6.15) that

$$\frac{d}{dx}(x^a J_a(x)) = x^a J_{a-1}(x) \text{ and } \frac{d}{dx}(x^{-a} J_a(x)) = -x^{-a} J_{a+1}(x)$$

and these identities imply

$$J_a'(x) + \frac{a}{x} J_a(x) = J_{a-1}(x) \text{ and } J_a'(x) + -\frac{a}{x} J_a(x) = -J_{a-1}(x)$$

In particular, $J_0'(x) = -J_1(x)$. These identities may be written in integral form as

$$\int x^a J_{a-1}(x)\,dx = x^a J_a(x) \text{ and } \int x^{-a} J_{a+1}(x)\,dx = x^{-a} J_a(x)\,. \tag{6.17}$$

Finally, we can show that

$$\int_0^\infty e^{-sx} J_0(x)\,dx = \frac{1}{\sqrt{s^2+1}} \tag{6.18}$$

which we regard as the formula for the Laplace transform of $J_0(x)$.

MODIFIED BESSEL FUNCTIONS

The general solution of the equation

$$x^2 y''(x) + xy'(x) + (p^2x^2 - a^2)\,y(x) = 0$$

(Bessel's equation with a parameter) can be written as

$$y(x) = C_1 J_a(px) + C_2 Y_a(px)$$

Then the general solution of the so-called *modified Bessel's equation*

$$x^2 y''(x) + xy'(x) - (p^2x^2 + a^2)\,y(x) = 0 \tag{6.19}$$

can be expressed as $y(x) = C_1 J_a(ipx) + C_2 Y_a(ipx)$. It is often convenient to introduce the *modified Bessel functions* of the first and second kinds

$$I_a(x) = i^{-a} J_a(ix) \quad \text{and} \quad K_a(x) = \frac{\pi}{2}\,\frac{I_{-a}(x) - I_a(x)}{\sin a\pi} \tag{6.20}$$

and express the general solution of (6.19) in the form

$$y(x) = C_1 I_a(x) + C_2 K_a(x)$$

The modified Bessel functions are not oscillatory and have no positive zeroes. The modified Bessel functions of the second kind grow without bound as x tends to zero.

SOLVED PROBLEMS

Boundary Value Problems

PROBLEM 6.1

An elastic beam of length L is subjected to a load per unit length $F(x)$ that increases linearly with x along the beam; i.e., $F(x) = Cx$ for positive constant C. Find $y(x)$, the deflection of the centerline of the beam if the end $x = 0$ is rigidly clamped while the end $x = L$ is free and unsupported.

SOLUTION 6.1

Consider an elastic beam with centerline lying along the x axis subjected to a transverse load distributed along the length of the beam. Let $y(x)$ denote the transverse displacement of a point on the beam's centerline at position x along the beam. Then if $y(x)$ and $y'(x)$ are each small, it is a reasonable approximation of reality to say that $M(x)$, the internal bending moment acting at position x in the beam, is proportional to $y''(x)$, the curvature in the beam centerline at the position x. More precisely,

$$M(x) = EIy''(x) \tag{1}$$

where E denotes a material dependent property known as *Young's modulus* and I denotes the moment of inertia of the beam cross section at x; i.e., I is a parameter depending on the shape of the beam cross section at the position x.

If $S(x)$ denotes the internal shear force acting at x, then $dM(x) = S(x)dx$; i.e.,

$$S(x) = M'(x) = EIy^{(3)}(x) \tag{2}$$

Similarly, if $F(x)$ denotes the applied transverse load per unit length, then $dS(x) = F(x)dx$; i.e., $S'(x) = F(x)$. Combining these two results with (1) then leads to

$$EIy^{(4)}(x) = F(x) \tag{3}$$

Equation (3) governs the (static) transverse deflection of an elastic beam subjected to a distributed transverse load given by $F(x)$. For $F(x) = Cx$, the deflection of the centerline, $y(x)$, satisfies

$$y^{(4)}(x) = \frac{Cx}{EI} \tag{4}$$

In addition, we express the fact that the beam is rigidly clamped at the end $x = 0$ by

$$y(0) = 0 \quad \text{and} \quad y'(0) = 0. \tag{5}$$

From (1) and (2), it follows that if the end $x = L$ is free (i.e., no shear and no moment), then

$$y''(L) = 0 \quad \text{and} \quad y^{(3)}(L) = 0 \tag{6}$$

The auxiliary equation associated with (1) has zero as a root of multiplicity 4. It follows from Theorem 4.3 that (4) has a homogeneous solution of the form

$$y_H(x) = A + Bx + Cx^2 + Dx^3$$

and a particular solution of the form $y_p(x) = Rx^4$. The method of undetermined coefficients leads to the result $R = \lambda/24EI$. Then the general solution of (4) is given by

$$y(x) = A + Bx + Cx^2 + Dx^3 + \frac{\lambda x^4}{24EI}.$$

The boundary conditions (5) imply $y(0) = A = 0$ and $y'(0) = B = 0$. Similarly, the boundary conditions (6) imply that

$$y''(L) = 2C + 6DL + (\lambda L^2)/(2EI) = 0$$

and

$$y^{(3)}(L) = 6D + (\lambda L)/(EI) = 0,$$

which leads to the solution

$$y(x) = \lambda x^2 \frac{(x^2 - 4xL + 6L^2)}{24EI} \tag{7}$$

Using (7), not only can we determine the deflection of the centerline of the beam, we can also compute the distribution of shear and moment in the beam as a function of x.

PROBLEM 6.2

Consider an elastic beam of length L loaded as in the previous problem. Find $y(x)$ for the case that both ends of the beam are simply supported.

SOLUTION 6.2

If the beam is simply supported at each end, then the ends of the beam are prevented from deflecting up or down but are free to rotate. The boundary conditions describing this state of affairs are

$y(0) = y(L) = 0$ (zero deflection at each end) and

$y''(0) = y''(L) = 0$ (zero moment at each end)

As before, the governing equation is (4) of the previous problem. It has a general solution of the form

$$y(x) = A + Bx + Cx^2 + Dx^3 + \frac{\lambda x^4}{24EI}$$

The boundary conditions at $x = 0$ imply that $y(0) = A = 0$, $y''(0) = 2C = 0$, and at $x = L$, we have $y(L) = BL + DL^3 + \lambda L^4/24EI = 0$, $y''(L) = 6DL + \lambda L^2/2EI = 0$. This leads easily to the result

$$y(x) = \lambda x \frac{(L^3 - 2Lx^2 + x^3)}{24EI}$$

Note that the point of maximum deflection of the beam may be found in this case by solving $y'(x) = 0$. In the previous problem, $y'(x)$ has no real zeroes and the point of maximum deflection there occurs at $x = L$.

PROBLEM 6.3

Consider a simply supported homogeneous elastic rod of uniform cross section and length L subjected to an axial compressive force P. Let the x axis lie along the beam and denote deflection in the direction normal to the axis of the beam by y. As P is increased from zero, the rod compresses with no deflection in the y direction until, at some value of P, the rod suddenly buckles. Show that the critical buckling load P can be found by solving an appropriate eigenvalue problem.

SOLUTION 6.3

As in the previous problems, the fundamental assumption of beam theory states that the bending moment producing deflection curve $y = y(x)$ is given by $M = EIy''(x)$. In addition, a simple geometric analysis shows that in this situation the moment is approximately equal to $M = -Py$. Equating these two expressions for M leads to the following differential equation:

$$y''(x) = -\lambda^2 y(x), \quad 0 < x < L \quad \text{where } \lambda^2 = \frac{P}{EI} \tag{1}$$

In addition, we have the boundary conditions reflecting that the ends of the rod are simply supported:

$$y(0) = y = L = 0 \tag{2}$$

The eigenvalue problem (1), (2) has nontrivial solutions only for certain values of λ called eigenvalues. We shall see that the eigenvalues of this problem are all positive and the smallest of these, λ_1, corresponds to the critical buckling load P_1; i.e., $P_1 = \lambda_1^2 EI$.

Equation (1) has as its general solution $y(x) = A \cos \lambda x + B \sin \lambda x$ and then the boundary conditions imply

$$y(0) = A = 0 \quad \text{and} \quad y(L) = B \sin \lambda L = 0$$

Clearly, nontrivial solutions exist only for λ such that $\sin \lambda L = 0$; i.e., for λL equal to an integer multiple of π. Then $\lambda_n = (n\pi)/L$ and $\lambda_1 = \pi/L$. Then the critical buckling load, corresponding to the smallest λ where the problem has a nontrivial solution, is $(\pi^2 EI)/L^2$. The corresponding eigenfunction $y(x;\lambda_1) = \sin(\pi x)/L$, indicates the deflection mode shape for the buckled rod.

Sturm-Liouville Problems

PROBLEM 6.4

Show that the following is a Sturm-Liouville problem and find all the eigenvalues and eigenfunctions.

$$-y''(x) = \lambda y(x), \quad y'(0) = 0, \quad y'(L) = 0 \tag{1}$$

SOLUTION 6.4

The problem (6.7) reduces to problem (1) when we choose: $p(x) = 1$, $r(x) = 1$, $q(x) = 0$, and $C_2 = C_4 = 0$ with $a = 0$ and $b = L$. Thus, (1) is a Sturm-Liouville problem and the conclusions of Theorem 6.3 apply. In particular, all the eigenvalues are real so that λ must be positive, zero, or negative.

If λ is negative, then $\lambda = -p^2$ for some nonzero real number p. Then the general solution of the differential equation (1) can be written as a linear combination of any two of the following six functions: e^{px}, e^{-px}, $\sinh px$, $\cosh px$, $\sinh p(L-x)$, $\cosh p(L-x)$. In fact, the last four of these six functions are just linear combinations of the first two. If we choose to write the general solution in the form
$y(x) = A \cosh px + B \cosh p(L-x)$ then the boundary conditions imply

$$y'(0) = -Bp \sinh pL = 0 \quad \text{and} \quad y'(L) = Ap \sinh pL = 0$$

Since $L > 0$ implies $\sinh pL$ is not zero, we conclude that $A = B = 0$ and, thus, (1) has only the trivial solutions corresponding to negative values of λ. Then there are no negative eigenvalues for (1). Note that if we had chosen to write the solution $y(x)$ as a linear combination of some other two functions from the list of six, say $y(x) = Ae^{px} + Be^{-px}$, then we would also find $A = B = 0$. However, we would be obliged in this case to solve a set of two algebraic equations in the two unknowns A and B to discover this fact (try it). Choosing the form of the general solution as we did makes finding A and B slightly more direct.

If $\lambda = 0$, then the general solution of the differential equation has the form $y(x) = Bx + A$. The boundary conditions require $y'(0) = B = 0$ and $y'(L) = B = 0$; i.e., $y(x) = A$ solves (1) for arbitrary nonzero constant A. Thus, $\lambda_0 = 0$ is an eigenvalue with corresponding eigenfunction $y_0(x) = A_0$.

Finally, if λ is positive, then $\lambda = p^2$ for some nonzero real number p. In this case, the general solution of the differential equation has the form

$$y(x) = A \cos px + B \sin px$$

and the boundary conditions now require

$$y'(0) = Bp = 0 \quad \text{and} \quad y'(L) = Ap \sin pL = 0$$

Then $B = 0$, but A need not be zero if we choose p such that $\sin pL = 0$. The function $\sin x$ has an infinite number of zeroes located at integer multiples of π. Then the eigenvalues of (1) are the positive numbers $\lambda_n = p_n^2 = ((n\pi)/L)^2$ for $n = 1, 2, \ldots$ (note that since p_n is squared, we need not include the negative values of p_n). Since $B = 0$, it follows that the corresponding eigenfunctions $y(x;\lambda_n)$ are given by $y(x;\lambda_n) = y_n(x) = A_n \cos(n\pi x/L)$, where A_n denotes an aribtrary nonzero constant.

Note that elementary integration shows that the family $\{y_0, y_1, \ldots\}$ of eigenfunctions is, in fact, an orthogonal family on $(0, L)$ with respect to the weight function $r(x) = 1$ as predicted by Theorem 6.4; i.e.,

$$\langle y_n, y_m \rangle = \int_0^L \cos\left(\frac{n\pi x}{L}\right) \cos\left(\frac{m\pi x}{L}\right) dx = 0 \quad \text{if } m \text{ does not equal } n$$

PROBLEM 6.5

Show that the following problem is a Sturm-Liouville problem

$$-y''(x) = \lambda y(x), \quad y(0) = 0, \quad y(L) - cy'(L) = 0 \tag{1}$$

and show that, for $c < 1$, there is one negative eigenvalue. Here c denotes a positive constant not equal to L.

SOLUTION 6.5

Clearly, (1) corresponds to (6.7) on the interval $(0, L)$ in the case that $p(x) = r(x) = 1$, $q(x) = 0$, and $C_1 = C_3 = 1$, $C_2 = -c$, $C_4 = 0$. Then this is a Sturm-Liouville problem and we shall now see whether there are any negative eigenvalues.

If λ is negative, then $\lambda = -p^2$ for some nonzero real number p. The general solution for (1) in this case can be written

$$y(x) = A \cosh px + B \sinh px$$

and the boundary conditions imply

$$y(0) = A = 0 \quad \text{and}$$
$$y(L) - ay'(L) = B(\sinh pL - cp \cosh pL) = 0$$

The coefficient B need not equal zero if there are nonzero roots p for the equation

$$\sinh pL = cp \cosh pL \tag{2}$$

Equation (2) is equivalent to $\tanh pL = cp$. The roots of this transcendental equation are not to be found by inspection and we resort to superimposing the graph of the function $F(p) = \tanh pL$ on the graph of $G(p) = cp$ (Figure 6.1). Clearly, the two graphs cross at $p = 0$, but it is only nonzero crossings that lead to negative eigenvalues. It is clear from Figure 6.1 that the two graphs meet at nonzero values of p if the slope of the tangent line to $F(p)$ at $p = 0$ exceeds the slope of $G(p) = cp$; i.e., equation (2) has nonzero roots, P_0 and $-P_0$ with a single corresponding negative eigenvalue $\lambda_0 = -P_0^2$. Since $A = 0$, the eigenfunction corresponding to this single negative eigenvalue is $y_0(x) = B_0 \sinh P_0 x$.

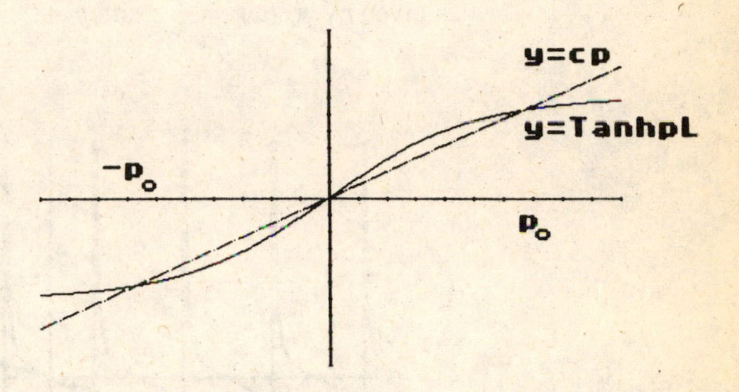

Figure 6.1
Negative Eigenvalues

PROBLEM 6.6

Find all the non-negative eigenvalues for the Sturm-Liouville problem

$$-y''(x) = \lambda y(x), \quad y(0) = 0, \quad y(L) - cy'(L) = 0 \tag{1}$$

SOLUTION 6.6

Since this is a Sturm-Liouville problem, the eigenvalues are all real and we are interested here in just the non-negative eigenvalues.

For $\lambda = 0$, the differential equation reduces to $y''(x) = 0$ for which the general solution is $y(x) = Bx + A$. It is easily checked that for c not equal to L, the boundary conditions imply $A = B = 0$. Then there is only the trivial solution corresponding to $\lambda = 0$ and zero is not an eigenvalue.

For positive λ, we can write $\lambda = p^2$ for p real and nonzero. Then the general solution for the differential equation can be written

$$y(x) = A \cos px + B \sin px.$$

The boundary conditions imply

$$y(0) = A = 0 \quad \text{and} \quad y(L) - ay'(L) = B(\sin pL - cp \cos pL) = 0$$

and we see that the coefficient B need not equal zero if there exists nonzero roots p for the equation

$$\sin pL = cp \cos pL. \tag{2}$$

This is a transcendental equation and the roots, therefore, cannot be found exactly from any algorithm of finitely many steps. However, the equation is equivalent to the equation $\tan pL = cp$ and by superimposing the graph of $F(p) = \tan pL$ on that of $G(p) = cp$ (see Figure 6.2) we see at once that (2) has infinitely many nonzero roots. For each positive root p_n, there is also a negative root $-p_n$ with the same magnitude, but each positive/negative pair produces just a single positive eigenvalue $\lambda_n = p_n^2, n = 1, 2, \ldots$ Since $A = 0$, the eigenfunction corresponding to the positive eigenvalue λ_n is given by $y_n(x) = B_n \sin p_n x$.

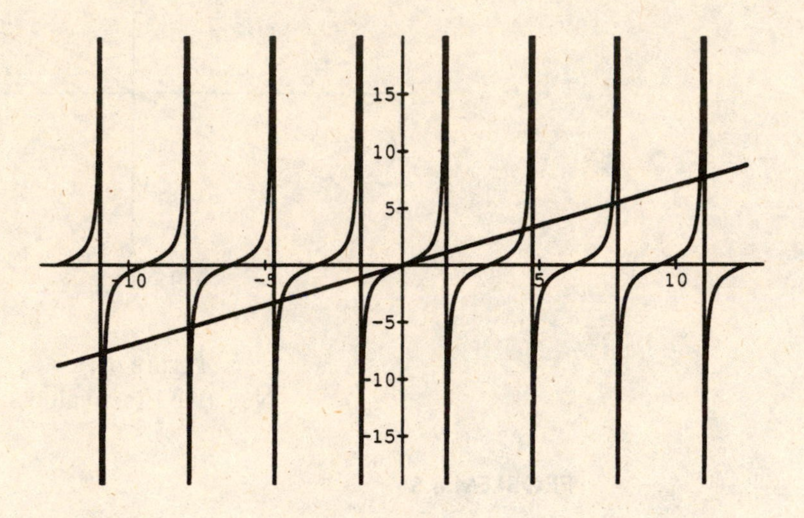

Figure 6.2
Positive Eigenvalues

PROBLEM 6.7

Find all the eigenvalues and eigenfunctions for the following problem:

$$-y''(x) = \lambda y(x) \quad -L < x < L \tag{1}$$

$$y(L) = y(-L) \quad \text{and} \quad y'(L) = y'(-L) \tag{2}$$

SOLUTION 6.7

Equation (1) has the Sturm-Liouville form, but the boundary conditions (2) do not have the same form as the boundary conditions in (6.7). The boundary conditions in (6.7) are referred to as *separated boundary condi-*

tions. The boundary conditions (2) are called *periodic boundary conditions* and even though they are not separated, all of the conclusions of Theorem 6.3 continue to hold for the problem (1), (2) with the exception of conclusion (b). As we shall see, the problem with periodic boundary conditions has two independent eigenfunctions for each nonzero eigenvalue.

We begin by considering negative values for λ. Then we write $\lambda = -p^2$ for a nonzero real number p and note that, in this case, (1) has the general solution $y(x) = A \cosh px + B \sinh px$. Then the boundary conditions imply

$$A \cosh pL + B \sinh pL = A \cosh pL - B \sinh pL$$
$$pA \sinh pL + B \cosh pL = p\text{–}A \sinh pL + B \cosh pL$$

The first equation leads to the conclusion $B = 0$ and, from the second, we see that $A = 0$. Then there are no nontrivial solutions corresponding to negative values for λ; i.e., there are no negative eigenvalues.

For $\lambda = 0$, equation (1) has the general solution $y(x) = Bx + A$. The boundary conditions imply that $B = 0$ but impose no constraints on A. Then $y(x) = A$ for arbitrary A is a nontrivial solution of (1), (2) corresponding to the value $\lambda = 0$. Then $\lambda_0 = 0$ is an eigenvalue with the corresponding eigenfunction $y_0(x) = A_0$.

When λ is positive, we write $\lambda = p^2$ for nonzero p. The general solution of (1) in this case is given by $y(x) = A \cos px + B \sin px$ and the boundary conditions lead to

$$y(L) - y(-L) = 2B \sin pL = 0$$
$$y'(L) - y'(-L) = 2A \sin pL = 0$$

Evidently, if p is such that $\sin pL = 0$, then the boundary conditions are both satisfied without either A or B having to vanish. Thus, for each value $\lambda = \lambda_n = (n\pi/L)^2$ $n = 1, 2, \ldots$, there are two independent functions $y_n(x)$ and $v_n(x)$.

Not only are all but one of the conclusions of Theorem 6.3 valid for this problem but the conclusion of Theorem 6.4 holds as well. The family of eigenfunctions $y_0, y_1, \ldots; v_1(x), v_2(x), \ldots$ is orthogonal in the sense that

$(y_m, v_n) = 0$ for all m and n

$(y_m, y_n) = 0$ and $(v_m, v_n) = 0$ for m not equal to n

Note that since $r(x) = 1$ here, the inner products (f, g) and $\langle f, g \rangle$ are identical.

Singular Sturm-Liouville Problems

PROBLEM 6.8

Show that the eigenvalue problem for Bessel's equation

$$x^2 y''(x) + xy'(x) + (\lambda x^2 - a^2)\, y(x) = 0, \quad 0 < x < b \qquad (1)$$

$y(x)$ tends to a finite limit as x tends to 0 and $y(b) = 0$ (2)

is a singular Sturm-Liouville problem and find the eigenvalues and eigenfunctions.

SOLUTION 6.8

We can rewrite (1) in the form $-(xy'(x))' + (a^2/x)y(x) = \lambda xy(x)$ from which it is clear that (1) is of Sturm-Liouville form with $p(x) = x = r(x)$ and $q(x) = a^2/x$. Then it follows from Theorem 6.3 that all eigenvalues are real.

If we consider the case $\lambda = 0$, then (1) reduces to the Euler equation,

$$x^2y''(x) + xy'(x) - a^2y(x) = 0$$

for which the general solution can be written as

$$y(x) = \begin{cases} C_1 + C_2 \ln x & \text{if } a = 0 \\ C_1x^a + C_2x^{-a} & \text{if } a \text{ is not zero} \end{cases}$$

The condition that $y(x)$ tend to a finite limit as x tends to zero forces $C_2 = 0$ and the condition $y(b) = 0$ leads to $C_1 = 0$. Then only trivial solutions result from the choice $\lambda = 0$ and zero is not an eigenvalue.

If λ is positive, we can write $\lambda = p^2$ for nonzero p, and then the general solution of (1) can be written

$$y(x) = C_1J_a(px) + C_2Y_a(px)$$

The condition at $x = 0$ forces $C_2 = 0$ and the condition $y(b) = 0$ is satisfied if

$$J_a(pb) = 0 \tag{3}$$

Recalling that $J_a(x)$ has infinitely many positive zeroes, we may label these zeroes $0 < \mu_1 < \mu_2 < \ldots$. Then (3) has an infinite number of roots p_n given by $p_n = \mu_n/b$ which leads to infinitely many positive eigenvalues $\lambda_n = (\mu_n/b)^2$. The corresponding eigenfunctions are given by

$$y_n(x) = J_a(p_nx).$$

For negative values of λ, $\lambda = -p^2$ and (1) becomes the modified Bessel's equation. Then the general solution must be expressed in terms of so-called modified Bessel functions (6.20). These functions have no positive zeroes and it follows that only the trivial combination of these modified Bessel functions can satisfy the boundary conditions. Hence, there are no negative eigenvalues.

Note that the family of eigenfunctions is an orthogonal family with respect to the weight function $r(x) = x$; i.e.

$$\int_0^b xJ_a\left(\frac{x\mu_n}{b}\right)J_a\left(\frac{x\mu_m}{b}\right)dx = 0 \quad \text{for } m \text{ not equal to } n$$

PROBLEM 6.9

Write the general solution of Airy's equation

$$y''(x) + xy(x) = 0, \quad x > 0 \tag{1}$$

in terms of Bessel functions.

SOLUTION 6.9

Airy's equation is one of a class of equations that can be reduced to Bessel's equation by suitable changes of variable. If we let $y(x) = \sqrt{x}z(x)$, then

$$y'(x) = \sqrt{x}z'(x) + \frac{z(x)}{2\sqrt{x}}$$

$$y''(x) = \sqrt{x}z''(x) + \frac{z'(x)}{\sqrt{x}} - \frac{z(x)}{(4\sqrt{x^3})}$$

Using this in (1) and multiplying the result by x^2 then leads to

$$x^2z''(x) + xz'(x) - \frac{1}{4}z(x) + x^3z(x) = 0 \tag{2}$$

Now introducing the new independent variable $t = x^{3/2}$, we compute

$$xz'(x) = \frac{3}{2}x^{3/2}\frac{dz}{dt} \quad \text{and} \quad x^2z''(x) = \frac{3}{4}x^{3/2}\frac{dz}{dt} + \frac{9}{4}x^3\frac{d^2z}{dt^2}$$

Then (2) becomes

$$t^2z''(t) + tz'(t) + (\frac{4}{9}t^2 - \frac{1}{9})z(t) = 0 \tag{3}$$

According to (6.16), the general solution of (3) is given by

$$z(t) = C_1J_{1/3}(\frac{2x^{3/2}}{3}) + C_2J_{-1/3}(\frac{2x^{3/2}}{3})$$

In general, any equation of the form

$$(x^ay'(x))' + bx^cy(x) = 0$$

has a general solution of the form

$$y(x) = x^{r/s}(C_1J_r(x^{1/s}s\sqrt{b}) + C_2y_r(x^{1/s}s\sqrt{b}))$$

where

$$r = \frac{1-a}{c-a+2} \quad \text{and} \quad s = \frac{2}{c-a+2}$$

For Airy's equation we had $a = 0$ and $b = c = 1$ so that $r = 1/3$ and $s = 2/3$.

The inhomogeneous linear boundary value problem (6.3) has a unique solution if and only if D, defined in (6.4), is not equal to zero. Note the similarilty with the statement that the $n \times n$ system of linear algebraic equations $AX = B$ has a unique solution if and only if det A is not zero and the homogeneous system where $B = 0$ has nontrivial solutions if and only if det A does equal zero.

If $p(x)$, $q(x)$ and $r(x)$ satisfy the conditions:

$p(x)$, $p'(x)$, $q(x)$ a d $r(x)$ are continuous on (a, b)

$p(x)$ and $r(x)$ are strictly positive on (a, b)

then the eigenvalue problem

$$-(p(x)y'(x))' + q(x)y(x) = \lambda r(x)y(x) \qquad a < x < b,$$

$$B[y;a] = C_1 y(a) + C_2 y'(a) = 0$$

$$B[y;b] = C_3 y(b) + C_4 y'(b) = 0$$

is called a regular Sturm-Liouville problem. Here, if neither C_1, C_2 nor C_3, C_4 both vanish, we say the problem has separated boundary conditions. The eigenvalues and eigenfunctions of a regular Sturm-Liouville problem with separated boundary conditions satisfy:

(a) the eigenvalues are all real,

(b) to each eigenvalue, there corresponds a single independent eigenfunction,

(c) the eigenvalues form an increasing infinite sequence, $\{\lambda_n\}$, with λ_n tending to plus infinity with increasing n;

(d) the family of eigenfunctions is an orthogonal family with respect to the weighted inner product

$$\langle f, g \rangle = \int_a^b f(x)\,g(x)\,r(x)\,dx$$

The following are examples of regular Sturm-Liouville problems with their eigenvalues and associated eigenfunctions:

1. $-y''(x) = \lambda y(x), \quad 0 < x < L; \quad y(0) = y(L) = 0$

$$\lambda_n = (n\pi/L)^2 \qquad y_n(x) = \sin(n\pi x/L) \qquad n = 1, 2, \ldots$$

2. $-y''(x) = \lambda y(x), \quad 0 < x < L; \quad y'(0) = y'(L) = 0$

$$\lambda_n = (n\pi/L)^2 \quad y_n(x) = \cos(n\pi x/L) \quad n = 0, 1, 2, \ldots$$

3. $-y''(x) = \lambda y(x), \quad 0 < x < L; \quad y'(0) = y(L) = 0$

$$\lambda_n = ((n+1/2)\pi/L)^2 \quad y_n(x) = \cos((n+1/2)\pi x/L), \quad n = 0, 1, 2, \ldots$$

4. $-y''(x) = \lambda y(x), \quad 0 < x < L; \quad y(0) = y'(L) = 0$

$$\lambda_n = ((n+1/2)\pi/L)^2 \quad y_n(x) = \sin((n+1/2)\pi x/L) \quad n = 0, 1, 2, \ldots$$

The following is a regular Sturm-Liouville problem having periodic boundary conditions. Properties (a), (c), and (d) apply to this problem, but for each nonzero eigenvalue there are two independent eigenfunctions.

5. $-y''(x) = \lambda y(x), \quad -L < x < L; \; y(-L) = y(L), \; y'(-L) = y'(L)$

$$\lambda_n = (n\pi/L)^2, u_n(x) = \sin(n\pi x/L), \; v_n(x) = \cos(n\pi x/L) \quad n = 0, 1, \ldots$$

The following are examples of singular Sturm-Liouville problems together with their eigenvalues and eigenfunctions.

6. $(1-x^2)y''(x) - 2xy'(x) + \lambda y(x) = 0, \quad -1 < x < 1$

$y(x)$ approaches finite limits as x tends to 1, −1

$$\lambda_n = n(n+1), \; y_n(x) = P_n(x) = n\text{th Legendre polynomial}, \quad n = 0, 1, \ldots$$

7. $x^2y''(x) + xy'(x) + (\lambda x^2 - a^2)y(x) = 0, \quad 0 < x < b$

$y(x)$ tends to a finite limit as x tends to 0 and $y(b) = 0$

$\lambda_n = (\mu_n/b)^2$ *where μ_n denotes the nth zero of $J_a(x)$,*

$$y_n(x) = J_a(\mu_n x/b)$$

7

Fourier Series and Eigenfunction Expansions

Power series representations of functions are often useful but they are limited to functions having derivatives of all orders on the interval where the representation applies. Fourier series representations on the other hand are less restricted and are even valid for some functions that are not continuous.

We show first how to generate the Fourier series for a given function $f(x)$ defined on a symmetric interval $[-L, L]$. A symmetric interval is used merely for convenience and since any bounded interval $[a, b]$ can always be transformed to such a symmetric interval by a linear change of variables, this restriction implies no loss of generality.

If the function $f(x)$ possesses certain symmetry, then it can be anticipated that some terms will be missing from the Fourier series for f. Conversely if the function f is defined only on the half range $[0, L]$, then we can extend it to the full range $[-L, L]$ so that f has the symmetry properties needed to cause certain terms to vanish from its Fourier series. These observations are the basis of the so-called half range Fourier sine and cosine series and are associated with the concept of the periodic extension of a function.

There are several useful ways in which convergence of infinite series may be defined. Here we define mean square, pointwise, and uniform convergence and state a convergence theorem for Fourier series that lists properties of the function $f(x)$ or its periodic extension sufficient to guarantee each of these modes of convergence.

Trigonometric Fourier series are a special case of the more general notion of generalized Fourier series. Given any orthonormal family, we can expand an arbitrary square integrable f in a generalized Fourier *series in terms of the given orthonormal family. A result known as Bessel's inequality guarantees that the series will converge but, in general, we cannot be sure that the series converges to f. However, if the orthonormal family is the family of eigenfunctions from some Sturm-Liouville problem, then the so-called eigenfunction expansion for f not only converges, it must converge to f. This fact is a powerful tool for solving certain problems in partial differential equations.*

FORMAL PROPERTIES OF FOURIER SERIES

We begin our discussion of Fourier series *formally*, without regard to questions of convergence. We will first show how to generate the Fourier series for an arbitrary given function $f(x)$. Later we will describe the properties $f(x)$ must have in order that this formal series actually converges to $f(x)$.

Trigonometric Infinite Series

An infinite series of the form

$$\frac{1}{2}a_0 + \sum_{n=1}^{\infty} a_n \cos\frac{n\pi x}{L} + b_n \sin\frac{n\pi x}{L} \tag{7.1}$$

is called a *trigonometric infinite series.* The constants $a_0, a_1, \ldots; b_1, b_2, \ldots$ are called the *coefficients* in the series.

ORTHOGONALITY RELATIONS

The trigonometric identities

$$\cos\alpha\cos\beta = \frac{1}{2}(\cos(\alpha+\beta) - \cos(\alpha-\beta))$$

$$\sin\alpha\sin\beta = \frac{1}{2}(\cos(\alpha-\beta) - \cos(\alpha+\beta))$$

$$\sin\alpha\cos\beta = \frac{1}{2}(\sin(\alpha+\beta) + \sin(\alpha-\beta))$$

lead at once to the following integration formulas. For all integers m and n:

$$\int_{-L}^{L} \cos\frac{n\pi x}{L}\cos\frac{m\pi x}{L}\,dx = \begin{cases} 0 & \text{if } m \neq n \\ 2L & \text{if } m = n = 0 \\ L & \text{if } m = n > 0 \end{cases} \tag{7.2}$$

$$\int_{-L}^{L} \sin\frac{n\pi x}{L} \sin\frac{m\pi x}{L} dx = \begin{cases} 0 & \text{if } m \neq n \\ L & \text{if } m = n \end{cases}$$

$$\int_{-L}^{L} \sin\frac{n\pi x}{L} \cos\frac{m\pi x}{L} dx = 0 \quad \text{for all integers } m \text{ and } n$$

These integral results are referred to as *orthogonality relations* for the family of functions

$$1, \cos\frac{\pi x}{L}, \cos\frac{2\pi x}{L}, \ldots; \sin\frac{\pi x}{L}, \sin\frac{2\pi x}{L}, \ldots$$

and the family of functions is sometimes referred to as an *orthogonal family* on the interval $[-L, L]$.

Fourier Coefficients

Suppose the series (7.1) is convergent on the interval $[-L, L]$ to a sum $f(x)$. If $f(x)$ is integrable on $[-L, L]$, the orthogonality relations (7.2) imply that the coefficients in the series must be related to $f(x)$ by

$$a_m = \frac{1}{L}\int_{-L}^{L} f(x) \cos\frac{m\pi x}{L} dx, \quad m = 0, 1, \ldots$$

$$b_m = \frac{1}{L}\int_{-L}^{L} f(x) \sin\frac{m\pi x}{L} dx, \quad m = 1, 2, \ldots \tag{7.3}$$

The constants a_m, b_m defined by (7.3) are called the *Fourier coefficients* for the function $f(x)$ and when the coefficients in the series (7.1) are given by (7.3), we say that (7.1) is the *Fourier series* for f.

Example 7.1 Fourier series

(a) For $L = \pi$, let $f(x)$ be defined on $[-\pi, \pi]$ by

$$f(x) = \begin{cases} 0 & \text{for } -\pi < x < 0 \\ 1 & \text{for } 0 \leq x \leq \pi \end{cases}$$

Then f is integrable on $[-\pi, \pi]$, but since f is piecewise defined, we must break the integral on $[-\pi, \pi]$ into an integral on $[-\pi, 0]$ and an integral on $[0, \pi]$ to accommodate the different definitions for f on these two subintervals. Thus

$$a_m = \frac{1}{\pi}\int_{-\pi}^{0} 0 \cos mx \, dx + \frac{1}{\pi} \cdot \int_{0}^{\pi} 1 \cos mx \, dx = 0 \text{ for } m = 1, 2, \ldots$$

$$a_0 = \frac{1}{\pi}\int_{0}^{\pi} 1 dx = 1 \quad (\text{recall } \cos mx = 1 \text{ for } m = 0)$$

$$b_m = \frac{1}{\pi}\int_{0}^{\pi} 1 \sin mx \, dx = -\frac{1}{m\pi}(\cos m\pi - 1) \quad \text{for } m = 1, 2, \ldots$$

Since $\cos m\pi = (-1)^m$ for all integers m, we get

$$b_1 = \frac{2}{\pi}, b_2 = 0, b_3 = \frac{2}{3\pi}, b_4 = 0, b_5 = \frac{2}{5\pi}, \ldots$$

so that the Fourier series for f is

$$f(x) = \frac{1}{2} + \frac{2}{\pi}\sum_{n=1}^{\infty} \frac{1}{2n-1} \sin(2n-1)x$$

We emphasize that this result is only formal at this point. We have not shown that the series converges at all much less that it converges to $f(x)$.

(b) For the function $f(x) = x$, $-\pi < x < \pi$, we compute

$$a_m = \frac{1}{\pi}\int_{-\pi}^{\pi} x \cos mx \, dx = \frac{1}{\pi}\left(\frac{x \sin mx}{m} - \frac{-\cos mx}{m^2}\right)\Bigg|_{-\pi}^{\pi}$$

$$= \frac{-2 \cos m\pi}{m} = (-1)^{m+1}\frac{2}{m}, \quad m = 1, 2, \ldots$$

Then, formally,

$$x = \sum_{n=1}^{\infty} (-1)^{n+1}\frac{2}{n} \sin nx$$

(c) Consider the function $f(x) = |x|$ for $-\pi < x < \pi$. Like the function in part (a), this function is piecewise defined with $f(x) = -x$ on $[-\pi, 0]$ and on $[0, \pi]$, $f(x) = x$. Then

$$a_0 = \frac{1}{\pi}\int_{-\pi}^{0} (-x dx) + \frac{1}{\pi}\int_{0}^{\pi} x dx = \pi$$

$$a_m = \frac{1}{\pi}\int_{-\pi}^{0} -x \cos mx \, dx + \frac{1}{\pi}\int_{0}^{\pi} \cos mx \, dx = \frac{2}{\pi}\frac{\cos m\pi - 1}{m^2}$$

$$b_m = \frac{1}{\pi}\int_{-\pi}^{0} -x \sin mx \, dx + \frac{1}{\pi}\int_{0}^{\pi} x \sin mx \, dx = 0$$

Thus, the formal Fourier series for $f(x) = |x|$ is given by

$$|x| = \frac{\pi}{2} - \frac{4}{\pi}\sum_{n=1}^{\infty} \frac{1}{(2n-1)^2} \cos (2n-1)x$$

PERIODIC FUNCTIONS

A function $f(x)$ defined for all values of x is said to be *periodic* with period P if

$$f(x) = f(x+P) \qquad \text{for all } x$$

Note that if $f(x)$ is periodic with period P, then it is also true that for each integer m, $f(x) = f(x + mP)$ for all x. Then for all integers n, $\sin nx$ and $\cos nx$ are periodic with period 2π. Similarly, $\sin (n\pi x/L)$ and $\cos (n\pi x/L)$ are periodic with period $2L$ for all integers n.

Finally if $u_1(x), \ldots, u_m(x)$ are all periodic functions having a common period P, then for all constants $c_1, \ldots, c_m$, the sum

$$\sum_{n=1}^{m} c_n u_n(x)$$

is periodic with period P.

PERIODIC EXTENSIONS

If $f(x)$ is defined on the closed bounded interval $[-L, L]$, then we can extend f to $(-\infty, \infty)$ as a periodic function of period $2L$. We refer to this extended function as the *2L-periodic extension* of f and we denote it by f_{2L}. Then f_{2L} is defined by

$$f_{2L}(x) = f(x) \qquad \text{if } -L \le x \le L$$

and

$$f_{2L}(x) = f_{2L}(x+2mL) \qquad \text{for all integers } m \text{ and all } x$$

Figure 7.1 shows the periodic extension of the functions from Example 7.1. Note that f continuous on $[-L, L]$ is not sufficient to imply that the extension f_{2L} is continuous on $\Re$. However, we have

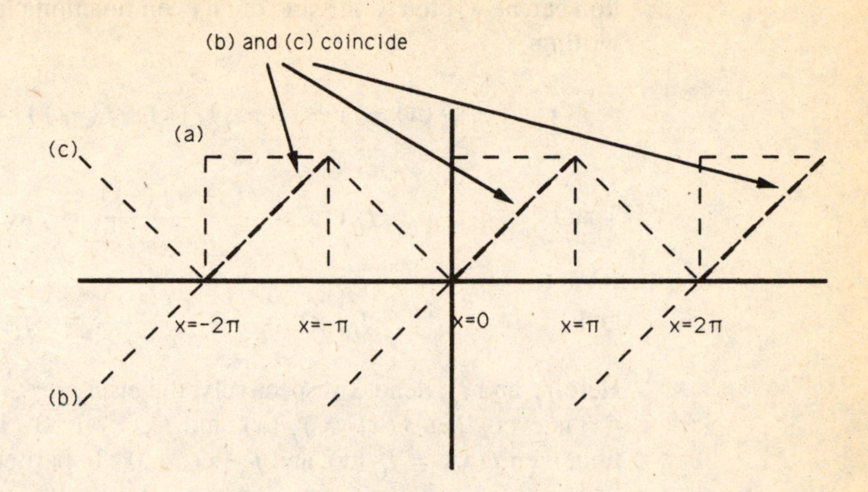

Figure 7.1
Periodic Extensions
(a) square wave (b) $f(x) = x$ (c) $f(x) = |x|$

Theorem 7.1

Theorem 7.1. For $f(x)$ defined on $[-L, L]$, the $2L$-periodic extension $f_{2L}(x)$ is everywhere continuous if and only if $f(x)$ is continuous on $[-L, L]$ *and* $f(L) = f(-L)$.

If the trigonometric series (7.1) is convergent, then the sum must be a $2L$-periodic function because each term of the series is $2L$-periodic. More to the point, suppose a series of the form (7.1) is the Fourier series for a function $f(x)$ defined on $[-L, L]$ and the series converges to f on $[-L, L]$. Then the series must be everywhere convergent to $f_{2L}(x)$, the $2L$-periodic extension of $f(x)$. As we shall see, the convergence properties of the Fourier series for $f(x)$ are determined by the smoothness properties of the extension, $f_{2L}(x)$.

Even and Odd Functions

Note that for each integer n,

$$f(x) = \cos\frac{n\pi x}{L} \quad \text{satisfies } f(x) = f(-x)$$

$$f(x) = \sin\frac{n\pi x}{L} \quad \text{satisfies } f(x) = -f(-x)$$

A function $f(x)$ is said to be *even* if $f(x) = f(-x)$, and if $f(x) = -f(-x)$, then f

is said to be *odd*. Most functions are neither even nor odd, but any function can be written as the sum of an even function plus an odd function by writing

$$f(x) = \frac{1}{2}(f(x)+f(-x)) + \frac{1}{2}(f(x)-f(-x)) = f_E(x) + f_O(x)$$

That is, $$f_E(x) = \frac{f(x)+f(-x)}{2} = f_E(-x)$$

and $$f_O(x) = \frac{f(x)-f(-x)}{2} = -f_O(-x)$$

Here f_E and f_O denote, respectively, the *even part* and *odd part* of $f(x)$. If $f(x)$ is even, then $f(x) = f_E(x)$ and $f_O(x) = 0$. If $f(x)$ is an odd function, then $f(x) = f_O(x)$ and $f_E(x) = 0$. In particular, if we have

$$f(x) = \frac{1}{2}a_0 + \sum_{n=1}^{\infty} a_n \cos\frac{n\pi x}{L} + b_n \sin\frac{n\pi x}{L}$$

then since $\cos(n\pi x/L)$ is even for every n and $\sin(n\pi x/L)$ is odd for each n, it follows that

$$f_E(x) = \frac{1}{2}a_0 + \sum_{n=1}^{\infty} a_n \cos\frac{n\pi x}{L} \tag{7.4}$$

and

$$f_O(x) = \sum_{n=1}^{\infty} b_n \sin\frac{n\pi x}{L} \tag{7.5}$$

This implies

for $f(x)$ even: $$a_m = \frac{2}{L}\int_0^L f(x)\cos\frac{m\pi x}{L}\,dx, \quad m = 0, 1, \ldots$$

$$b_m = 0, \quad m = 1, 2, \ldots \tag{7.6}$$

for $f(x)$ odd: $$a_m = 0, \quad m = 0, 1, \ldots$$

$$b_m = \frac{2}{L}\int_0^L f(x)\sin\frac{m\pi x}{L}\,dx, \quad m = 1, 2, \ldots \tag{7.7}$$

SYMMETRIES

When a function $f(x)$ defined on $[-L, L]$ is even, then it is symmetric about the vertical axis and it follows, without need for any computation, that $b_m = 0$ for every m. Similarly, when $f(x)$ is odd, then it (or more precisely, its graph) is symmetric about the origin and it follows that $a_m = 0$ for every m. Other types of symmetry lead to the vanishing of additional Fourier coefficients. Some of these symmetries are explored in the solved problems.

Half Range Series

We may also apply the preceding remarks regarding odd and even functions to the situation in which f(x) is defined only on the half range $[0, L]$. In this case we may find it convenient to extend $f(x)$ to the interval [–L, 0] as an even function. This leads to the

Half Range Cosine Series. We define the Fourier coefficients for $f(x)$ from (7.6). Then the Fourier series for $f(x)$ has the form

$$\frac{1}{2}a_0 + \sum_{n=1}^{\infty} a_n \cos\frac{n\pi x}{L}$$

We refer to this as the *half range Fourier cosine series* for $f(x)$. If this series converges to $f(x)$ for x in $[0, L]$, then it converges everywhere to $f_{2L,\text{ even}}$; the even $2L$-periodic extension of $f(x)$.

In other cases, we may find it more convenient to extend $f(x)$ to the interval $[-L, 0]$ as an odd function. Then we obtain:

Half Range Sine Series. We define the Fourier coefficients for $f(x)$ from (7.7). Then the Fourier series for $f(x)$ has the form

$$\sum_{n=1}^{\infty} b_n \sin\frac{n\pi x}{L}$$

We refer to this as the *half range Fourier sine series* for $f(x)$. If this series converges to $f(x)$ in $[0, L]$, then it converges everywhere to $f_{2L,\text{ odd}}$; the odd $2L$-periodic extension of $f(x)$.

When $f(x)$ is defined only on $[0, L]$, the choice for the extension of $f(x)$ to $[-L, 0]$ is generally dictated by the application at hand.

Example 7.2 Half Range Fourier Series

(a) Let $f(x) = x$ for $0 < x < \pi$. Then using (7.7) to compute the sine series coefficients, we find

$$b_m = \frac{2}{m}(-1)^{m+1}, \quad m = 1, 2, \ldots$$

These are the Fourier coefficients we computed in Example 7.1(b) for the odd function $f(x) = x$ on $[-\pi, \pi]$. Thus, when we write a half range sine series for a function f defined on the half range, we are in fact writing the full Fourier series for the odd extension of f to the full range.

(b) Similarly, if we use (7.6) to compute the cosine series coefficients for the half range function $f(x) = x$, we find

$$a_0 = \pi \quad \text{and} \quad a_m = \frac{2}{\pi}\,\frac{\cos m\pi - 1}{m^2}, \qquad m = 1, 2, \ldots$$

These are the coefficients we obtained in Example 7.1(c) for the even function $f(x) = |x|$ on $[-\pi, \pi]$. This is consistent with the observation that when we write a half range cosine series for the half range function f, we are in fact writing the Fourier series for the even extension of f to the full range.

EXPONENTIAL FORM OF THE FOURIER SERIES

The Euler identity

$$e^{ix} = \cos x + i \sin x \qquad i = \sqrt{-1}$$

implies

$$\cos x = \frac{e^{ix} + e^{-ix}}{2} \quad \text{and} \quad \sin x = \frac{e^{ix} - e^{-ix}}{2i}$$

This suggests that the series (7.1) can be written in the form

$$\sum_{m=-\infty}^{\infty} c_m e^{im\pi x/L} \tag{7.8}$$

where the coefficients c_m can be expressed in terms of the a_m and b_m. In fact, the c_m can be obtained directly by noting the following orthogonality relation for the family $\{e^{im\pi x/L}\}$

$$\int_{-L}^{L} e^{im\pi x/L} e^{-in\pi x/L}\, dx = \begin{cases} 0 & \text{if } m \neq n \\ 2L & \text{if } m = n \end{cases} \tag{7.9}$$

It follows easily that

$$c_m = \frac{1}{2L}\int_{-L}^{L} f(x)\, e^{-im\pi x/L}\, dx, \qquad m = \text{integer} \tag{7.10}$$

For some applications the exponential form of the Fourier series is more convenient than the trigonometric form.

Example 7.3 Exponential Form of the Fourier Series

Consider the function defined on $[-\pi, \pi]$ by

$$f(x) = \begin{cases} 1 & \text{if } |x| \le a \\ 0 & \text{if } a < |x| \le \pi \end{cases} \qquad \text{for } 0 < a < \pi.$$

Then

$$c_m = \frac{1}{2\pi}\int_{-a}^{a} e^{-imx}dx = \frac{-1}{i2m\pi}e^{-imx}\Big|_{x=-a}^{x=a}$$

$$= \frac{\sin ma}{m\pi} \quad \text{for } m^2 = 1, 2, \ldots$$

where

$$c_0 = \frac{a}{\pi}$$

Thus, we have (formally) the following exponential series for $f(x)$:

$$f(x) = \frac{a}{\pi} + \sum_{m<0} \frac{\sin ma}{m\pi}e^{imx} + \sum_{m>0} \frac{\sin ma}{m\pi}e^{imx}$$

Since

$$\frac{\sin ma}{m\pi} = \frac{\sin(-ma)}{-m\pi} \quad \text{for every integer } m$$

we can write

$$f(x) = \frac{a}{\pi} + \sum_{m>0} \frac{\sin ma}{m\pi}(e^{imx} + e^{-imx})$$

$$= \frac{a}{\pi} + \sum_{m>0} \frac{2}{m\pi} \sin ma \cos(mx)$$

This last series is the trigonometric form of the Fourier series for $f(x)$, [note that since $f(x)$ is even, it is a cosine series]. This example illustrates the equivalence of the exponential and trigonometric forms of the series.

CONVERGENCE OF FOURIER SERIES

We are going to state a Fourier convergence theorem prescribing conditions sufficient to imply that the Fourier series for $f(x)$ converges to the sum $f(x)$. First, we define the sense in which the series is to converge. It will be convenient to define more than a single mode of convergence.

POINTWISE CONVERGENCE

For positive integer m, let $S_m(x)$ denote the mth partial sum of the Fourier series for $f(x)$:

$$S_m(x) = \frac{1}{2}a_0 + \sum_{n=1}^{m} a_n \cos\frac{n\pi x}{L} + \sum_{n=1}^{m} b_n \sin\frac{n\pi x}{L}$$

We say that the Fourier series for $f(x)$ is *pointwise convergent* on $I = [-L, L]$ to the sum $f(x)$ if, for each x in I, the difference $|S_m(x) - f(x)|$ tends to zero as m tends to infinity.

UNIFORM CONVERGENCE

For m a positive integer let M_m denote the maximum value for all x in I of the difference $|S_m(x) - f(x)|$. If M_m tends to zero as m tends to infinity then we say the Fourier series for $f(x)$ *converges uniformly* on I to the sum $f(x)$. Clearly uniform convergence implies pointwise convergence but the converse is false.

MEAN SQUARE CONVERGENCE

For m a positive integer let J_m denote the value of the integral

$$J_m = \int_I |S_m(x) - f(x)|^2 dx$$

We say that the Fourier series for $f(x)$ converges in the *mean square sense* to $f(x)$ if J_m tends to zero as m tends to infinity. Uniform convergence implies mean square convergence, but the converse is false. Mean square convergence does not imply pointwise convergence nor does pointwise convergence imply mean square convergence.
We will now define properties of $f(x)$ sufficient to imply these various modes of convergence for the Fourier series. Recall that a function is said to be square integrable on interval I if the integral of the square of $f(x)$ over I is finite.

SIMPLE DISCONTINUITIES

The function $f(x)$ is said to have a *simple discontinuity* at the point $x = a$ if the following one sided limits exist but are unequal:

$$f(a+) = \lim_{h \to 0} f(a+h) = f(a+) \neq f(a-) = \lim_{h \to 0} f(a-h), \quad h > 0$$

Sometimes a simple discontinuity is called a *finite jump discontinuity* because $f(x)$ jumps from one finite value to another at the point $x = a$. At a point $x = a$ where $f(x)$ is continuous, we have $f(a+) = f(a-)$.

PIECEWISE CONTINUOUS FUNCTIONS

A function $f(x)$ is said to be *piecewise continuous* on the closed bounded interval I if $f(x)$ is continuous on I except for a finite number of simple discontinuities. We denote the class of functions that are piecewise continuous on I by $PC[I]$ and write $f \in PC[I]$ to indicate that f is piecewise continuous on I. If f is continuous at each point of I, we will write $f \in C[I]$. Clearly, if f is continuous, then it is piecewise continuous, and if f is piecewise continuous on I, then f is square integrable on I.

Example 7.4

(a) The function $f(x) = |x|^{-1/4}$ is square integrable on $[-L, L]$ since

$$\int_{-L}^{L} |f(x)|^2 dx = \int_{-L}^{L} |x|^{-1/2} dx = 2\int_0^L |x|^{-1/2} dx$$

$$= 4|x|^{-1/2}\Big|_0^L = 4\sqrt{L}$$

However, $f(x)$ is not piecewise continuous because the one sided limits at $x = 0$ do not exist.

(b) The function $g(x) = |x|/x$ satisfies $g(x) = -1$ if $x < 0$ and $g(x) = 1$ for $x > 0$. Then, however, $g(x)$ is defined at $x = 0$; g is piecewise continuous on $[-L, L]$ but g is not continuous. We say that the everywhere defined function $f(x)$ is *piecewise continuous* and write $f \in PC$ if $f \in PC[I]$ for every closed bounded interval I. Note that while $f(x)$ continuous on $[-L, L]$ is not sufficient to imply that $f_{2L}(x)$ is continuous everywhere, it is sufficient to have $f(x)$ piecewise continuous on $[-L, L]$ in order to have $f_{2L} \in PC$. Note that the derivative of a piecewise continuous function is not defined at points where f has a simple discontinuity. However, if the one sided limits of $f'(x)$ exist at each point, then we will say that both f and f' are piecewise continuous.

Theorem 7.2 Fourier Convergence Theorem

Theorem 7.2. Suppose $f(x)$ is defined on $I = [-L, L]$.

(a) If f is square integrable on I, then the Fourier series for f converges to f in the mean square sense on I.

(b) If the periodic extensions f_{2L} and f'_{2L} are piecewise continuous, then the Fourier series for $f(x)$ converges pointwise to the value:

$$\frac{1}{2}\left(f_{2L}(x+) + f_{2L}(x-)\right) \qquad \text{at each point } x \text{ in } \Re^1$$

(c) In order for the Fourier series for f to converge uniformly to f on I:

(i) it is *necessary* that the extension f_{2L} is everywhere continuous

(ii) it is *sufficient* that $f_{2L} \in C$ and $f'_{2L} \in PC$

Note that conclusion (b) implies that at each point x in $[-L, L]$, the Fourier series for $f(x)$ converges to $(f(x+) + f(x-))/2$. In particular, at each point x in $[-L, L]$ where $f(x)$ is continuous, the Fourier series converges pointwise to $f(x)$.

Example 7.5

(a) For $f(x)$ as in Example 7.1(a), we have $f \in PC[-\pi, \pi]$ and $f_{2\pi}$ is piecewise continuous but not continuous. In addition, $f'(x) = 0$ for $x \neq 0$ and $f'(x)$ has one sided limits at $x = 0$ as follows: $f'(0+) = 0, f'(0-) = 0$. Thus, $f' \in PC[-\pi, \pi]$ and $f'_{2\pi} \in PC$. Then part (b) of Theorem 7.2

implies that the Fourier series for f converges pointwise to the average of the left and right hand one sided limits for f at each x in $[-\pi, \pi]$. In particular, when x equals an integer multiple of π, all of the sine terms in the series vanish and the series sums to the value 1/2. But this is just the average of the right and left hand one sided limits for f_{2L} at such points. Because f_{2L} is not continuous, part (c) of the theorem implies the convergence of the Fourier series is not uniform.

(b) For $f(x)$ as in Example 7.1(b) we have $f \in C[-\pi, \pi]$ hence $f_{2\pi} \in PC$. Note that we do not have $f_{2\pi} \in C$, however, since $f_{2\pi}$ has a simple discontinuity at each integer multiple of π at such points we have

$$f_{2\pi}(x+) = -\pi \quad \text{and} \quad f_{2\pi}(x-) = \pi.$$

In addition, $f'_{2\pi}(x) = 1$ at all points where $f_{2\pi}$ is continuous and at the simple discontinuities, the one sided limits of $f'_{2\pi}$ exist (and are, in fact, equal). Then $f'_{2\pi} \in PC$ and it follows that the Fourier series for $f(x)$ converges pointwise to the sum

$$\frac{1}{2}\left(f_{2\pi}(x+) + f_{2\pi}(x-)\right) \quad \text{at each point } x \text{ in } \Re^1$$

Then, at integer multiples of π, the Fourier series for $f(x)$ converges to zero. At all other points, it converges to $f_{2\pi}(x)$. Again, because $f_{2\pi}$ is not continuous, the convergence of the Fourier series is not uniform.

(c) For $f(x)$ as in Example 7.1(c), we have $f \in C[-\pi, \pi]$, and since $f(\pi) = f(-\pi)$, we also have $f_{2\pi} \in C$. Note that

$$f'_{2\pi}(x) = \begin{cases} -1 & \text{if } (2k-1)\pi < x < 2k\pi \\ 1 & \text{if } 2k\pi x < (2k-1)\pi \end{cases}$$

Thus, $f'_{2\pi}(x)$ is continuous except at integer multiples of π. At these points the left and right hand one sided limits for $f'_{2\pi}$ exist and it follows that $f'_{2\pi} \in PC$. Then, by part (c) of Theorem 7.2, the Fourier series for $f(x)$ converges uniformly to $f_{2\pi}(x)$.

Note that term-by-term differentiation of the series for $f(x)$,

$$|x| = \frac{\pi}{2} - \frac{4}{\pi} \sum_{n=1}^{\infty} \frac{1}{(2n-1)^2} \cos(2n-1)x$$

leads to

$$\frac{4}{\pi} \sum_{n=1}^{\infty} \frac{1}{(2n-1)} \sin(2n-1)x$$

and it is easy to show that this is the Fourier series for

and it is easy to show that this is the Fourier series for

$$f'_{2\pi}(x) = \begin{cases} -1 & \text{if } -\pi < x < 0 \\ 1 & \text{if } 0 < x < \pi \end{cases}$$

It remains to be seen under what general conditions on f such a result must hold.

(d) Consider the function $f(x) = \sqrt{|x|}$ on $I = [-1, 1]$. Then $f \in C[-1, 1]$ and $f_2 \in C$. However,

$$f'(x) = \begin{cases} -|x|^{-1/2} & \text{if } -1 < x < 0 \\ |x|^{-1/2} & \text{if } 0 < x < 1 \end{cases}$$

and it is easy to see that the one sided limits for $f'(x)$ fail to exist at $x = 0$. Then $f'(x)$ does not belong to $PC[-1, 1]$, so parts (b) and (c) of Theorem 7.2 fail to apply to this function. However, f is clearly square integrable on I and thus, by part (a) of the theorem, the Fourier series must converge in the mean square sense on I.

UNIQUENESS OF FOURIER SERIES

Suppose f and g denote two functions that are defined and piecewise continuous with piecewise continuous derivatives on $[-L, L]$. Suppose also that $f(x) = g(x)$ at all but finitely many points x in $[-L, L]$. Because altering the integrand at a finite set of points does not affect the value of a definite integral, f and g have the same Fourier coefficients and, thus, the same Fourier series on $[-L, L]$. Then it is not the case that functions having the same Fourier series on $[-L, L]$ are necessarily equal at all points of the interval. However, we have:

Theorem 7.3

Theorem 7.3. If $f, g \in C[-L, L]$ have the same Fourier coefficients on $[-L, L]$, then $f(x) = g(x)$ at all points x in $[-L, L]$.

Operations on Fourier Series

It can be shown that uniform convergence is sufficient to permit term-by-term integration of an infinite series and term-by-term differentiation of a series is permitted if the differentiated series converges uniformly. These operations are valid for Fourier series under less restrictive conditions.

Theorem 7.4 Integration of Fourier Series

Theorem 7.4. Suppose that $f \in PC[-L, L]$. Then term-by-term integration of the Fourier series for $f(x)$ from $-L$ to x leads to the series

$$\frac{1}{2}a_0(x+L) + \sum_{n=1}^{\infty} \frac{1}{n}\left(a_n \sin\frac{n\pi x}{L} - b_n\left(\cos\frac{n\pi x}{L} - (-1)^n\right)\right)$$

This series converges uniformly on $[-L, L]$ to the sum

$$F(x) = \int_{-L}^{x} f$$

regardless of whether the Fourier series for $f(x)$ converges.

Theorem 7.5 Differentiation of Fourier Series

Theorem 7.5. Suppose $f(x)$ is defined on $[-L, L]$ such that $f_{2L} \in C$. Suppose also that f'_{2L} and $f''_{2L} \in PC$. Then term-by-term differentiation of the Fourier series for $f(x)$ leads to the series

$$\sum_{n=1}^{\infty} \frac{n\pi}{L}\left(b_n \cos\frac{n\pi x}{L} - a_n \sin\frac{n\pi x}{L}\right)$$

This series converges pointwise on $[-L, L]$ to the limit $(f'(x+) + f'(x-))/2$ at each x in $[-L, L]$.

GENERALIZED FOURIER SERIES

In the previous chapter, we introduced the notion of orthogonal and orthonormal families of functions in the vector space of square integrable functions, $R^2(I)$, and in connection with orthonormal families we defined the *generalized Fourier series* of a square integrable function f. The trigonometric Fourier series of this chapter are but a special case of this more general concept.

Example 7.5 An Orthogonal Family

For $I = [-L, L]$, the orthogonality relations (7.2) imply that the family:

$$\begin{cases} h_{2n}(x) = \cos\dfrac{n\pi x}{L} \\ \\ h_{2n+1}(x) = \sin\dfrac{n\pi x}{L} \end{cases} \quad n = 0, 1, \ldots$$

i.e.,

$$1, \sin\frac{\pi x}{L}, \cos\frac{\pi x}{L}, \sin\frac{2\pi x}{L}, \cos\frac{2\pi x}{L}, \ldots$$

is an orthogonal family in $R^2[-L, L]$. In fact, this family is just the orthogonal family generated by the Sturm-Liouville problem of solved Problem 6.7.

GENERALIZED FOURIER SERIES

Let $\{g_1, g_2, \ldots\}$ be a given orthonormal family in $R^2(I)$. Then, for arbitrary f in $R^2(I)$ we define the *generalized Fourier series* for f with respect to the family $\{g_n\}$ to be the series

$$\sum_{n=1}^{\infty} (f, g_n)\, g_n(x)$$

In the previous chapter we were not in a position to discuss the convergence of generalized Fourier series. We consider this question now.

BESSEL'S INEQUALITY, RIEMANN-LEBESGUE LEMMA

Theorem 7.6

Theorem 7.6. Let $\{g_1, g_2, \ldots\}$ be an orthonormal family in $R^2(I)$. Then for arbitrary f in $R^2(I)$, let $f_n = (f, g_n)$ for $n = 1, 2, \ldots$. For each positive integer N and any choice of constants $a_1, a_2, \ldots, a_N$, we have

(a) $$\left\| f(x) - \sum_{n=1}^{N} a_n g_n(x) \right\|^2 = \|f\|^2 - \sum_{n=1}^{N} f_n^2 + \sum_{n=1}^{N} (f_n - a_n)^2$$

(b) $$\left\| f(x) - \sum_{n=1}^{N} a_n g_n(x) \right\| \geq \left\| f(x) - \sum_{n=1}^{N} f_n g_n(x) \right\|$$

Corollary 7.7 For an orthonormal family $\{g_1, g_2, \ldots\}$ in $R^2(I)$ and for arbitrary f in $R^2(I)$, let $f_n = (f, g_n)$. Then

$$\sum_{n=1}^{\infty} |f_n|^2 \leq \|f\|^2$$

Corollary 7.8 For an orthonormal family $\{g_1, g_2, \ldots\}$ in $R^2(I)$ and arbitrary f in $R^2(I)$, let $f_n = (f, g_n)$. Then $f_n \to 0$ as $n \to \infty$.

Corollary 7.8 is a version of the *Riemann-Lebesgue lemma* and Corollary 7.7 is known as *Bessel's inequality*.

Complete Orthonormal Families

The Bessel's inequality implies that the generalized Fourier series for f converges in the mean square sense. However, it does not imply that the series converges to f. We say that the orthonormal family $\{g_1, g_2, \ldots\}$ in $R^2(I)$ is *complete* if, for every f in $R^2(I)$, the series

$$\sum_{n=1}^{\infty} (f, g_n)\, g_n(x)$$

converges in the mean square sense to f.

Theorem 7.9

Theorem 7.9. The following statements are equivalent:

(a) The orthonormal family $\{g_1, g_2, \ldots\}$ in $R^2(I)$ is generated by a Sturm-Liouville problem.

(b) The orthonormal family $\{g_1, g_2, \ldots\}$ in $R^2(I)$ is complete.

(c) For every f in $R^2(I)$, $\sum_{n=1}^{\infty} f^2 = \|f\|^2$, where $f_n = (f, g_n)$.

(d) For every f in $R^2(I)$, $f_n = (f, g_n) = 0$ for all n if and only if $\|f\| = 0$.

(e) For f, g in $R^2(I)$, $(f, g) = \sum_{n=1}^{\infty} f_n g_n$.

It follows from this theorem that the eigenfunctions of any Sturm-Liouville problem are a complete orthogonal family and can be used for writing generalized Fourier series. Such series are often referred to as eigenfunction expansions. We have the following convergence theorem for eigenfunction expansions.

Theorem 7.10

Theorem 7.10. Let $g_1, g_2, \ldots$ denote the normalized eigenfunctions of the Sturm-Liouville problem

$$-(p(x)y'(x))' + q(x)y(x) = \lambda r(x)y(x),$$
$$B[y; a] = 0,\ B[y; b] = 0$$

i.e., $\langle g_m, g_n\rangle$ equals 1 if m equals n and is zero otherwise. Then, for f defined on $I = [a, b]$, the eigenfunction expansion of f in terms of $g_n(x)$

$$\sum_{n=1}^{\infty} \langle f, g_n\rangle g_n(x)$$

converges as follows:

(a) in the mean square sense if f is square integrable on I,

(b) pointwise to $(f(x+) + f(x-))/2$ at each x if f and $f' \in PC(I)$,

(c) uniformly to $f(x)$ on I if $f \in C(I)$, if $f' \in PC(I)$ and if $BC[f;a] = 0$, $BC[f;b] = 0$.

Example 7.8

(a) $1, \sin(\pi x)/\lambda, \cos(\pi x)/L, \sin(2\pi x)/L, \cos(2\pi x)/L, \ldots$ The orthogonal family associated with the full range Fourier series is reduced to an orthonormal family by dividing each function in the family by its norm. Then

$$\frac{1}{\sqrt{2L}}, \frac{1}{\sqrt{L}} \sin\frac{\pi x}{L}, \frac{1}{\sqrt{L}} \cos\frac{\pi x}{L}, \ldots$$

is an orthonormal family. Since these are the eigenfunctions of a Sturm-Liouville problem, it follows from Theorem 7.9 that the family is complete and has the convergence properties of Theorem 7.10.

(b) The family $\{\sin(n\pi x/L): n = 1, 2, \ldots\}$, orthogonal on the interval $(0, L)$, is associated with the half range Fourier sine series. Each function has norm equal to $\sqrt{L}$; thus $\{L^{-1/2} \sin(n\pi x/L): n = 1, 2, \ldots\}$ is the corresponding orthonormal family. These are the eigenfunctions of the Sturm-Liouville problem $-y''(x) = \lambda y(x), y(0) = y(L) = 0$, and thus, by Theorem 7.9, they are complete and have the convergence properties of Theorem 7.10.

(c) The family $\{J_a(\mu_n x): n = 1, 2, \ldots\}$ of Bessel functions are the eigenfunctions of Problem 6.8. Then these functions are orthogonal on $(0, b)$ with respect to the weight function $r(x) = x$ and

$$\langle J_a(\mu_n x), J_a(\mu_m x)\rangle = \int_0^b xJ_a(\mu_n x) J_a(\mu_m x)\, dx \tag{7.11}$$

$$= \begin{cases} 0 & \text{if } m \text{ does not equal } n \\ N(a, n) & \text{if } m = n \end{cases}$$

Then for $f(x)$ defined on $(0, b)$ we can write the generalized Fourier series in terms of the orthogonal family of eigenfunctions

$$f(x) = \sum_{n=1}^{\infty} c_n J_a(\mu_n x)$$

and use (7.11) to solve for the coefficients c_n,

$$c_n = \frac{\langle f, J_a(\mu_n x)\rangle}{N(a, n)}, \quad n = 1, 2, \ldots$$

This series converges as prescribed by Theorem 7.10.

SOLVED PROBLEMS

Formal Properties of Fourier Series

PROBLEM 7.1

Derive the formulas (7.3) for the Fourier coefficients of $f(x)$ defined on $[-L, L]$.

SOLUTION 7.1

Suppose that for f defined on $I = [-L, L]$,

$$f(x) = \frac{1}{2}a_0 + \sum_{n=1}^{\infty} a_n \cos\frac{n\pi x}{L} + b_n \sin\frac{n\pi x}{L} \tag{1}$$

and we assume for convenience that the convergence of this series is uniform on I. In order to find a_m for any fixed integer m, we multiply both sides of the equation by $\cos(m\pi x/L)$ and integrate over I. Thus,

$$\int_I f(x) \cos\frac{m\pi x}{L} = \frac{1}{2}a_0 \int_I \cos\frac{m\pi x}{L} + \sum_{n=1}^{\infty} a_n \int_I \cos\frac{m\pi x}{L} \cos\frac{n\pi x}{L} dx \tag{2}$$

$$+ \sum_{n=1}^{\infty} b_n \int_I \cos\frac{m\pi x}{L} \sin\frac{n\pi x}{L} dx$$

The orthogonality relations (7.2) imply that for any choice of the integer m,

$$\int_I \cos\frac{m\pi x}{L} \sin\frac{n\pi x}{L} dx = 0 \quad \text{for all } n.$$

If $m = 0$ then

$$\int_I \cos\frac{m\pi x}{L} \cos\frac{n\pi x}{L} dx = 0 \quad \text{for } n = 1, 2, \ldots$$

and

$$\int_I \cos\frac{m\pi x}{L} dx = \int_I 1\, dx = 2L$$

Then (2) reduces to

$$\int_I f dx = \frac{1}{2}a_0 2L = a_0 L \tag{3}$$

If $m > 0$ then

$$\int_I \cos\frac{m\pi x}{L} \cos\frac{n\pi x}{L} dx = 0 \quad \text{for } n = 1, 2, \ldots\ n \neq m$$

$$\int_I \cos\frac{m\pi x}{L} dx = 0$$

$$\int_I \cos\frac{m\pi x}{L} \cos\frac{n\pi x}{L} dx = L \quad \text{for } n = m$$

Then (2) reduces to

$$\int_I f(x) \cos\frac{m\pi x}{L} dx = a_m L. \tag{4}$$

Equations (3) and (4) are the formula (7.3) for the coefficients a_m. The formula for the b_m coefficients is obtained in a similar way.

PROBLEM 7.2

For N a positive integer, let

$$S_N(x) = \frac{1}{2}a_0 + \sum_{n=1}^{N} a_n \cos\frac{n\pi x}{L} + b_n \sin\frac{n\pi x}{L}$$

denote the Nth partial sum of the Fourier series for $f(x)$. For $N = 1, 2, 4,$ and 20 sketch the graphs of the partial sums for the three functions of Example 7.1.

SOLUTION 7.2

Parts (a), (b), and (c) of Figure 7.1 show the periodic extensions for these functions over the range $(-2\pi, 2\pi)$. In Figure 7.2, we have sketched the graphs of $S_N(x)$ for $N = 1, 2, 4,$ and 20 on the range $(0, 2\pi)$. Figure 7.2(a) shows the partial sums for the square wave function of Example 7.1(a). Note that as N increases the graph of $S_N(x)$ approaches the graph of $f(x)$ more closely, but in a neighborhood of the discontinuities for $f(x)$ $(x = 0, \pi, 2\pi)$, the distance between $f(x)$ and $S_N(x)$ is greatest. This is also evident in Figure 7.2(b), showing the graphs of the partial sums for the discontinuous sawtooth wave $f(x) = x$. Near $x = \pi$ where the periodic extension of this function has a discontinuity, the discrepancy between the graphs of $S_N(x)$ and $f_{2\pi}(x)$ is most pronounced. On the other hand, Figure 7.1(c) shows the periodic extension of $f(x) = |x|$ to be everywhere continuous. Correspondingly, in Figure 7.2(c) we see that the graphs of

Generate the Fourier series for the function

$$f(x) = \begin{cases} -\pi - x & \text{if } -\pi < x < -\pi/2 \\ x & \text{if } -\pi/2 < x < \pi/2 \\ \pi - x & \text{if } \pi/2 < x < \pi \end{cases}$$

Sketch the graph of $f_{2\pi}$ on $[-2\pi, 2\pi]$. Use the notion of symmetry to explain any missing terms in the resulting Fourier series.

SOLUTION 7.3

The graph of $f_{2\pi}$ on $[-2\pi, 2\pi]$ is sketched in Figure 7.3 where it can be seen that $f(x)$ is an odd function. Then $a_m = 0$ for every m and it only remains to compute the coefficients b_m. We have

$$b_m = \frac{2}{\pi}\int_0^{\pi} f(x) \sin xmx \, dx$$

$$= \frac{2}{\pi}\int_0^{\pi/2} x\sin mx \, dx + \frac{2}{\pi}\int_{\pi/2}^{\pi} (\pi - x) \sin mx \, dx$$

$$= \frac{2}{\pi}\left(-\frac{x \cos mx}{m}\right)\Bigg|_{x=0}^{x=\pi/2} + \frac{2}{\pi m}\int_0^{\pi/2} \sin mx \, dx$$

$$+ \frac{2}{\pi}\left(-\frac{(\pi - x) \cos mx}{m}\right)\Bigg|_{x=\pi/2}^{x=\pi} - \frac{2}{\pi m}\int_{\pi/2}^{\pi} \sin mx \, dx$$

$$= \frac{4}{\pi m^2} \sin\frac{m\pi}{2}$$

Then

$$f(x) = \frac{4}{\pi}\left(\sin x - \frac{1}{9}\sin 3x + \frac{1}{25}\sin 5x - \dots\right)$$

$$= \frac{4}{\pi}\sum_{n=1}^{\infty} \frac{(-1)^{n+1}}{(2n-1)^2} \sin (2n-1) x$$

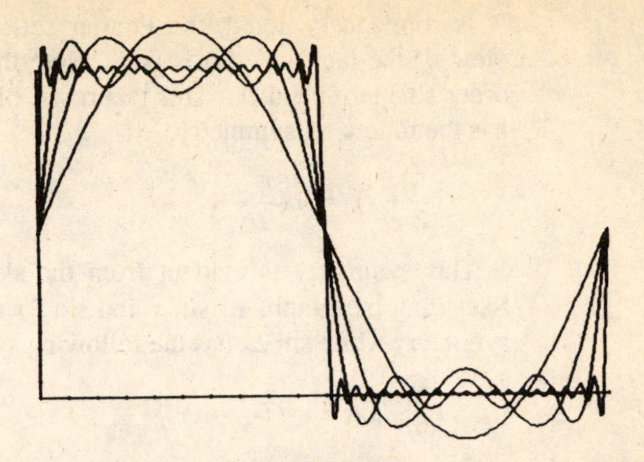

Figure 7.2(a)
Partial Sums for Square Wave

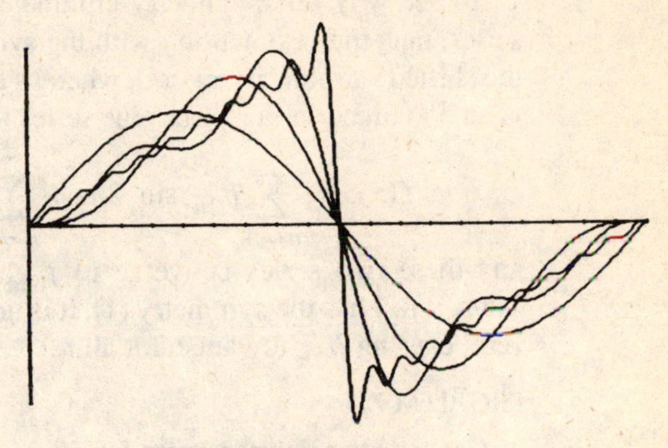

Figure 7.2(b)
Partial Sums for $f(x) = x$

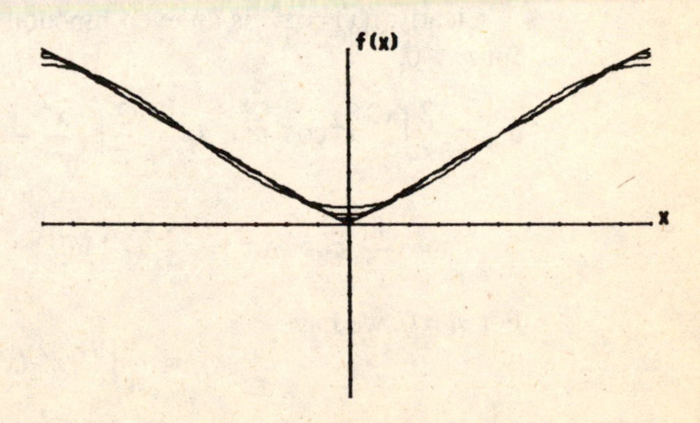

Figure 7.2(c)
Partial Sums for $f(x) = |x|$

As previously noted, the Fourier series contains no cosine terms in view of the fact that $f(x)$ is odd. Note that the series contains no even order sine terms either. This is a result of the fact that this function $f(x)$ has the following symmetry:

$$f\left(\frac{\pi}{2}+x\right) = f\left(\frac{\pi}{2}-x\right) \tag{1}$$

This symmetry is evident from the sketch of the graph of $f_{2\pi}(x)$. Sketching the graphs of sin x and sin $2x$ shows that sin x has this same symmetry while sin $2x$ has the following symmetry

$$f\left(\frac{\pi}{2}+x\right) = -f\left(\frac{\pi}{2}-x\right) \tag{2}$$

It is not hard to show that for each integer n, $\sin(2n-1)x$ has the symmetry (1) and sin $2nx$ has the symmetry (2).

To see why this symmetry eliminates certain terms from the Fourier series, note that a function f with the symmetry (1) is an even function of the shifted variable $z = x - \pi/2$, whereas a function f with the symmetry (2) is an odd function of z. The sine series for $f(x)$ can be written

$$f(x) = \sum_{n=1}^{\infty} b_{2n} \sin 2nx + \sum_{n=1}^{\infty} b_{2n-1} \sin(2n-1)x$$

and these two series converge to $f_{\text{odd}}(z)$ and $f_{\text{even}}(z)$, respectively. Thus, if $f(x)$ has the symmetry (1), it is an even function of z so $f_{\text{odd}}(z)$ is zero, causing b_{2n} to vanish for all n.

PROBLEM 7.4

Generate the Fourier series for $f(x) = x^2, -\pi < x < \pi$. S

SOLUTION 7.4

Clearly, $f(x) = x^2$ is an even function and so $b_m = 0$ for all m. Then for $m > 0$,

$$a_m = \frac{2}{\pi}\int_0^{\pi} x^2 \cos mx\, dx = \frac{2}{\pi}\left(\left(\frac{x^2}{m} - \frac{2}{m^3}\right)\sin mx + \frac{2x}{m^2}\cos mx\right)\Bigg|_{x=0}^{x=\pi}$$

$$= \frac{4}{m^2}\cos m\pi = \frac{4}{m^2}(-1)^m$$

For $m = 0$, we have

$$a_0 = \frac{2}{\pi}\int_0^{\pi} x^2 dx = \frac{2\pi^2}{3}$$

thus,

$$x^2 = \frac{\pi^2}{3} - 4\left(\cos x - \frac{1}{4}\cos 2x + \frac{1}{9}\cos 3x - \cdots\right)$$

$$= \frac{\pi^2}{3} + 4\sum_{n=1}^{\infty} \frac{(-1)^n}{n^2} \cos nx$$

Note that for $x = \pi$ this leads to the result

$$\frac{\pi^2}{6} = \sum_{n=1}^{\infty} \frac{1}{n^2}$$

PROBLEM 7.5

Generate the Fourier series for

$$f(x) = \begin{cases} 0 & \text{for } -L < x < 0 \\ x & \text{for } 0 < x < L \end{cases}$$

SOLUTION 7.5

If one sketches the graph of $f(x)$, it is apparent that f_{2L} is neither even nor odd. Then

$$a_0 = \frac{1}{L}\int_0^L x\,dx = \frac{L}{2}$$

$$a_m = \frac{1}{L}\int_0^L x\cos\frac{m\pi x}{L}dx = \frac{L(\cos m\pi - 1)}{(m\pi)^2}, \quad m = 1, 2, \ldots$$

$$b_m = \frac{1}{L}\int_0^L x\sin\frac{m\pi x}{L}dx = -\frac{L\cos m\pi}{m\pi}, \quad m = 1, 2, \ldots$$

and

$$f(x) = \frac{L}{4} - \frac{2L}{\pi^2}\sum_{n=1}^{\infty}\frac{1}{(2n-1)^2}\cos(2n-1)\frac{\pi x}{L} + \frac{L}{\pi}\sum_{n=1}^{\infty}\frac{(-1)^{n+1}}{n}\sin\frac{n\pi x}{L}$$

Note that $f(x) = (|x| + x)/2$ and that this Fourier series is just the sum of the Fourier series for $|x|/2$ and the Fourier series for $x/2$.

PROBLEM 7.6

Write a Fourier series on $I = [-\pi, \pi]$ for the functions

$$f(x) = \sin x \quad \text{and} \quad g(x) = \sin(\mu x) \quad \mu \neq \text{integer}$$

SOLUTION 7.6

It follows from the orthogonality relations (7.2) that the Fourier coefficients for $f(x)$ satisfy $a_m = 0$ for all m, $b_1 = 0$, $b_m = 0$ for $m > 1$.

Thus, the Fourier series for $\sin x$ on I contains just one term, $\sin x$. One may interpret this to mean that since $\sin x$ is a member of the orthogonal family $\{1, \cos x, \ldots; \sin x, \ldots\}$ the expression of $\sin x$ as a linear combination of functions in the family requires only one term.

Since $g(x)$ is an odd function, its Fourier series contains no cosine terms. That is, $a_m = 0$ for all m. Using (7.7) we compute

$$b_m = \frac{2}{\pi}\int_0^{\pi} \sin(\mu x)\sin mx dx = \frac{1}{\pi}\int_0^{\pi} (\cos(m-\mu)x - \cos(m+\mu)x)\,dx$$

$$= \frac{1}{\pi}\left(\frac{\sin(m-\mu)x}{m-\mu} - \frac{\sin(m+\mu)x}{m+\mu}\right)\Bigg|_{x=0}^{x=\pi}$$

$$= \frac{2m}{\pi}\frac{(-1)^{m+1}\sin\pi\mu}{m^2-\mu^2}, \qquad m = 1, 2, \ldots$$

Then the Fourier series for $g(x)$ has infinitely many terms. This is a result of the fact that for μ not an integer, $\sin \mu x$ is not a member of the orthogonal family $\{1, \cos x, \ldots; \sin x, \ldots\}$. Not only is $g(x)$ not one of the functions in the family, but also no finitely long linear combination of functions in the family can be equal to $g(x)$.

PROBLEM 7.7

Consider the function $f(x) = \pi - x$.

(a) Write a Fourier series for $f(x)$ on $[-\pi, \pi]$.

(b) Write a cosine series for $f(x)$ on $[0, \pi]$.

(c) Write a sine series for $f(x)$ on $[0, \pi]$.

SOLUTION 7.7

Part (a) Using (7.2) with $L = \pi$, we find

$$a_0 = \frac{1}{\pi}\int_{-\pi}^{\pi} (\pi - x)\,dx = 2\pi$$

$$_m = \frac{1}{\pi}\int_{-\pi}^{\pi} (\pi - x)\cos mx\,dx = \qquad \text{for } m = 1, 2, \ldots$$

$$_m = \frac{1}{\pi}\int_{-\pi}^{\pi} (\pi - x)\sin mx\,dx = \frac{2(-1)'}{m} \qquad \text{for } m = 1, 2, \ldots$$

Then

$$f(x) = \pi + 2 \sum_{m=1}^{\infty} \frac{(-1)^m}{m} \sin mx$$

Part (b) Using (7.6), we find

$$a_0 = \frac{2}{\pi}\int_0^{\pi} (\pi - x)\, dx = \pi$$

$$a_m = \frac{2}{\pi}\int_0^{\pi} (\pi - x) \cos mx dx = \frac{2}{\pi m^2}(1 - (-1)^m) \text{ for } m = 1, 2, \ldots$$

Then

$$f_{\text{even}}(x) = \frac{\pi}{2} + \frac{4}{\pi} \sum_{m=1}^{\infty} \frac{1}{(2m-1)^2} \cos (2m-1)x$$

Part (c) Using (7.7) we compute the sine series coefficients

$$b_m = \frac{2}{\pi} \cdot \int_0^{\pi} (\pi - x) \sin mx\, dx = \frac{2}{m} \quad \text{for } m = 1, 2, \ldots$$

thus,

$$f_{\text{odd}}(x) = 2 \sum_{m=1}^{\infty} \frac{1}{m} \sin mx$$

Note that the functions have the same definitions on the interval $[0, \pi]$ but each of the three differs from the other two on $[-\pi, 0]$. In particular, we have

$$f(x) = \pi - x \qquad -\pi \le x \le \pi \qquad f \,\varepsilon\, C[-\pi, \pi] \text{ and } f_{2\pi} \in PC$$

$$f_{\text{even}}(x) = \begin{cases} \pi + x & -\pi \le x < 0 \\ \pi - x & 0 \le x \le \pi \end{cases} \qquad f \in C[-\pi, \pi] \text{ and } f_{2\pi} \in C$$

$$f_{\text{odd}}(x) = \begin{cases} -\pi - x & -\pi \le x < 0 \\ \pi - x & 0 < x \le \pi \end{cases} \qquad f \in PC[-\pi, \pi] \text{ and } f_{2\pi} \in PC$$

Convergence of Fourier Series

PROBLEM 7.8

Discuss the convergence of the Fourier series generated in Problem 7.3

SOLUTION 7.8

The periodic extension $f_{2\pi}$ of the function from Problem 7.3 is shown in Figure 7.3 and is clearly continuous for all x. The derivative of this function can be obtained by inspection from the graph of the piecewise linear function $f_{2\pi}(x)$ or it can be calculated from the definition of $f(x)$:

$$f(x) = \begin{cases} -\pi-x & \text{if } -\pi < x < -\pi/2 \\ x & \text{if } -\pi/2 < x < \pi/2 \\ \pi-x & \text{if } \pi/2 < x < \pi \end{cases}$$

$$f'(x) = \begin{cases} -1 & \text{if } -\pi < x < -\pi/2 \\ 1 & \text{if } -\pi/2 < x < \pi/2 \\ -1 & \text{if } \pi/2 < x < \pi \end{cases}$$

Then $f'(x)$ is discontinuous at $\pi/2$ and $-\pi/2$ but the one sided limits of $f'(x)$ at these points clearly exist. Then $f'_{2\pi}$ is piecewise continuous. In addition, the second derivative $f''(x)$ exists and equals zero at each point where $f'(x)$ is continuous, and at points of discontinuity for $f'(x)$, the one sided limits for f'' exists (and equal zero from both sides).

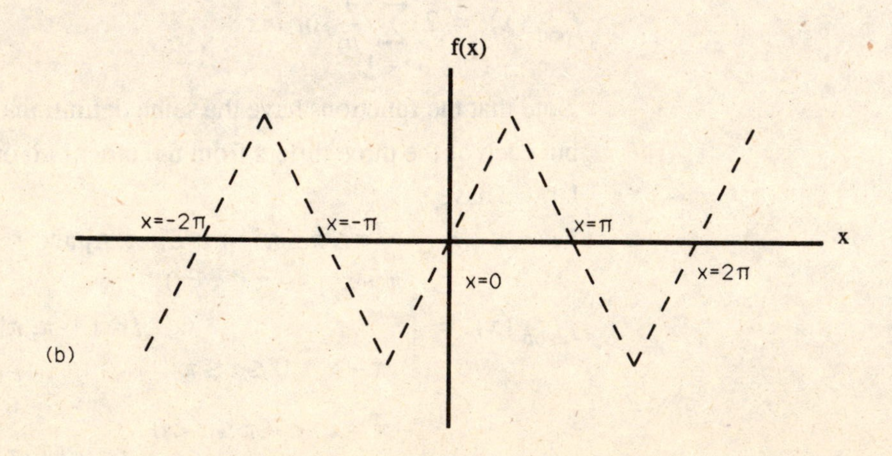

Figure 7.3
Periodic Extension Showing Symmetry about $\pi/2$

Then it follows from Theorem 7.2 that the Fourier series for $f(x)$ converges uniformly to $f_{2\pi}(x)$. In addition, Theorem 7.5 implies that this series can be differentiated term by term and the series obtained converges pointwise to the average of the left and right hand limits of the derivative $f'_{2\pi}(x)$. That is, the differentiated series

$$\frac{4}{\pi}\sum_{n=1}^{\infty}\frac{(-1)^{n+1}}{(2n-1)}\cos(2n-1)x$$

converges pointwise to the square wave function $f'_{2\pi}(x)$. In particular, at the points $-\pi/2$ and $\pi/2$ where $f'(x)$ is discontinuous, the differentiated series is convergent to the value zero since $(f'(-\pi/2-)+f'(-\pi/2+))/2 = (+1-1)/2 = 0$ and $(f'(\pi/2-)+f'(\pi/2+))/2 = (-1+1)/2 = 0$.

PROBLEM 7.9

Discuss the convergence of the Fourier series generated in Problem 7.4.

SOLUTION 7.9

The function $f(x) = x^2$ on $[-\pi, \pi]$ has a continuous 2π-periodic extension. The derivative of this function is $f'(x) = 2x$, $-\pi \leq x \leq \pi$, and this function has a 2π-periodic extension that is piecewise continuous but not continuous. That is, the derivative has finite jump discontinuities at integer multiple of π, but the one sided limits of the derivative exist. This derivative is just two times the function pictured in Figure 7.1(b). The second derivative $f''(x)$ exists and equals zero at each point where $f'(x)$ is continuous and thus $f''_{2\pi}(x)$ is piecewise continuous as well.

Theorem 7.2 implies the uniform convergence to $f_{2\pi}$ of the Fourier series for f and Theorem 7.5 implies that this series can be differentiated term by term. The resulting series converges pointwise to $(f'(x+) + f'(x-))/2$ at each x. The differentiated series

$$-4\sum_{n=1}^{\infty}\frac{(-1)^{n}}{n}\sin nx$$

is twice the series obtained in Example 7.1(b) for $f(x) = x$.

Generalized Fourier Series

PROBLEM 7.10

For fixed positive integer n let $x_j = j/n$, $j = 0, 1, \ldots n$, and consider the family $\{g_1, \ldots, g_n\}$ of square integrable functions on $I = [0, 1]$ defined by

$$g_j(x) = \begin{cases} 1 & \text{if } x_{j-1} < x < x_j \\ 0 & \text{otherwise} \end{cases} \qquad j = 1, 2, \ldots, n$$

Show that this family of piecewise continuous functions is orthogonal but not orthonormal nor complete. For an arbitrary $f(x)$ in $R^2(I)$ what is the relationship between f and the sum

$$n \sum_{j=1}^{n} (f, g_j)\, g_j(x) \tag{1}$$

SOLUTION 7.10

Note that I is divided into n equal subintervals of width $w = 1/n$ by the points x_1 to x_n. Each of the functions g_j is equal to 1 on one of the subintervals and is zero on all the others. Then the functions $g_j(x)$ are piecewise continuous but not continuous, and the product $g_j(x)\, g_k(x)$ is zero for all x in I if j is different from k. Then the inner product (g_j, g_k), which is just the integral over I of the product $g_j(x)\, g_k(x)$, is zero if j is different from k. If j equals k, then

$$(g_j, g_j) = \int_I g_j(x)^2 dx = \int_{x_{j-1}}^{x_j} 1\, dx = w$$

Thus, the functions are mutually orthogonal but not orthonormal. To see that the family is not complete we need only find a nontrivial function $f(x)$ that is orthogonal to all of the functions in the family. Then part (d) of Theorem 7.9 will imply the family is not complete. The function $f(x)$ that equals 1 on the interval $(0, x_1/2)$, equals -1 on $(x_1/2, x_1)$ and is zero otherwise is one example of a nontrivial function such that $(f, g_j) = 0$ for each j.

The sum in (1) equals a piecewise continuous function $F(x)$ that is constant on each of the subintervals $I_j = (x_{j-1}, x_j)$. The value of $F(x)$ for x in I_j is equal to

$$(f, g_j) = n \int_{x_{j-1}}^{x_j} f(x)\, dx = \frac{1}{w} \int_{x_{j-1}}^{x_j} f(x)\, d.$$

which is just the average value for $f(x)$ on the interval (x_{j-1}, x_j) of width w.

PROBLEM 7.11

For fixed positive integer n, let $x_j = j/n, j = 0, 1, \ldots n$, and consider the family $\{h_0, h_1, \ldots, h_n\}$ of square integrable functions on $I = [0, 1]$ defined by

$$h_j(x) = \begin{cases} (x - x_{j-1})/w & \text{if } x_{j-1} < x < x_j \\ (x_{j+1} - x)/w & \text{if } x_j < x < x_{j+1} \quad j = 0, 1, \ldots, n \\ 0 & \text{otherwise} \end{cases}$$

Show that this family of continuous piecewise linear functions is neither orthonormal nor orthogonal on $I = [0, 1]$ but for arbitrary f in $R^2(I)$ the expression

$$Q(c_0, c_1, \ldots, c_n) = \left\| f - \sum_{j=0}^{n} c_j h_j \right\|^2$$

is minimized if the vector of coefficients, $C = [c_0, c_1, \ldots, c_n]^T$, solves $AC = F$, where A_{ij}, the ij entry of the $n+1 \times n+1$ matrix A, is equal to (h_i, h_j) and F_i, the ith entry of the $n+1$ vector F, is equal to (f, h_i).

SOLUTION 7.11

The functions $\{h_0, h_1, \ldots, h_n\}$ are seen to be continuous on I and piecewise linear; i.e., h_0 decreases linearly from 1 to 0 on I_1, h_n increases linearly from 0 to 1 on I_n, and for $j = 1, \ldots, n-1$, h_j increases linearly from 0 to 1 on I_j, and then decreases in a linear manner from 1 to 0 on I_{j+1}. Note that the product $h_i(x)\,h_j(x)$ is equal to zero for all x in I if the difference $|i - j|$ is greater than one. In this case the inner product (h_i, h_j) is zero. If the difference $|i - j|$ is less than or equal to one (i.e., if i equals j or if i equals j plus or minus one), then the inner product (h_i, h_j) is not zero. Thus, the functions are not mutually orthogonal nor are they normalized since $(h_i, h_j) = h^2$ is not equal to one.

Note that for arbitrary f in $R^2(I)$

$$\begin{aligned} Q(c_0, c_1, \ldots, c_n) &= (f - \textstyle\sum c_j h_j, f - \sum c_i h_i) \\ &= \|f\|^2 - 2\textstyle\sum c_j (f, h_j) + \sum c_j \sum c_j (h_j, h_i) \\ &= \|f\|^2 - 2C^\mathrm{T} F + C^\mathrm{T} A C \end{aligned}$$

and in Problem 1.16 we showed that **grad** $Q = 2(AC - F)$. Then, clearly, the quadratic functional Q is minimized by $C = [c_0, c_1, \ldots, c_n]^\mathrm{T}$ such that $AC - F$. The sum $\sum c_j h_j(x)$ with this choice of constants c_j is then the *least squares approximation* of f by a linear combination of the functions $\{h_0, h_1, \ldots, h_n\}$. We shall encounter this concept again in connection with the finite element method for solving differential equations.

PROBLEM 7.12

Show that each function $f(x)$ that is square integrable on $I = (-1, 1)$ has a convergent generalized Fourier series expansion of the form

$$f(x) = \sum_{n=0}^{\infty} c_n P_n(x) \tag{1}$$

where the coefficients c_n are given by

$$c_n = \frac{2n+1}{2}\int_{-1}^{1} f(x)\,P_n(x)\,dx \qquad (2)$$

SOLUTION 7.12

Since the Legendre polynomials $P_n(x)$ are the eigenfunctions of a Sturm-Liouville problem, Theorem 7.9 implies they are a complete orthogonal family. The functions $P_n(x)$ are orthogonal on I with respect to the weight function $r(x) = 1$ so that $(P_n, P_m) = 0$ for m not equal to n. We can also show that

$$(P_n, P_n) = \frac{2}{2n+1} \qquad \text{for } n = 0, 1, \ldots \qquad (3)$$

and thus if we multiply both sides of (1) by $P_m(x)$ for m a fixed but arbitrary nonnegative integer and integrate over I, we obtain

$$\int_{-1}^{1} f(x)\,P_m(x)\,dx = \sum_{n=0}^{\infty} c_n (P_n, P_m) = c_m \frac{2}{2m+1}$$

We solve this for c_m, and since we can do so for each integer m, (2) follows. According to Theorem 7.10 the expansion (1) converges in the mean square sense to $f(x)$ if f is square integrable. Note that $P_n(x)$ is an even (odd) function when n is an even (odd) integer and, thus, (2) implies that $c_{2n} = 0$ if $f(x)$ is odd and $c_{2n+1} = 0$ if $f(x)$ is an even function of x on I.

PROBLEM 7.13

Write an eigenfunction expansion for the function $f(x) = 3x^3 + 1$ in terms of the Legendre polynomials.

SOLUTION 7.13

Since $P_n(x)$ is a polynomial in x of degree n and $f(x)$ is a polynomial of degree 3 we can show that the functions $f(x)$, $P_0(x)$, $P_1(x)$, $P_2(x)$, $P_3(x)$ form a linearly dependent set; i.e., $f(x)$ can be written as a linear combination of $P_0(x)$ through $P_3(x)$

$$\begin{aligned} 3x^3 + 1 &= c_0 P_0(x) + c_1 P_1(x) + c_2 P_2(x) + c_3 P_3(x) \\ &= c_0 1 + c_1 x + c_2 \frac{3x^2 - 1}{2} + c_3 \frac{5x^3 - 3x}{2} \\ &= (c_0 - c_2/2) + (c_1 - 3c_3/2)\,x + (3c_2/2)\,x^2 + (5c_3/2)\,x^3 \end{aligned}$$

equating coefficients of like powers of x, we see that this is an identity in x if

$$c_0 - c_2/2 = 1, \quad c_1 - 3c_3/2 = 0, \quad 3c_2/2 = 0, \quad 5c_3/2 = 3$$

i.e., $c_0 = 1$, $c_2 = 0$, $c_3 = 6/5$, and $c_1 = 9/5$. Thus, $f(x) = (5P_0(x) + 9P_1(x) + 6P_3(x))/5$.

PROBLEM 7.14

Compute the Legendre polynomial expansion for the function $f(x)$ such that $f(x) = -1$ when $-1 < x < 0$ and $f(x) = 1$ when $0 < x < 1$.

SOLUTION 7.14

Since $f(x)$ is an odd function, it follows that $c_{2n} = 0$ for $n = 0, 1, \ldots$ and, for odd integers n, we have

$$c_n = \frac{2n+1}{2}\int_{-1}^{1} f(x)\,P_n(x)\,dx = (2n+1)\int_0^1 P_n(x)\,dx$$

Then $c_1 = 3/2$, $c_3 = -7/8$, $c_5 = 11/16 \ldots$ and

$$f(x) = (24P_1(x) - 14P_3(x) + 11P_5(x) + \cdots)/16$$

PROBLEM 7.15

Expand the function $F(x, y) = 6 - 2x - 3y$, $0 < x < 3$, $0 < y < 2$, in a series of the form

$$F(x, y) = \sum_{m=1}^{\infty}\sum_{n=1}^{\infty} C_{mn} \sin(m\pi x/3)\sin(n\pi y/2) \tag{1}$$

SOLUTION 7.15

We know that the family $\{\sin(n\pi x/L)$: $n = 1, 2, \ldots\}$ is a complete orthogonal family on the interval $(0, L)$. Thus, $\{\sin(m\pi x/3)\}$ and $\{\sin(n\pi y/2)\}$ are each complete orthogonal families on the intervals (0, 3) and (0, 2), respectively. Then the two-parameter family $\{\sin(m\pi x/3)\sin(n\pi y/2)\}$ can be shown to be a complete orthogonal family in the space of functions which are square integrable on the 2-dimensional set $\Omega = \{0 < x < 3, 0 < y < 2\}$. Let p and q denote fixed positive integers, multiply both sides of (1) by $\sin(p\pi x/3)\sin(q\pi y/2)$ and integrate over Ω. Then

$$\int_0^3\int_0^2 F(x, y)\,X_p(x)\,Y_q(y)\,dxdy = \sum_{m=1}^{\infty}\sum_{n=1}^{\infty} C_{mn}(X_p, X_m)(Y_q, Y_n) \tag{2}$$

where $X_p(x) = \sin(p\pi x/3)$ and $Y_q(y) = \sin(q\pi y/2)$. Since $\{X_p(x)\}$ and $\{Y_q(y)\}$ are each mutually orthogonal families of eigenfunctions, the right hand side of (2) reduces to $C_{pq}(X_p, X_p)(Y_q, Y_q) = C_{pq}(3/2)(1)$. In addition, we can compute

$$\int_0^3\int_0^2 F(x, y)\,X_p(x)\,Y_q(y)\,dxdy = \int_0^2 Y_q(y)\int_0^3 (6 - 2x - 3y)\,X_p(x)\,dxdy$$

$$= \int_0^2 \frac{3}{p\pi} (6 - (1 - (-1)^p 3y) Y_q(y)) \, dy$$

$$= \frac{36}{\pi^2} \frac{1 - (-1)^{p+q}}{pq}$$

Then

$$F(x, y) = \frac{24}{\pi^2} \sum_{p=1}^{\infty} \sum_{q=1}^{\infty} \frac{1 - (-1)^{p+q}}{pq} X_p(x) Y_q(y)$$

and this series has the convergence properties described in Theorem 7.10.

A *function $f = f(x)$ has a Fourier expansion of the form (7.1) if $f(x)$ is everywhere defined and 2L-periodic or if $f(x)$ is defined on the interval $[-L, L]$ (only). The Fourier coefficients are computed from (7.3). If the graph of f has certain symmetries, then some of the Fourier coefficients must vanish:*

(a) if $f(x) = f(-x)$, i.e., f is even, then $b_m = 0$ for all m if in addition:
(i) $f(L/2 + x) = f(L/2 - x)$ then $a_{2m-1} = 0$ for all m,
(ii) $f(L/2 + x) = -f(L/2 - x)$then $a_{2m} = 0$ for all m.

(b) if $f(x) = -f(-x)$, i.e., f is odd, then $a_m = 0$ for all m if in addition,
(i) $f(L/2 + x) = f(L/2 + x) = f(L/2 - x)$ then $a_{2m} = 0$ for all m,
(ii) $f(L/2 + x) = -f(L/2 - x)$ then $a_{2m-1} = 0$ for all m.

For a function $f(x)$ to have a convergent power series expansion on an interval I, $f(x)$ must have derivatives of all orders on I. For the Fourier series for $f(x)$ to converge to f on I much less is needed. For $f(x)$ defined on $I = [-L, L]$:

(a) If f is square integrable on I, then the Fourier series for f converges to f in the mean square sense on I.

(b) If the periodic extensions f_{2L} and f'_{2L} are piecewise continuous, then the Fourier series for $f(x)$ converges pointwise to $(f_{2L}(x+) + f_{2L}(x-))/2$ at each point x.

(c) For the Fourier series for f to converge uniformly to f on I, it is necessary that the extension f_{2L} is everywhere continuous and it is sufficient that $f_{2L} \in C$ and $f'_{2L} \in PC$.

The trigonometric Fourier series is a special case of a more general form of infinite series representation. Any family of functions $\{g_n(x)\}$

defined on an interval I with the property

$$(g_n, g_m) = \int_I g_n(x)\, g_m(x)\, dx = \begin{cases} 1 & \text{if } m = n \\ 0 & \text{otherwise} \end{cases}$$

is an orthonormal family on I. Given an orthonormal family we can write a generalized Fourier series

$$f(x) = \sum_m (f, g_m)\, g_m(x)$$

for any f(x) that is square integrable on I. Bessel's inequality

$$\sum_m (f, g_m)^2 \leq \|f\|^2$$

always holds and implies that the generalized Fourier series converges in the mean square sense. However, the sum of the series equals f(x) for every square integrable f if and only if the orthonormal family is complete. In general, it is not easy to show that a given orthonormal family is complete, but if the orthonormal family is generated by a Sturm-Liouville problem over an interval I, then the family is guaranteed to be complete in the space $R^2(I)$ of square integrable functions on I.

8

Integral Transforms

Fourier series and generalized Fourier series can be used to represent a function that is everywhere defined and periodic and they work equally well for a function whose domain is an interval of finite length. In this latter case, the series converges to a periodic function that is a periodic extension of the given function.

For functions that are everywhere defined but are not periodic, we must use a different mode of representation, one that is based on integral transforms rather than infinite series. Both representations, eigenfunction expansion and integral transform, will be used in the next chapter for solving problems in partial differential equations.

In this chapter we will discuss two of the most commonly used integral transforms, the Fourier and Laplace transforms. The Fourier transform is defined for functions whose domain is the whole real line. In particular, a function that is square integrable on the real line has a Fourier transform that is a square integrable function of the transform variable. We develop several properties of the Fourier transform that will be of use in solving partial differential equations in which at least one of the independent variables involved ranges over the whole real line.

The Laplace transform is similar to the Fourier transform but is defined for functions whose domain is the positive real axis. Elementary properties of the Laplace transform have been derived in chapter 4 where they were used to solve problems in ordinary differential equations. Here we present additional properties of the Laplace transform that will prove useful in the next chapter.

THE FOURIER TRANSFORM

Let $f(x)$ be defined for all real x. Then we define the Fourier transform of f to be the function

$$F(\alpha) = (2\pi)^{-1}\int_{-\infty}^{\infty} f(x)\, e^{-ix\alpha} dx \tag{8.1}$$

The integral in (8.1) is an improper integral since the domain of integration is an unbounded interval. The convergence or divergence of this integral depends entirely on the function $f(x)$ because $e^{-ix\alpha}$ is a complex valued function with modulus equal to one; i.e.,

$$|e^{-ix\alpha}| = |\cos x\alpha - i \sin x\alpha| = (\cos^2 x\alpha + \sin^2 x\alpha)^{1/2} = 1$$

There are various conditions we may place on the class of functions whose Fourier transforms we want to compute and each choice leads to a somewhat different theory of the Fourier transform. For our purposes it will be most convenient to work in the class of square integrable functions.

Square Integrable Functions

If $f(x)$ is defined for all real values x and if the integral of $f(x)^2$ from negative infinity to plus infinity is finite then we will say f is *square integrable* and write $f \in R^2(-\infty, \infty)$.

Example 8.1

(a) Let p denote a fixed real number and consider the function

$$f(x) = \begin{cases} x^p & \text{if } |x| > 1 \\ 0 & \text{if } |x| < 1 \end{cases}$$

Then

$$\int_{-\infty}^{\infty} f(x)^2 dx = 2\int_1^{\infty} x^{2p} dx = \begin{cases} 2/(2p+1) & \text{if } 2p+1 < 0 \\ \infty & \text{if } 2p+1 \geq 0 \end{cases}$$

Evidently in order for $f(x) = x^p$ to be square integrable, it is necessary for the function to tend to zero sufficiently fast as $|x|$ tends to infinity. Here "sufficiently fast" means p must be less than $-1/2$.

(b) Consider the function $g(x)$ that is equal to 1 if $n < x < n + 1/n^2$ for $n = 1, 2, \ldots$ and $g(x)$ is zero otherwise. Then

$$\int_{-\infty}^{\infty} g(x)^2 dx = \sum_{n=1}^{\infty} 1\frac{1}{n^2} < \infty$$

Thus, g is square integrable but g does not tend to zero as x tends to infinity. In fact, the graph of $g(x)$ consists of a sequence of square pulses beginning at $x = n$ for each positive integer n with each pulse of height one and width equal to $1/n^2$. Thus, $g(x)$ is bounded and piecewise continuous but not continuous.

(c) For $n = 1, 2, \ldots$, let $h(x)$ be equal to $n^{1/2}$ if $n < x < n + 1/n^3$ and let $h(x)$ be zero otherwise. Then

$$\int_{-\infty}^{\infty} h(x)^2 dx = \sum_{n=1}^{\infty} n\frac{1}{n^3} < \infty$$

so $h(x)$ is square integrable but not only does h not tend to zero as x tends to infinity, but $h(x)$ grows without bound as x tends to infinity.

In the two examples, g and h are intended to illustrate the fact that it is not necessary for a function to tend to zero as x tends to infinity in order for the function to be square integrable over the real line. However, we do have the following result.

Theorem 8.1

Theorem 8.1. Suppose $f(x)$ is square integrable. Suppose also that f is differentiable almost everywhere and that f' is also square integrable. Then $f(x)$ tends to zero as x tends to plus or minus infinity.

Here f is differentiable almost everywhere means that f may be only piecewise continuous but $f'(x)$ has left and right hand limits at the isolated points where f is discontinuous.

Fourier Transform Formulas

We can now compute the Fourier transforms of a few simple functions. We will occasionally use the alternative notation $F\{f\}$ to denote the Fourier transform of a function f.

Example 8.2

(a) Let p denote a fixed positive constant and consider the function $I_p(x)$ that is equal to one if $|x| < p$ and is equal to zero otherwise. Then it is easy to check that $I_p(x)$ is square integrable and has Fourier transform given by

$$F\{I_p\}(\alpha) = (2\pi)^{-1}\int_{-p}^{p} 1e^{-i\alpha x}dx = \left.\frac{-e^{-i\alpha x}}{2\pi i\alpha}\right|_{-p}^{p} = \frac{\sin p\alpha}{\pi\alpha}$$

(b) For b a fixed positive constant, the function $f(x) = e^{-b|x|}$ is square integrable and we can compute

$$F(\alpha) = (2\pi)^{-1}\int_{-\infty}^{\infty} e^{-b|x|}e^{-i\alpha x}dx$$

$$= (2\pi)^{-1}\int_{-\infty}^{0} e^{bx-i\alpha x}dx + (2\pi)^{-1}\int_{0}^{\infty} e^{-bx-i\alpha x}dx$$

$$= (2\pi)^{-1}\left.\frac{e^{bx-i\alpha x}}{b-i\alpha}\right|_{-\infty}^{0} + (2\pi)^{-1}\left.\frac{e^{-bx-i\alpha x}}{-b-i\alpha}\right|_{0}^{\infty}$$

But $|e^{bx-i\alpha x}| = |e^{bx}|$ tends to zero as x tends to negative infinity, and as x tends to plus infinity, it is also clear that $|e^{-bx-i\alpha x}| = |e^{-bx}|$ tends to zero. Thus,

$$F(\alpha) = (2\pi)^{-1}\left[\frac{1}{b-i\alpha}+\frac{1}{b+i\alpha}\right] = \frac{b}{\pi}\,\frac{1}{b^2+\alpha^2}$$

Inverse Fourier Transform

The formula (8.1) defining the Fourier transform of the square integrable function $f(x)$ is analogous to the following formula for the exponential form of the Fourier coefficients of a function $f(x)$ that is square integrable on $I = [-\pi, \pi]$ given by

$$F_n = (2\pi)^{-1}\int_{-\pi}^{\pi} f(x)e^{-inx}dx$$

In the case of the Fourier series, we may recover the function $f(x)$ by summing the series

$$f(x) = \sum_{n=-\infty}^{\infty} F_n e^{inx}$$

In the case of the Fourier transform it can be shown that $f(x)$ is recovered from its Fourier transform via the following integration

$$f(x) = \int_{-\infty}^{\infty} F(\alpha)e^{ix\alpha}d\alpha \tag{8.2}$$

However, for most functions f, the integration in (8.2) cannot be carried out by elementary means. Therefore, we shall use an alternate means for recovering f when we are given its Fourier transform F. In particular we will use a dictionary of Fourier transform formulas, Table 8.1, together with a number of operational properties of the Fourier transform. The operational properties of the Fourier transform are very similar to the operational properties of the Laplace transform developed in an earlier chapter.

	$f(x)$	$F(\alpha)$
1.	$I_p(x) \quad p>0$	$\dfrac{\sin p\alpha}{\pi\alpha}$
2.	$e^{-b\|\alpha\|} \quad b>0$	$\dfrac{b/\pi}{b^2+\alpha^2}$
3.	$e^{-cx^2} \quad c>0$	$(4\pi c)^{-1/2}e^{-\alpha^2/4c}$
4.	$2\dfrac{\sin ax}{x} \quad a>0$	$I_a(\alpha)$
5.	$\dfrac{2b}{b^2+x^2} \quad b>0$	$e^{-b\|\alpha\|}$

Table 8.1
Fourier Transform Formulas

Operational Properties of the Fourier Transform

The following properties hold for arbitrary square integrable functions f and g and their Fourier transforms.

ALGEBRAIC PROPERTIES

1. For arbitrary constants a and b: $F\{af+bg\} = aF(\alpha)+bG(\alpha)$

2. For nonzero constant a, $F\{f(ax)\} = \dfrac{1}{|a|}F(\dfrac{\alpha}{a})$

DIFFERENTIATION

3. If f and f' are both square integrable, then $F\{f'\} = i\alpha F(\alpha)$.
   If f, f', and f'' are all square integrable, then $F\{f''\} = -\alpha^2 F(\alpha)$.

4. If $xf(x)$ is square integrable, then $F\{xf(x)\} = iF'(\alpha)$.

SHIFTING PROPERTIES

5. For any real constant c, $F\{f(x-c)\} = e^{-i\alpha c}F(\alpha)$.

6. For any real constant c, $F\{e^{icx}f(x)\} = F(\alpha - c)$.

REPEATED TRANSFORMATION

7. If f is square integrable, then $F(\alpha)$ is square integrable as well and $F\{F(x)\} = f(-\alpha)/2\pi$.

PRODUCTS OF TRANSFORMS

If we define the convolution product of two functions by

$$f*g(x) = \int_{-\infty}^{\infty} f(x-y)\,g(y)\,dy \tag{8.3}$$

then we have

8. $F\{f*g(x)\} = 2\pi F(\alpha)\,G(\alpha)$.

The Fourier transform, like the Laplace transform, may be used to reduce linear constant coefficient ordinary differential equations to algebraic equations in the transform space. These transforms also have applications to problems in partial differential equations on unbounded domains. Such applications are discussed in chapter 9.

LAPLACE TRANSFORM

The Fourier transform is defined for functions $f(x)$ whose domain of definition is the whole real line, $-\infty < x < \infty$. We have further restricted our attention to functions that are square integrable over the real line. Consider then a function $f(t)$ defined for $t > 0$ and satisfying, for some constants M and b,

$$|f(t)| \le Me^{bt} \quad \text{for } t > 0 \tag{8.4}$$

Then for such an f, we can define the Laplace transform of f as follows:

$$\mathcal{L}\{f(t)\} = F(s) = \int_0^{\infty} f(t)\,e^{-st}\,dt \tag{8.5}$$

Then $F(s)$ is a well defined function of s, at least on $s > b$, provided $f(t)$ is at least regular enough for the integral (8.5) to exist. It is sufficient for $f(t)$ to be piecewise continuous. We denote the independent variable by t instead of x in order to emphasize that the domain is now the half line rather than the whole real line. In applications it is often natural to restrict the time variable to the positive real axis and, for that reason, we denote the half range variable by t.

ADDITIONAL LAPLACE TRANSFORM FORMULAS

Elementary properties of the Laplace transform have already been discussed in chapter 4. However, we will need additional formulas to be used in applications to problems in partial differential equations. The following auxiliary result will be useful.

Theorem 8.2

Theorem 8.2. If $L\{f(t)\} = F(s)$, then

$$L^{-1}\{F(\sqrt{s})\} = (4\pi t^3)^{-1/2} \int_0^\infty z e^{-z^2/4t} f(z)\, dz \tag{8.6}$$

Using Theorem 8.2 we can derive the following additions to Table 4.1:

For $a > 0$: $$L\{a(4\pi t^3)^{-1/2} e^{-a^2/4t}\} = e^{-a\sqrt{s}} \tag{8.7}$$

For $a \geq 0$: $$L\{(\pi t)^{-1/2} e^{-a^2/4t}\} = s^{-1/2} e^{-a\sqrt{s}} \tag{8.8}$$

For $a \geq 0$: $$L\{\operatorname{erfc}(a/\sqrt{4t})\} = s^{-1} e^{-a\sqrt{s}} \tag{8.9}$$

where erfc(x) denotes the *complementary error function*; i.e.,

$$\operatorname{erfc} x = \pi^{-1/2} \int_x^\infty e^{-z^2} dz \tag{8.10}$$

SOLVED PROBLEMS

Fourier Transform Formulas

PROBLEM 8.1

For positive constant c, $f(x) = e^{-cx^2}$ is square integrable. Compute the Fourier transform.

SOLUTION 8.1

By definition (8.1) we have

$$(\alpha) = (2\pi)^{-1}\int_{-\infty}^{\infty} e^{-cx^2}e^{-ix\alpha}dx$$

$$= (2\pi)^{-1}\int_{-\infty}^{\infty} e^{-c(x^2+ix\alpha/c-\alpha^2/4c^2)}dx\, e^{-\alpha^2/4c}$$

$$= (2\pi)^{-1}e^{-\alpha^2/4c}\int_{-\infty}^{\infty} e^{-c(x+i\alpha)^2}dx$$

But it is well known that

$$\int_{-\infty}^{\infty} e^{-c(x+i\alpha)^2}dx = \int_{-\infty}^{\infty} e^{-cz^2}dz = \sqrt{\pi/c} \quad \text{for } c>0$$

Hence

$$F\{e^{-cx^2}\} = (4\pi c)^{-1/2}e^{-\alpha^2/4c}$$

i.e.,

$$F\{e^{-cx^2}\} = Ae^{-b\alpha^2} \quad \text{for } A = (4\pi c)^{-1/2} \quad \text{and} \quad b = 1/4c$$

Thus, the Fourier transform of the Gaussian function e^{-cx^2} is another Gaussian function.

PROBLEM 8.2

Show that for any positive constant b,

$$F\{2\sin bx/x\} = I_b(\alpha) \tag{1}$$

and

$$F\{2b/(b^2+x^2)\} = e^{-b|\alpha|} \tag{2}$$

SOLUTION 8.2

By property 7 of the Fourier transform and the Fourier transform formula

$$F\{I_b(x)\} = \frac{\sin b\alpha}{\alpha\pi} = F(\alpha)$$

we have

$$F\{F(x)\} = F\{\frac{\sin b\alpha}{\alpha\pi}\} = \frac{f(-\alpha)}{2\pi} = \frac{I_b(-\alpha)}{2\pi}$$

and since $I_b(-\alpha) = I_b(\alpha)$, (1) follows after multiplying by 2π. In the same way,

$$F\{e^{-b|x|}\} = \frac{b}{\pi}\frac{1}{b^2+\alpha^2} = F(\alpha)$$

thus

$$F\{\frac{b}{\pi}\frac{1}{b^2+\alpha^2}\} = \frac{f(-\alpha)}{2\pi} = \frac{e^{-b|\alpha|}}{2\pi}$$

Multiplying both sides by 2π leads to (2).

Inverse Fourier Transforms

PROBLEM 8.3

If $f(x)$ is a square integrable function with Fourier transform $F(\alpha)$, then find the function whose Fourier transform is:

(a) $\cos p\alpha\ F(\alpha)$ (b) $\frac{\sin p\alpha}{\alpha}F(\alpha)$, $p = \text{constant}$

SOLUTION 8.3

If we write

$$\cos p\alpha\ F(\alpha) = (e^{ip\alpha}F(\alpha) + e^{-ip\alpha}F(\alpha))/2$$

then it follows from the first shifting property, property 5, of the Fourier transform that

$$F\{(f(x+p) + f(x-p))/2\} = \cos p\alpha\ F(\alpha)$$

i.e.,

$$F^{-1}\{\cos p\alpha\ F(\alpha)\} = \frac{f(x+p) + f(x-p)}{2}$$

Note also that

$$\frac{\sin p\alpha}{\alpha}F(\alpha) = \frac{1}{2}(e^{ip\alpha} - e^{-ip\alpha})\frac{F(\alpha)}{i\alpha}$$

Now let $g(x)$ denote an antiderivative of $f(x)$; i.e., $g'(x) = f(x)$. Then by property 3 of the Fourier transform, $F\{g'(x)\} = i\alpha G(\alpha) = F(\alpha)$ from which we get

$$\frac{\sin p\alpha}{\alpha}F(\alpha) = \frac{1}{2}(e^{ip\alpha} - e^{-ip\alpha})G(\alpha)$$

and, thus,

$$F^{-1}\{\frac{\sin p\alpha}{\alpha}F(\alpha)\} = \frac{g(x+p) - g(x-p)}{2}$$

PROBLEM 8.4

Find the function whose Fourier transform is equal to $F(\alpha) = \sinh p\alpha / \sinh \alpha$ for p a real constant.

SOLUTION 8.4

Note first that $F(\alpha) = F(|\alpha|)$ and then write

$$\frac{\sinh p|\alpha|}{\sinh|\alpha|} = \frac{e^{p|\alpha|} - e^{-p|\alpha|}}{e^{|\alpha|} - e^{-|\alpha|}} = \left(e^{(p-1)|\alpha|} - e^{-(p+1)|\alpha|}\right)\frac{1}{1-e^{-2|\alpha|}}$$

But

$$(1-e^{-2|\alpha|})^{-1} = \sum_{n=0}^{\infty} e^{-2n|\alpha|}$$

hence,

$$F(\alpha) = \sum_{n=0}^{\infty} e^{(p-2n-1)|\alpha|} - \sum_{n=0}^{\infty} e^{-(p+2n-1)|\alpha|}$$

Then by the result (2) of Problem 8.2

$$f(x) = \sum_{n=0}^{\infty} \frac{2(p-2n-1)}{(p-2n-1)^2+x^2} - \sum_{n=0}^{\infty} \frac{2(p+2n+1)}{(p+2n+1)^2+x^2}$$

Properties of the Fourier Transform

PROBLEM 8.5

Suppose $f(x) = 0$ for $|x| > A$ for some positive constant A. Then show that for any L > A

$$F(n\pi/L) = \frac{L}{\pi} f_n \quad \text{for all integers n} \tag{1}$$

where

$$f(x) = \sum_{n=-\infty}^{\infty} f_n e^{in\pi x/l} \tag{2}$$

SOLUTION 8.5

If $f(x) = 0$ for $|x| > A$, then f has both a Fourier transform and a Fourier series. For any $L > A$, we can expand f in a Fourier series of the form (2) for f_n given by

$$f_n = \frac{1}{2L}\int_{-A}^{A} f(x)\, e^{-in\pi x/L} dx, \quad n = \text{integer} \tag{3}$$

Then, comparing (3) with (8.1) leads to (1) which says that the coefficients in the exponential form of the Fourier series for f are proportional to the values at discrete points of the Fourier transform for f.

PROBLEM 8.6

Suppose f is such that $F(\alpha) = 0$ for $|\alpha| > A$ for some positive constant A. Then show that for all real x, we have

$$f(x) = \sum_{n=-\infty}^{\infty} f(n\pi/A) \frac{\sin(A(x-n\pi))}{A(x-n\pi)} \tag{1}$$

i.e., by sampling $f(x)$ at the discrete points $x_n = n\pi/A$, we can recapture the value of $f(x)$ at any real value x.

SOLUTION 8.6

A function $f(x)$ with the property that $F(\alpha) = 0$ for $|\alpha| > A$ is said to be *band limited* with *cutoff frequency* equal to A. Then $F(\alpha)$ can be expanded in a Fourier series of the form

$$F(\alpha) = \sum_{-\infty}^{\infty} F_n e^{in\pi\alpha/A} \tag{2}$$

where

$$F_n = (2\pi)^{-1} \int_{-A}^{A} F(\alpha) e^{-in\pi\alpha/A} d\alpha \tag{3}$$

Comparing (3) with (8.2) leads to the result

$$F_n = (2\pi)^{-1} f(-n\pi/A), \quad n = \text{integer} \tag{4}$$

and

$$F(\alpha) = (2\pi)^{-1} \sum_{n=-\infty}^{\infty} f(n\pi/A) e^{-in\pi\alpha/A} \tag{5}$$

where we have replaced n with $-n$. According to (5), it is sufficient to know $f(x)$ at the discrete values $x_n = n\pi/A$ in order to have $F(\alpha)$ for all α. Now using (5) in (8.2) and interchanging the order of integration and summation, we get (1).

PROBLEM 8.7

Give an example of two square integrable functions $f(x)$, $g(x)$ such that $f*g(x)$ is zero for all x, but neither f nor g is zero for all x.

SOLUTION 8.7

We use here the fact that $F\{f*g\} = F(\alpha)G(\alpha)$. Then choose f and g such that the set where $F(\alpha)$ is different from zero is disjoint from the set where $G(\alpha)$ is different from zero. For example

$h(x) = (2\sin x)/x$ has for its Fourier transform $H(\alpha) = I_1(\alpha)$. Then using the second shifting property, we find that for the choice

$$f(x) = e^{i2x}h(x) \quad \text{and} \quad g(x) = e^{-i2x}h(x)$$

we have

$$F(\alpha) = I_1(\alpha-2) \quad \text{and} \quad G(\alpha) = I_1(\alpha+2)$$

hence,

$$F(\alpha)G(\alpha) = 0 \text{ for all } \alpha \text{ and } f*g(x) = 0 \text{ for all } x$$

Laplace Transform Formulas

PROBLEM 8.8

Show that if $L\{f(t)\} = F(s)$, then

$$L^{-1}\{F(\sqrt{s})\} = (4\pi t^3)^{-1/2}\int_0^\infty ze^{-z^2/4t}f(z)\,dz \tag{1}$$

SOLUTION 8.8

If the result holds, then

$$F(\sqrt{s}) = \int_0^\infty e^{-st}(4\pi t^3)^{-1/2}\int_0^\infty ze^{-z^2/4t}f(z)\,dzdt \tag{2}$$

$$= \int_0^\infty \pi^{-1/2}f(z)\int_0^\infty e^{-st}ze^{-z^2/4t}(4t^3)^{-1/2}\,dtdz$$

If we introduce a new variable $\tau = z/\sqrt{4t}$, then $d\tau = -(z/2)(4t^3)^{-1/2}dt$ and

$$F(\sqrt{s}) = \int_0^\infty \pi^{-1/2}f(z)\int_0^\infty 2e^{-z^2s/4t^2}e^{-\tau^2}\,dtdz$$

$$= 2\pi^{-1/2}\int_0^\infty f(z)\frac{\sqrt{\pi}}{2}e^{-z\sqrt{s}}dz$$

i.e.,

$$F(\sqrt{s}) = \int_0^\infty f(z)e^{-z\sqrt{s}}dz \tag{3}$$

where we have used

$$\int_0^\infty 2e^{-z^2s/4t^2}e^{-\tau^2}d\tau = \frac{\sqrt{\pi}}{2}e^{-z\sqrt{s}}$$

Then reversing the steps from (3) to (2) proves (1).

PROBLEM 8.9

Use Theorem 8.2 to prove (8.7).

SOLUTION 8.9

According to Problem 4.9, we have $F(s) = e^{-as}$ for $f(t) = \delta(t-a)$. We apply the theorem in this case to obtain

$$L^{-1}\{e^{-a\sqrt{s}}\} = (4\pi t^3)^{-1/2} \int_0^\infty ze^{-z^2/4t}\delta(z-a)\,dz$$

But for any function $g(z)$ that is continuous for $z > 0$, the δ-function has the property that

$$\int_0^\infty g(z)\,\delta(z-a)\,dz = g(a) \qquad \text{for } a > 0$$

Then

$$L^{-1}\{e^{-a\sqrt{s}}\} = (4\pi t^3)^{-1/2} ae^{-a^2/4t} \qquad \text{for } a > 0$$

PROBLEM 8.10

Prove the Laplace transform result (8.8).

SOLUTION 8.10

We begin with the result (8.7),

$$L^{-1}\{e^{-a\sqrt{s}}\} = (4\pi t^3)^{-1/2} ae^{-a^2/4t}$$

and integrate both sides with respect to the parameter a from $b \geq 0$ to ∞

$$L^{-1}\{\int_b^\infty e^{-a\sqrt{s}}da\} = (4\pi t^3)^{-1/2} \int_b^\infty ae^{-a^2/4t}da$$

Then

$$L^{-1}\{\int_b^\infty e^{-a\sqrt{s}}da\} = L^{-1}\{s^{-1/2}e^{-b\sqrt{s}}\}$$

and

$$(4\pi t^3)^{-1/2} \int_b^\infty e^{-a\sqrt{s}}da = (\pi t)^{-1/2}e^{-b^2/4t}$$

proving (8.8). Note that, for $b = 0$, this result gives

$$L\{(\pi t)^{-1/2}\} = s^{-1/2}$$

PROBLEM 8.11

Derive the Laplace transform formula (8.9).

SOLUTION 8.11

We begin with the result (8.8)

$$L^{-1}\{s^{-1/2}e^{-b\sqrt{s}}\} = (\pi t)^{-1/2}e^{-b^2/4t} \quad \text{for } b \geq 0$$

If we integrate both sides with respect to the parameter b from $a \geq 0$ to ∞, we get

$$L^{-1}\{\int_a^\infty s^{-1/2}e^{-b\sqrt{s}}db\} = (\pi t)^{-1/2}\int_a^\infty e^{-b^2/4t}db \quad \text{for } b \geq 0$$

Then

$$L^{-1}\{\int_a^\infty s^{-1/2}e^{-b\sqrt{s}}db\} = L^{-1}\{s^{-1}e^{-a\sqrt{s}}\}$$

and the change of variable $z = b/\sqrt{4t}$ shows that

$$(\pi t)^{-1/2}\int_a^\infty e^{-b^2/4t}db = 2\pi^{-1/2}\int_{a/\sqrt{4t}}^\infty e^{-z^2}dz = \operatorname{erfc}(a/\sqrt{4t})$$

Using these in (1) proves the result.

PROBLEM 8.12

Show that for c not equal to zero,

$$L^{-1}\left\{\frac{e^{-a\sqrt{s}}}{\sqrt{s}(\sqrt{s}+c)}\right\} = e^{ac+c^2t}\operatorname{erfc}(c\sqrt{t}+a/\sqrt{4t}) \tag{1}$$

SOLUTION 8.12

Note first that partial fractions leads to the result

$$L^{-1}\{\frac{1}{s(s+c)}\} = L^{-1}\{\frac{1}{c}(\frac{1}{s}-\frac{1}{s+c})\} = \frac{1}{c}(1-e^{-ct}) \tag{2}$$

and then the shifting property of the Laplace transform implies

$$L^{-1}\{\frac{e^{-as}}{s(s+c)}\} = \frac{1}{c}(1-e^{-c(t-a)})H(t-a) = f(t;a) \tag{3}$$

For convenience, we denote the right side of (3) by $f(t; a)$. Now we apply Theorem 8.2 to (3)

$$L^{-1}\left\{\frac{e^{-a\sqrt{s}}}{\sqrt{s}(\sqrt{s}+c)}\right\} = (4\pi t^3)^{-1/2}\int_0^\infty ze^{-z^2/4t}f(z;a)\,dz$$

$$= (4\pi t^3)^{-1/2}\int_0^{\infty} ze^{-z^2/4t}\frac{1}{c}(1-e^{-c(t-a)})\,dz$$

Integration by parts reduces the integral on the right to

$$I = e^{ac}(\pi t)^{-1/2}\int_a^{\infty} e^{-cz}e^{-z^2/4t}dz$$

$$= e^{ac}(\pi t)^{-1/2}\int_a^{\infty} e^{-(z^2+4zct+4c^2t^2)/4t}e^{c^2t}dz$$

After completing the square we have

$$I = e^{ac+c^2t}(\pi t)^{-1/2}\int_a^{\infty} e^{-(z+2ct)^2/4t}dz$$

$$= 2e^{ac+c^2t}\pi^{-1/2}\int_{c\sqrt{t}+a/\sqrt{4t}}^{\infty} e^{-y^2}dy$$

$$= e^{ac+c^2t}\operatorname{erfc}(c\sqrt{t}+a\sqrt{4t})$$

Here we used the change of variable $y = (z + 2ct)/\sqrt{4t}$ to reduce I to an expression in terms of the complementary error function.

If f(x) is defined for all real x then the Fourier transform of f is defined to be

$$F(\alpha) = (2\pi)^{-1}\int_{-\infty}^{\infty} f(x)e^{-ix\alpha}dx$$

If f(x) is a square integrable function of x, then $F(\alpha)$ *can be shown to be a square integrable function of* α *and, theoretically, f can be recovered from F by means of the inversion formula*

$$f(x) = \int_{-\infty}^{\infty} F(\alpha)e^{ix\alpha}d\alpha$$

For practical purposes, the inversion of an integral transform is generally accomplished by means of a dictionary of transform formulas used in conjunction with the operational properties of the transform. Table 8.1 lists several Fourier transform.

For f(t) defined on $\{t \geq 0\}$, *we have the Laplace transform, defined by*

$$L\{f(t)\} = F(s) = \int_0^{\infty} f(t)e^{-st}dt$$

A piecewise continuous function f(t) satisfying (8.4) is said to be of exponential type. F(s) is a continuous function of s, s > b, if f is of exponential type. Elementary Laplace transform formulas and operational properties can be found in chapter 4. Some additional formulas include:

For $a > 0$: $$L\{a(4\pi t^3)^{-1/2} e^{-a^2/4t}\} = e^{-a\sqrt{s}}$$

For $a \geq 0$: $$L\{(\pi t)^{-1/2} e^{-a^2/4t}\} = s^{-1/2} e^{-a\sqrt{s}}$$

For $a \geq 0$: $$L\{\operatorname{erfc}(a/\sqrt{rt})\} = s^{-1} e^{-a\sqrt{s}}$$

where erfc(x) *denotes the complementary error function.*

9

Linear Problems in Partial Differential Equations

In this chapter we will consider linear problems in partial differential equations. We will focus for the most part on second order equations in two independent variables. Not only do such equations include most of the so-called equation of mathematical physics but each such equation falls into one of the following three classes of equations: elliptic, parabolic, and hyperbolic. Each of these three classes is represented by a prototypical equation in the sense that what is true for the prototype equation is generally true of all equations in that class. For elliptic equations, the prototype is Laplace's equation; for parabolic and hyperbolic, the prototypes are the heat and wave equations, respectively. We will consider a variety of problems for each of the prototype equations.

We will consider problems where the independent variables range over unbounded sets as well as problems where they are restricted to bounded sets. We shall see that an integral transform is the method of choice for a problem on an unbounded interval whereas eigenfunction expansions are the indicated solution method for problems on bounded domains. In either case, the so-called principle of superposition is the underlying principle for either approach. The principle of superposition is valid only in the case of linear equations so the solution methods of this chapter are restricted to linear problems.

In the solved problems, we begin with a number of examples of problems on unbounded regions and solve these using either the Laplace or the Fourier transform. Following this, we use the method of eigenfunction expansions to solve several problems on bounded domains. Finally, we solve several examples in which there are more than two independent variables. Here it can be seen that while the conceptual difficulty of a problem does not increase as the number of variables in the problem increases, the computational difficulty increases very rapidly.

PARTIAL DIFFERENTIAL EQUATIONS

An equation involving one or more of the partial derivatives of an unknown function of two or more independent variables is called a *partial differential equation*. The equation is said to be *linear* if it is linear in the unknown function and its derivatives. We will consider only linear equations in this chapter.

Auxiliary Conditions

The general solution of a linear ordinary differential equation of order N contains N arbitrary constants of integration. The general solution of a linear partial differential equation contains arbitrary functions. In applications, however, we will rarely be interested in the general solution of an equation. Instead, we will seek particular solutions subject to certain auxiliary conditions descriptive of the particular physical system being modeled. These auxiliary conditions will consist mainly of two types of conditions.

INITIAL CONDITIONS

In a partial differential equation in which one of the variables can be interpreted as time, it is customary to specify the state of the system at the instant when the equation commences to apply. The conditions which specify the initial state are called *initial conditions*. The number of conditions required to specify the state of the system depends on the order of the time derivatives that appear in the equation. Generally, if the equation contains time derivatives of order n and no time derivatives of order higher than n, then the initial state is characterized by n initial conditions specifying the unknown function and its derivatives up to order $n - 1$ throughout the spatial domain where the partial differential equation applies.

BOUNDARY CONDITIONS

Auxiliary conditions that are satisfied by the particular solution at all points on the boundary S of the spatial domain D in which the equation applies are called *boundary conditions*. In physical applications, boundary conditions frequently occur in one of the following forms:

Dirichlet Conditions. The value of the solution u is specified on S.

Neumann Conditions. The normal derivative of the solution u is specified on S; i.e., the value of the function grad $u \cdot N$ is specified on S, where N denotes the unit outward normal to S.

Robin Conditions. A combination of solution values and normal derivative values are specified on S. For example, $pu + q$ **grad** $u \cdot N$ is specified for p and q given functions defined on S.

Robin conditions are also called boundary conditions of the third kind. If the spatial domain D is a bounded interval, then the boundary S consists of the two endpoints. If D is a simple bounded region in the plane then S is a simple closed curve and S is a simple closed surface if D is a simple region in 3-space.

BOUNDARY CONDITIONS AT INFINITY

If the domain D is unbounded then in some cases we impose *behavior at infinity conditions* on the solution. These conditions act, in effect, as boundary conditions at infinity. For example, we may require that the solution remain bounded for all values of the spatial variables or we may require the solution to have specified growth or decay rates at infinity. Some partial differential equations on unbounded domains do not require that behavior at infinity be specified.

WELL POSED PROBLEMS

A linear partial differential equation will generally have infinitely many solutions. By imposing auxiliary conditions, we eliminate some of the solutions. If we impose too many auxiliary conditions, we may eliminate all of the solutions. When the number of auxiliary conditions is such that there is one and only one function satisfying the differential equation and the auxiliary conditions, then we say that the problem is *algebraically well posed.* If small changes in the data specified by the auxiliary conditions results in correspondingly small changes in the solution, then we say that the solution *depends continuously on the data.* If the problem is algebraically well posed and the solution depends continuously on the data, then we say the problem is *well posed.* The solution to any ordinary differential equation problem that is algebraically well posed automatically will depend continuously on the data. However, there are algebraically well posed problems for partial differential equations where the solution does not depend continuously on the data. There is no general method for distinguishing a well posed problem from one that is not well posed. Here we will only consider examples that are well posed.

Classification

Partial differential equations is an enormous subject containing so much material that in order to grasp even a part of it we must be able to divide the problems into classes. One way to do this is to separate the partial differential equations into distinct classes.

ORDER

If the partial differential equation contains derivatives of order n and no derivatives of order higher than n, then we say the equation is of *order n.*

The most general linear partial differential equation of order two in two independent variables has the form

$$a(x,y)\,\partial_{xx}u(x,y) + 2b(x,y)\,\partial_{xy}u(x,y) + c(x,y)\,\partial_{yy}u(x,y) + \cdots = F(x,y)$$

We have denoted the independent variables by x and y and the unknown function by $u(x, y)$. We use the notation $\partial_{xx}u$, $\partial_{xy}u$, and $\partial_{yy}u$ to denote the various partial derivatives of order two of the function $u = u(x, y)$. If $F(x, y)$ is the zero function, then we say equation (9.1) is *homogeneous*; if F is not the zero function, then (9.1) is said to be *inhomogeneous*.

CLASSIFICATION BY TYPE

We have shown only the terms involving derivatives of order two in (9.1) since the *lower order terms* have no bearing on the classification of the equation. The highest order terms, the *principal* part of the equation (9.1), can also be written using matrix notation as follows:

$$[\partial_x, \partial_y]^{\mathsf{T}} \begin{bmatrix} a & b \\ b & c \end{bmatrix} \begin{bmatrix} \partial_x \\ \partial_y \end{bmatrix} u(x,y) + \cdots = F(x,y) \qquad (9.1)$$

Then the classification of the equation (9.1) is based on the 2×2 coefficient matrix which we denote by A. We say that equation (9.1) is:

Elliptic if the eigenvalues of A are of one sign and zero is not an eigenvalue;

Parabolic if zero is an eigenvalue for A;

Hyperbolic if the eigenvalues of A are of opposite sign and zero is not an eigenvalue.

Every equation of order two in two independent variables falls into one of these three classes; i.e., the classification is *exhaustive*. This classification scheme can be generalized to apply to equations of higher order and equations in more than two variables, but then the classification is no longer exhaustive; there are equations that are none of these three types.

Example 9.1 We list three equation of order two in two variables, show the corresponding matrix A, and list the type of the equation:

Laplace's equation: $\partial_{xx}u + \partial_{yy}u = 0;\ A = \begin{bmatrix} 1 & 0 \\ 0 & 1 \end{bmatrix}$; elliptic

Heat equation: $\partial_{x}u - \partial_{yy}u = 0;\ A = \begin{bmatrix} 0 & 0 \\ 0 & 1 \end{bmatrix}$; parabolic

Wave equation: $-\partial_{xx}u + \partial_{yy}u = 0;\ A = \begin{bmatrix} -1 & 0 \\ 0 & 1 \end{bmatrix}$; hyperbolic

CHARACTERISTIC BEHAVIOR

Each of the three types of partial differential equation models a distinct kind of physical behavior. For example

(a) Elliptic equations model: Steady-state conduction of heat; steady-state diffusion. Static electric and magnetic fields. Potential flow of fluids. Saturated flow in a porous medium.

(b) Parabolic equations model: Nonsteady heat conduction; nonsteady diffusion. Transport of electrons and holes in semiconductors. Unsaturated flow in porous media.

(c) Hyperbolic equations model: Transverse vibration of strings and membranes; torsional and longitudinal vibration of rods. Propagation of acoustic, electric, and magnetic waves.

It is fair to make the generalization that elliptic equations describe physical systems that are in a state of equilibrium whose behavior is independent of time, and that parabolic and hyperbolic equations describe the time evolution of physical systems that are not in a state of equilibrium. Typically, parabolic equations describe irreversible processes like conduction and diffusion, whereas hyperbolic equations are associated with wavelike evolution.

We shall see that each type of equation is similarly associated with a characteristic mathematical behavior of its solution. In the course of solving problems we will point out characteristic solution behavior and relate it to the physical behavior of the system being modeled.

We shall also see that the Laplace equation can be viewed as the prototype for elliptic equations in the sense that what is true about solutions of Laplace's equation is usually true about the solutions of elliptic equations in general. Similarly the heat and wave equations are prototypes for parabolic and hyperbolic equations, respectively.

Linearity and Superposition

Let $L[u]$ denote a linear partial differential operator and $B[u]$ denote a linear boundary operator; e.g.,

$$L[u] = a(x,y)\,\partial_{xx}u(x,y) + 2b(x,y)\,\partial_{xy}u(x,y) + c(x,y)\,\partial_{yy}u(x,y)$$

$$B[u] = (pu(x,y) + q\ \mathbf{grad}\ u(x,y) \bullet N)\,|_S$$

Here $f|_S$ denotes the restriction of f to S. Then the *principle of superposition* states that for smooth functions $u_1, \ldots, u_n$ and arbitrary constants $C_1, \ldots, C_n$

$$L[C_1u_1 + \cdots + C_nu_n] = C_1L[u_1] + \cdots + C_nL[u_n] \tag{9.2}$$

$$B[C_1u_1 + \cdots + C_nu_n] = C_1B[u_1] + \cdots + C_nB[u_n]$$

It is this version of the principle that permits a problem to be split into simpler subproblems.

More generally, if the infinite series

$$\sum_{n=1}^{\infty} C_nu_n \quad \text{and} \quad \sum_{n=1}^{\infty} C_nL[u_n]$$

are both convergent then we have

$$L\left[\sum_{n=1}^{\infty} C_nu_n\right] = \sum_{n=1}^{\infty} C_nL[u_n] \tag{9.3}$$

This version of the principle permits the use of eigenfunction expansions to solve problems in partial differential equations. Finally, the use of integral transforms for solving problems is supported by the following version of the principle of superposition. If $G(x, y, z)$ and $g(z)$ are integrable functions of z for z in the (not necessarily bounded) interval I and if the integrals

$$\int_I G(x, y, z)\, g(z)\, dz \quad \text{and} \quad \int_I L[G(x, y, z)]\, g(z)\, dz$$

both exist, then

$$L\left[\int_I G(x, y, z)\, g(z)\, dz\right] = \int_I L[G(x, y, z)]\, g(z)\, dz$$

SOLVED PROBLEMS

Boundary Value Problems on Infinite Sets

PROBLEM 9.1

Find the bounded solution $u = u(x, y)$ of the Dirichlet problem for the half space

$$\partial_{xx}u(x, y) + \partial_{yy}u(x, y) = 0 \quad \text{for } -\infty < x < \infty,\ y > 0, \tag{1}$$

$$u(x, 0) = f(x) \quad \text{for } -\infty < x < \infty \tag{2}$$

where $f(x) = bI_a(x)$ for some $a > 0$ and $b > 0$.

SOLUTION 9.1

The half space $H = \{(x, y) : -\infty < x < \infty, y > 0\}$ is the simplest example of a set with a boundary. The Dirichlet boundary condition (2) is imposed over the x axis which forms the boundary of the half space. The fact that we are seeking the bounded solution acts as a boundary condition at infinity for this problem.

Since x ranges over the whole real line, we will use the Fourier transform in the variable x to solve this problem. If we let $U(\alpha, y)$ denote the Fourier transform with respect to x of the unknown function $u(x, y)$, then

$$F\{\partial_{xx}u(x,y)\} = -\alpha^2 U(\alpha,y) \text{ and } F\{\partial_{yy}u(x,y)\} = U''(\alpha,y) \quad (3)$$

Since the Fourier transform is taken in the x variable, the derivative $\partial_{xx}u(x,y)$ interacts with the transform as prescribed by operational property 3 for the Fourier transform. However, the variable y plays the role of a parameter with respect to the Fourier transform in x with the result that the y derivative commutes with the transform.

In view of (3), application of the Fourier transform reduces the boundary value problem (1), (2) to the problem of finding the bounded solution for

$$-\alpha^2 U(\alpha,y) + U''(\alpha,y) = 0 \text{ for } y > 0 \text{ and } U(\alpha,0) = F(\alpha) \quad (4)$$

The general solution of the ordinary differential equation can be written

$$U(\alpha,y) = C_1 e^{-y|\alpha|} + C_2 e^{y|\alpha|} \qquad \text{for } y > 0$$

and since we want the bounded solution, we choose $C_2 = 0$. Then the condition $U(\alpha, o) = F(\alpha)$ leads to $C_1 = F(\alpha)$. Here $F(\alpha)$ denotes the Fourier transform of $f(x)$. Then the solution of the transformed problem (4) is

$$U(\alpha,y) = F(\alpha)\, e^{-y|\alpha|} \quad (5)$$

and it follows from operational property 8 together with formula (5) in Table 8.1 that

$$u(x,y) = (2\pi)^{-1}\int_{-\infty}^{\infty} \frac{2y}{y^2 + (x-z)^2} f(z)\,dz \quad (6)$$

For $f(x) = bI_a(x)$, this reduces to

$$u(x,y) = \frac{y}{\pi}\int_{-a}^{a} \frac{b\,dz}{y^2 + (x-z)^2} = \frac{b}{\pi}\left(\arctan\left(\frac{x+a}{y}\right) - \arctan\left(\frac{x-a}{y}\right)\right)$$

We could interpret $u(x, y)$ physically as the hydraulic head in a saturated field. The x axis is the surface boundary of the field and y is positive in the downward direction. We assume complete uniformity in the direction normal to the xy plane so that the problem can be treated as 2-dimensional. The boundary condition can be viewed as approximating a condition of ponded water in a channel of width $2a$ and depth b centered at $x = 0$.

Note that although y ranges over the positive half line, use of the Laplace transform with respect to y is not suggested here. The Laplace transform of the derivative $\partial_{yy}u(x, y)$ involves both $u(x, 0)$ and $\partial_y u(x, 0)$. Since only one of these is given, use of the Laplace transform would be awkward.

Finally, note that since $b > 0$ and $a > 0$, it follows from (6) that for each $y > 0$ we have $u(x, y) > 0$ for all x. This means that no matter how small the values a and b (as long as they are positive), the boundary function $f(x) = bI_a(x)$ will produce a positive $u(x, y)$ at every point inside the half space, H. This is sometimes referred to as *organic behavior* of the solution and is typical of systems in equilibrium; i.e., if a system is in equilibrium, then even a small change in the conditions on the boundary must result in changes everywhere inside as the system assumes a new equilibrium state.

PROBLEM 9.2

Find the bounded solution of the Neumann problem on the half plane

$$\partial_{xx}u(x, y) + \partial_{yy}u(x, y) = 0 \quad \text{in } H \tag{1}$$

$$\partial_y u(x, 0) = g(x) \quad \text{for } -\infty < x < \infty \tag{2}$$

SOLUTION 9.2

Rather than solve this problem directly, note that if we differentiate (1) with respect to y, we obtain $\partial_{xx}(\partial_y u) + \partial_{yy}(\partial_y u) = 0$. Letting $w(x, y) = \partial_y u(x, y)$, it then follows that w satisfies

$$\partial_{xx}w(x, y) + \partial_{yy}w(x, y) = 0 \quad \text{and } w(x, 0) = g(x)$$

The results of the previous problem imply that the bounded solution of this problem is

$$w(x, y) = \partial_y u(x, y) = \frac{1}{2\pi}\int_{-\infty}^{\infty} \frac{2y}{y^2 + (x-z)^2}\, g(z)\, dz \tag{3}$$

We integrate this with respect to y to recover $u(x, y)$

$$u(x, y) = \frac{1}{2\pi}\int_{-\infty}^{\infty} \log\left(y^2 + (x-z)^2\right) g(z)\, dz + c \tag{4}$$

Here C denotes an indeterminate constant of integration. Physically, the Neumann boundary condition could be used in the context of the saturated porous medium to model a pumping well or an injection well.

PROBLEM 9.3

Find the bounded solution of the following *quarter plane problem*

$$\partial_{xx}u(x, y) + \partial_{yy}u(x, y) = 0 \quad \text{for } x > 0,\ y > 0 \tag{1}$$

$$u(x,0) = f(x),\quad x > 0,\qquad \partial_x u(0,y) = g(y),\ y > 0 \tag{2}$$

SOLUTION 9.3

Here we seek a solution $u = u(x, y)$ for equation (1) that is bounded in the quarter plane $Q = \{(x, y)\colon\ x > 0, y > 0\}$ and satisfies mixed Dirichlet and Neumann boundary conditions (2) on the boundary of Q. Our strategy in this problem will be to break the problem (1), (2) into two subproblems, each of which can then be reduced to one or the other of the previous two problems.

Subproblem A. Let $v = v(x, y)$ denote the bounded solution of

$$\partial_{xx} v(x,y) + \partial_{yy} v(x,y) = 0 \quad \text{in } Q \tag{3}$$

$$v(x,0) = f(x),\quad x > 0,\qquad \partial_x v(0,y) = 0,\ y > 0 \tag{4}$$

Subproblem B. Let $w = w(x, y)$ denote the bounded solution of

$$\partial_{xx} w(x,y) + \partial_{yy} w(x,y) = 0 \quad \text{in } Q \tag{5}$$

$$w(x,0) = f(x),\quad x > 0,\qquad \partial_x w(0,y) = g(y),\ y > 0 \tag{6}$$

The principle of superposition implies that $u(x, y) = v(x, y) + w(x, y)$.

To solve subproblem A, replace the quarter plane problem (3), (4) by the half plane problem:

$$\partial_{xx} v(x,y) + \partial_{yy} v(x,y) = 0 \quad \text{for all } x \text{ and } y > 0 \tag{7}$$

$$v(x,0) = f_e(x) \quad \text{for all } x \tag{8}$$

Here $f_e(x)$ denotes the even extension of $f(x)$; i.e., $f_e(x) = f(x)$ for $x > 0$ and $f_e(x) = f(-x)$ for $x < 0$. Then according to the result of Problem 9.1

$$v(x,y) = \frac{y}{\pi}\int_{-\infty}^{\infty} \frac{1}{y^2 + (x-z)^2} f_e(z)\,dz \tag{9}$$

and this can be expressed in terms of f rather than f_e by writing

$$v(x,y) = \frac{y}{\pi}\int_{-\infty}^{0} \frac{1}{y^2 + (x-z)^2} f_e(z)\,dz + \frac{y}{\pi}\int_{0}^{\infty} \frac{1}{y^2 + (x-z)^2} f_e(z)\,dz$$

$$= \frac{y}{\pi}\int_{-\infty}^{0} \frac{1}{y^2 + (x-z)^2} f(-z)\,dz + \frac{y}{\pi}\int_{0}^{\infty} \frac{1}{y^2 + (x-z)^2} f(z)\,dz$$

$$= \frac{y}{\pi}\int_0^\infty \frac{1}{y^2+(x+z)^2} f(z)\,dz + \frac{y}{\pi}\int_0^\infty \frac{1}{y^2+(x-z)^2} f(z)\,dz$$

Finally,

$$v(x,y) = \frac{y}{\pi}\int_0^\infty \left(\frac{1}{y^2+(x+z)^2} + \frac{1}{y^2+(x-z)^2}\right) f(z)\,dz \tag{10}$$

It is easy to check from either (9) or (10) that $\partial_x v(0,y) = 0$. Thus, $v(x, y)$ solves subproblem A.

We can solve subproblem B in a similar manner by replacing the quarter plane problem (5), (6) by the half plane problem:

$$\partial_{xx} w(x,y) + \partial_{yy} w(x,y) = 0 \quad \text{for all } y \text{ and } x > 0 \tag{11}$$

$$\partial_x w(0,y) = g_0(y) \quad \text{for all } y \tag{12}$$

where $g_0(y)$ denotes the odd extension of the data function g in (6). Extending g in this fashion will automatically cause $w(x, 0)$ to equal zero for $x > 0$ just as extending f as an even function in subproblem A produced the effect $\partial_x v(0,y) = 0$ for $y > 0$. The half plane Neumann problem was solved in Problem 9.2 (with the roles of x and y interchanged). Thus,

$$w(x,y) = \frac{1}{2\pi}\int_{-\infty}^\infty \log(x^2+(y-z)^2)\, g_0(z)\,dz + C \tag{13}$$

We can express the solution in terms of g rather than the odd extension g_0 by writing

$$w(x,y) = \frac{1}{2\pi}\int_{-\infty}^0 \log(x^2+(y-z)^2)\, g_0(z)\,dz$$

$$+ \frac{1}{2\pi}\int_0^\infty \log(x^2+(y-z)^2)\, g_0(z)\,dz + C$$

Using the definition of the odd extension and changing variables in the first integral reduces this to

$$w(x,y) = \frac{1}{2\pi}\int_0^\infty \log\frac{x^2+(y-z)^2}{x^2+(y+z)^2}\, g(z)\,dz \tag{14}$$

Adding (14) and (10) gives the solution to the original problem. Note that as the complexity of the boundary increases, so too does the solution increase in complexity.

PROBLEM 9.4

Find a bounded function $u = u(x, y)$ that satisfies

$$\partial_{xx}u(x,y) + \partial_{yy}u(x,y) = 0 \quad \text{for all } x \text{ and } 0<y<1 \tag{1}$$

$$u(x,0), \quad u(x,1) = g(x) \quad \text{for all } x \tag{2}$$

SOLUTION 9.4

This problem could arise in connection with finding the electrostatic potential $u(x, y)$ in the region bounded by the planes $y = 0$ and $y = 1$, assuming that the plane at $y = 0$ is grounded and the plane at y = *1 carries a charge distribution that varies with x* (only) according to the prescribed function $g(x)$.

Since x varies over the whole real line, we apply the Fourier transform in x to this Dirichlet problem. The transformed problem is then

$$-\alpha^2 U(\alpha,y) + U''(\alpha,y) = 0, \quad 0<y<1 \tag{3}$$

$$U(\alpha,0) = 0 \quad \text{and} \quad U(\alpha,1) = G(\alpha) \tag{4}$$

If we write the general solution of (3) in the form

$$U(\alpha,y) = C_1 \sinh \alpha y + C_2 \sinh \alpha(1-y)$$

then the homogeneous boundary condition in (4) implies $C_2 = 0$ and the other boundary condition leads to $U(\alpha,1) = C_1 \sinh \alpha = G(\alpha)$. Therefore,

$$U(\alpha,y) = G(\alpha)\frac{\sinh \alpha y}{\sinh \alpha}, \quad 0<y<1 \tag{5}$$

To find the inverse transform of (5) we recall Problem 8.4 where we showed that

$$\frac{\sinh \alpha y}{\sinh \alpha} = \sum_{n=0}^{\infty} e^{-(2n+1-y)|\alpha|} - \sum_{n=0}^{\infty} e^{-(2n+1+y)|\alpha|}$$

Then the convolution property of the Fourier transform, together with entry 5 in Table 8.1 leads to the result

$$u(x,y) = \frac{1}{\pi}\sum_{n=0}^{\infty}\int_0^{\infty} \frac{(2n+1-y)}{(2n+1-y)^2+(x-z)^2}\, g(z)\,dz$$
$$-\frac{1}{\pi}\sum_{n=0}^{\infty}\int_0^{\infty} \frac{(2n+1+y)}{(2n+1+y)^2+(x-z)^2}\, g(z)\,dz \tag{6}$$

Initial Value Problems on Infinite Domains

PROBLEM 9.5 THE DIFFUSION EQUATION

Find the bounded solution for the following initial value problem for the 1-dimensional diffusion equation

$$\partial_t u(x,t) = D\partial_{xx}u(x,t) \quad \text{for all } x \text{ and } t>0 \tag{1}$$
$$u(x,0) = f(x) \quad \text{for all } x. \tag{2}$$

Here D is a positive constant.

SOLUTION 9.5

Here $u(x, t)$ could represent the time dependent distribution of contaminant in an infinitely long thin tube, given that the distribution is initially described by the function $f(x)$. Of course saying that the tube is infinitely long simply means we are ignoring the influence of conditions at the ends of the tube. This means, in turn, that the solution $u(x, t)$ is valid only far away from the ends and only for a limited amount of time. The precise meaning of the terms "far away" and "limited amount" depends on the constant D and on $f(x)$ among other things.

In this problem we could apply either the Fourier transform in x or the Laplace transform in t. The latter approach would leave us with an inhomogeneous second order ordinary differential equation in x to solve in the transform space. Using the Fourier transform, however, leads to a homogeneous first order equation in t in the transform space. We therefore choose to apply the Fourier transform in x to the initial value problem (1), (2) to obtain an ordinary differential equation in t for the transform $U(\alpha, t) = F\{u(x, t)\}$ of the unknown solution $u(x, t)$:

$$U'(\alpha, t) = -D\alpha^2 U(\alpha, t), \quad U(\alpha, 0) = F(\alpha) \tag{3}$$

This first order initial value problem is easily solved to find

$$U(\alpha, t) = F(\alpha) e^{-Dt\alpha^2} \quad \text{for } t > 0 \tag{4}$$

Now we use the convolution property of the Fourier transform together with entry 6 from Table 8.1 for $c = Dt$. This leads to

$$u(x, t) = (4\pi Dt)^{-1/2} \int_{-\infty}^{\infty} e^{-(x-z)^2/4Dt} f(z)\, dz \tag{5}$$

The change of variable $s = (x-z)/\sqrt{4Dt}$ reduces (5) to

$$u(x, t) = \pi^{-1/2} \int_{-\infty}^{\infty} e^{-s^2} f(x - s(4Dt)^{1/2})\, ds \tag{6}$$

PROBLEM 9.6 INFLUENCE OF LOWER ORDER TERMS

Find the bounded solution for the initial value problem for the 1-dimensional diffusion equation with lower order terms

$$\partial_t v(x, t) = D\partial_{xx} v(x, t) + B\partial_x v(x, t) + Cv(x, t) \quad \text{for all } x \text{ and } t > 0 \tag{1}$$

$$v(x, 0) = f(x) \quad \text{for all } x \tag{2}$$

Here D, B, and C are all constant with D positive.

SOLUTION 9.6

As in the previous problem, we apply the Fourier transform in x to transform the initial value problem (1), (2) into an ordinary differential equation in t for the unknown transform $V(\alpha, t) = F\{v(x, t)\}$,

$$V'(\alpha, t) = -\alpha^2 DV(\alpha, t) + i\alpha BV(\alpha, t) + CV(\alpha, t) \quad V(\alpha, 0) = F(\alpha) \tag{3}$$

The differential equation $V'(t) = (-\alpha^2 D + i\alpha B + C)\,V(\alpha, t)$ is again easily solved and we find

$$V(\alpha, t) = F(\alpha)\,e^{(-\alpha^2 D + i\alpha B + C)\,t} = F(\alpha)\,e^{-Dt\alpha^2} e^{iBt\alpha} e^{Ct} \tag{4}$$

Thus, $V(\alpha, t)$ is equal to $U(\alpha, t)$ from the previous problem multiplied by a complex exponential [which, according to operational property 5, acts as a shift] and multiplied also by an exponential function of t (which commutes with the inverse Fourier transform). Then it follows that

$$v(x, t) = (4\pi Dt)^{-1/2} e^{Ct} \int_{-\infty}^{\infty} e^{-(x + Bt - z)^2/4Dt} f(z)\,dz \tag{5}$$

$$= \pi^{-1/2} e^{Ct} \int_{-\infty}^{\infty} e^{-s^2} f(x + Bt - s\,(4Dt)^{1/2})\,ds \tag{6}$$

The zero order term $Cv(x, t)$ in equation (1) could, in the context of diffusion, be interpreted as a leakage term representing leakage of contaminant into $(C > 0)$ or out of $(C < 0)$ the diffusion tube through the lateral surface. It is evident from the solution that causes the concentration of contaminant to grow or decay exponentially with time. The first order term $B\partial_x v(x, t)$ can be shown to arise when the medium in which the contaminant diffuses is moving with velocity equal to B. The sign of B is then related to the direction of movement of the diffusion medium.

PROBLEM 9.7

Show that the influence of the initial conditions in Problem 9.5 could be said to propagate with infinite speed. Show further that this is a misleading interpretation of what could better be described as "instantaneous smoothing" of the initial data.

SOLUTION 9.7

Consider the solution to the initial value Problem 9.5 expressed in the form (6) in the case that $f(x) = I_a(x)$ for some small positive value for a. Then

$$u(x, t) = \pi^{-1/2} \int_{-\infty}^{\infty} e^{-s^2} I_a(x - s\,(4Dt)^{1/2})\,ds$$

$$= \pi^{-1/2} \int_{(x-a)/(4Dt)^2}^{(x+a)/(4Dt)^2} e^{-s^2}\,ds \tag{1}$$

Since the exponential function $\exp[-s^2]$ is positive for all s, it is clear that for each positive t, no matter how small, $u(x, t)$ is positive for all values of x. Thus, even though the initial state is zero everywhere except in the interval $(-a, a)$ near the origin, $u(x, t)$ is immediately positive everywhere. This is often referred to as the infinite speed of propagation associated with the diffusion equation (and parabolic equations in general).

Using the form (5) for the solution $u(x, t)$ with $f(x) = I_a(x)$

$$u(x, t) = (4\pi Dt)^{-1/2} \int_{-a}^{a} e^{-(x-z)^2/4Dt} dz \tag{2}$$

we see that for each $t > 0$, $u(x, t)$ can be differentiated arbitrarily often with respect to either x or t and the result is still a smooth function of x and t; i.e., the solution is an infinitely smooth function of x and t in spite of the fact that $u(x, 0)$ is only piecewise continuous. Also, by the mean value theorem for integrals, there exists some p between $x - a$ and $x + a$ such that

$$0 \le u(x, t) \le a(\pi Dt)^{-1/2} e^{-p^2/4Dt}$$

Thus, even though the influence of the initial state propagates with infinite speed, the strength of the influence diminishes exponentially as the distance from $(-a, a)$ increases.

A more realistic interpretation of the speed of propagation for effects that are governed by the diffusion equation can be obtained by considering an example in which the initial state is given by $f(x) = H(x)$; i.e., $f(x) = 1$ if $x > 0$ and $f(x) = 0$ if $x < 0$. Then the solution in the form (6) for Problem 9.5 reduces to

$$u(x, t) = \pi^{-1/2} \int_{-\infty}^{x/(4Dt)^{1/2}} e^{-s^2} ds \tag{3}$$

Since

$$\pi^{-1/2} \int_{-\infty}^{\infty} e^{-s^2} ds = 1$$

it follows that for each positive t and all values of x, $u(x, t)$ assumes a value between 0 and 1. In fact, $u(x, t)$ is clearly constant along parabolic curves of the form

$$x = C\sqrt{4Dt} \quad \text{for } C > 0 \tag{4}$$

For a fixed positive C, let u_c denote the constant value of u along the curve. Then, for each $t > 0$, there is a single point $x = X(t)$ such that $u(X(t), t) = u_c$ and it follows from (4) that the speed of this point is given by

$$X'(t) = C\sqrt{D/t} \quad \text{for } t > 0 \tag{5}$$

For each $t > 0$, this speed is finite and represents the speed at which the value $u = u_c$ propagates.

PROBLEM 9.8 A TRAVELLING WAVE

Find a function $u = u(x, t)$ that solves the initial value problem

$$\partial_t u(x, t) = \partial_x u(x, t) \qquad \text{for all } x \text{ and for } t > 0 \tag{1}$$

$$u(x, 0) = f(x) \qquad \text{for all } x \tag{2}$$

SOLUTION 9.8

This problem could be solved by applying the Fourier transform in x or by applying the Laplace transform in t. Application of the Laplace transform in t leads to an inhomogeneous ordinary differential equation in x in which the transform of the unknown function $u(x, t)$ is the unknown. Using the Fourier transform gives a somewhat simpler problem.

If we apply the Fourier transform in x, letting $U(\alpha, t)$ denote the transform of $u(x, t)$, then the problem (1), (2) becomes

$$U'(\alpha, t) = i\alpha U(\alpha, t), \qquad U(\alpha, 0) = F(\alpha) \tag{3}$$

The prime here denotes differentiation with respect to t and $F(\alpha)$ is the transform of $f(x)$. The initial value problem (3) is easily solved with the result

$$U(\alpha, t) = F(\alpha)e^{i\alpha t} \tag{4}$$

Then by the shifting property of the Fourier transform, it follows that

$$u(x, t) = f(x + t). \tag{5}$$

The solution (5) is often called a *travelling wave* solution since it consists of the waveform $f(x)$ travelling from left to right with speed equal to 1. More generally, the function that satisfies (2) and the equation

$$\partial_t u(x, t) = A\partial_x u(x, t) \qquad \text{for all } x \text{ and } t > 0$$

is found to be $u(x, t) = f(x + At)$ which is a travelling wave moving with speed equal to A. Sometimes the speed A is referred to as the *group velocity* or the *phase velocity* for the travelling wave. Note that the direction of the wave is determined by the sign of A.

PROBLEM 9.9 THE WAVE EQUATION

Find the solution of the initial value problem for the wave equation

$$\partial_{tt} u(x, t) = a^2 \partial_{xx} u(x, t) \qquad \text{for all } x \text{ and } t > 0 \tag{1}$$

$$u(x, 0) = f(x), \quad \partial_t u(x, 0) = g(x), \qquad \text{for all } x \tag{2}$$

This initial value problem could describe the propagation of acoustic waves in a long thin pipe where $u(x, t)$ equals the deviation at position x and time t from an equilibrium pressure state. The initial functions f and g

could be chosen to model some initial disturbance whose propagation we wish to study.

SOLUTION 9.9

This problem, like the last, could be solved by applying the Laplace transform in t. However, it is more convenient to apply the Fourier transform in x. This reduces the initial value problem to

$$U''(\alpha, t) = -a^2\alpha^2 U(\alpha, t), \; U(\alpha, 0) = F(\alpha) \quad U'(\alpha, 0) = G(\alpha) \tag{3}$$

where the primes denote differentiation with respect to t of the transform $U(\alpha, t)$ of $u(x, t)$. We choose to write the general solution of (3) in the form

$$U(a, t) = C_1 \cos a\alpha t + C_2 \sin a\alpha t$$

and then the initial conditions lead to the result

$$U(\alpha, t) = F(\alpha) \cos a\alpha t + \frac{G(\alpha)}{a\alpha} \sin a\alpha t$$

In order to find the inverse transform from our table of transforms, it will be helpful to rewrite this in the form

$$U(\alpha, t) = \frac{1}{2} F(\alpha)\,(e^{iat\alpha} + e^{-iat\alpha}) + \frac{1}{2a}\frac{G(\alpha)}{i\alpha}\,(e^{iat\alpha} - e^{-iat\alpha}) \tag{4}$$

Note that if $h(x)$ is a function such that $h'(x) = g(x)$, then $H(\alpha) = G(\alpha)/i\alpha$. Then using the shifting property of the Fourier transform to invert expression (4), we find

$$u(xt) = \frac{1}{2}(f(x+at) + f(x-at)) + \frac{1}{2a}(h(x+at) - h(x-at)) \tag{5}$$

Since $h'(x) = g(x)$ is equivalent to $h(x) = \int^x g$ we can write (5) in the form

$$u(x, t) = \frac{1}{2}(f(x+at) + f(x-at)) + \frac{1}{2a}\int_{x-at}^{x+at} g(s)\,ds \tag{6}$$

When the solution is written in this form, it is referred to as the *D'Alembert form* of the solution to the initial value problem for the wave equation. It is apparent from (5) that the solution consists of two travelling waves; one moves from left to right and the other from right to left, each travelling with speed $|a|$. Recall that the first order wave equation of the previous problem permitted propagation in only one direction.

Note that for fixed values of x and t, the value of $u(x, t)$ does not depend on all of the data but only on those values in the closed interval $[x - at, x + at]$. We refer to this interval as the *domain of dependence* for the point (x, t). The data values $f(s)$ and $g(s)$ for s not in this interval do not affect the value of u at the point (x, t). We interpret this to mean that the domain of dependence for (x, t) contains all the data values that will reach the point x by the time t when travelling at speed $|a|$ from time $t = 0$. Finite speed of propagation is characteristic of solutions to hyperbolic partial differential equations like the wave equation. One consequence of the finite propagation speed is that the solution behavior at infinity can have no bearing on the solution behavior on any bounded set. Therefore, it is neither necessary nor proper to impose boundary conditions at infinity for a hyperbolic problem on an infinite set.

Initial-Boundary Value Problems

PROBLEM 9.10

Find the bounded function $u = u(x, t)$ satisfying the diffusion equation

$$\partial_t u(x, t) = D\partial_{xx} u(x, t) \quad \text{for } x > 0 \text{ and } t > 0 \tag{1}$$

and the auxiliary conditions

$$u(x, 0) = f(x) \quad \text{for } x > 0 \tag{2}$$

$$u(0, t) = p(t) \quad \text{for } t > 0 \tag{3}$$

The problem (1), (2), (3) is called an *initial-boundary value problem* since the auxiliary conditions consist of an initial condition (2) and a boundary condition (3).

SOLUTION 9.10

It will be convenient to split this problem, as we did Problem 9.3, into two subproblems. The Fourier transform will turn out to be the better method for solving the subproblem having the inhomogeneous initial condition and zero boundary condition, whereas the Laplace transform will be more suitable for the problem where the initial condition is homogeneous and the boundary condition is not zero.

Subproblem A. Let $v = v(x, t)$ denote the bounded solution of

$$\partial_t v(x, t) = D\partial_{xx} v(x, t) \quad \text{for } x > 0, \text{ and } t > 0 \tag{4}$$

$$v(x, 0) = f(x), \quad x > 0 \quad \text{and} \quad v(0, t) = 0 \quad \text{for } t > 0 \tag{5}$$

Subproblem B. Let $w = w(x, t)$ denote the bounded solution of

$$\partial_t w(x, t) = D\partial_{xx} w(x, t) \quad \text{for } x > 0 \text{ and } t > 0 \tag{6}$$

$$w(x, 0) = 0 \quad \text{for } x > 0 \quad \text{and} \quad w(0, t) = p(t) \quad \text{for } t > 0 \tag{7}$$

Then superposition implies that $u(x, t) = v(x, t) + w(x, t)$.

To solve subproblem A we will replace it with an equivalent problem in which x ranges over the whole real line. Note that if $v(x, t)$ solves the problem

$$\partial_t v(x,t) = D\partial_{xx} v(x,t) \quad \text{for all } x \text{ and } t>0 \tag{8}$$

$$v(x,0) = f_0(x) \quad \text{for all } x \tag{9}$$

where $f_0(x)$ denotes the odd extension of $f(x)$ to the whole real line, then based on the result of Problem 9.5

$$v(x,t) = \int_{-\infty}^{\infty} K(x-z, Dt) f_0(z)\, dz \tag{10}$$

where

$$K(x,t) = (4\pi t)^{-1/2} e^{-x^2/4t} \tag{11}$$

Thus,

$$v(x,t) = \int_{-\infty}^{0} K(x-z, Dt) f_0(z)\, dz + \int_{0}^{\infty} K(x-z, Dt) f_0(z)\, dz$$

$$= \int_0^{\infty} K(x+z, Dt) f_0(-z)\, dz + \int_0^{\infty} K(x-z, Dt) f_0(z)\, dz$$

and then, by the definition of the odd extension, we find

$$v(x,t) = \int_0^{\infty} (K(x-z, Dt) - K(x+z, Dt))\, f(z)\, dz \tag{12}$$

It is obvious from (12) that $v(0,t) = 0$ and (10) implies that $v(x,t)$ solves (8), (9). Then $v(x,t)$ given by (12) solves the initial-boundary value problem of subproblem A. Observe that extending $f(x)$ as an odd function ensures that the solution of (8), (9) will vanish at $x = 0$. Using the even extension instead would have led to a solution whose derivative was equal to zero at $x = 0$.

To solve subproblem B, let $W(x,s)$ denote the Laplace transform of $w(x,t)$ with respect to t. Then (6) is transformed to

$$sW(x,s) - 0 = DW''(x,s) \quad \text{for } x>0 \tag{13}$$

where the primes denote differentiation with respect to x. The boundary condition (7) becomes

$$W(0,s) = P(s) \quad \text{where } P(s) = L\{p(t)\} \tag{14}$$

The general solution of (13) can be written as

$$W(x,s) = C_1 e^{-x\sqrt{s/D}} + C_2 e^{x\sqrt{s/D}}$$

To obtain the bounded solution, we choose $C_2 = 0$ and then (14) implies $C_1 = P(s)$. Then

$$W(x,s) = P(s)\, e^{-x\sqrt{s/D}} \tag{15}$$

The inversion is accomplished by using the convolution property for the Laplace transform together with formula (8.7) with the parameter a set equal to $x/\sqrt{D}$. This leads to

$$w(x,t) = \int_0^t E(x, D(t-r))\, p(r)\, dr \tag{16}$$

where

$$E(x,t) = \frac{x}{t} K(x,t) \tag{17}$$

Then $u(x, t) = v(x, t) + w(x, t)$, where $v(x, t)$ can be interpreted as the response to the initial state, $f(x)$, and $w(x, t)$ can be thought of as the part of the solution that is due to the boundary forcing, $p(t)$.

PROBLEM 9.11

Find the solution to Problem 9.10 in the special case that $f(x) = 0$ and $p(t)$ is an increasing function of t such that $p(0) = 0$. Show that $g(t) = \partial_x u(0, t)$ is then a nonpositive function of t given by

$$g(t) = -(\pi D)^{-1/2} \int_0^t (t-r)^{-1/2} p'(r)\, dr \tag{1}$$

SOLUTION 9.11

If $f(x) = 0$, the $u(x, t) = w(x, t)$; i.e., the solution is composed only of the response to the boundary forcing. Now in the context of diffusion, $g(t) = \partial_x u(0, t)$ represents the contaminant flux at the boundary. In view of (15) from the previous problem,

$$g(t) = L^{-1}\{\partial_x W(x,s)|_{x=0}\} = L^{-1}\{-\sqrt{s/D}\; P(s)\}$$

$$= L^{-1}\{-(sD)^{-1/2} sP(s)\}$$

The formula for the Laplace transform of the derivative implies that if $p(0) = 0$, then $sP(s) = L\{p'(t)\}$ and formula (8.8) with $a = 0$ gives $(sD)^{-1/2} = L\{(\pi D t)^{-1/2}\}$. Then the convolution property for the Laplace transform leads to (1). Note that if $p(t)$ is increasing then $p'(t) > 0$ and (1) implies $g(t) < 0$ for $t > 0$. For example, if $p(t) = At$ for a positive constant A, then $g(t) = -A(t/\pi D)^{-1/2}$.

PROBLEM 9.12

Find the solution to Problem 9.10 in the case $f(x) = 0$ and $p(t) = A \cos \Omega t$, for given constants A and Ω. Identify the transient and steady-state components of the solution and give a physical interpretation in the context of heat conduction.

SOLUTION 9.12

In the case that $f(x) = 0$ and $p(t) = A \cos \Omega t$, the solution to Problem

9.10 is given by

$$u(x, t) = \int_0^t E(x, D(t-r)) A \cos \Omega r \, dr$$

The change of variable $z = x/(rD(t-r))^{1/2}$ reduces this integral to

$$u(x, t) = 2\pi^{-1/2} \int_{\lambda(x,t)}^{\infty} e^{-z^2} p(t - x^2/(4Dz^2)) \, dz \tag{1}$$

for $\lambda(x, t) = x(4Dt)^{-1/2}$. We can write this as

$$u(x, t) = 2\pi^{-1/2} \int_0^{\infty} e^{-z^2} p(t - x^2/(4Dz^2)) \, dz$$
$$- 2\pi^{-1/2} \int_0^{\lambda(x,t)} e^{-z^2} p(t - x^2/(4Dz^2)) \, dz \tag{2}$$

Since $\lambda(x, t)$ tends to zero as t tends to infinity it follows that the second of the two integrals in (2) tends to zero with increasing t. Then this part of the solution is the transient portion of the solution and the first integral in (2) is the steady-state component of the solution. Now it can be shown that

$$2\pi^{-1/2} \int_0^{\infty} e^{-z^2} \cos \Omega (t - x^2/(4Dz^2)) \, dz = e^{-\gamma x} \cos(\Omega t - \gamma x)$$

where $\gamma = (\Omega/2D)^{1/2}$. Then as t tends to infinity, $u(x, t)$ approaches the steady-state solution

$$u_{ss}(x, t) = Ae^{-\gamma x} \cos(\Omega t - \gamma x)$$

Thus, the steady-state response to a periodic input on the boundary is a periodic function whose amplitude decreases exponentially with increasing x and whose phase shift increases linearly with x. In the context of heat conduction, this means that periodically varying the temperature at the surface of a semi-infinite heat conducting medium eventually produces a response in the subsurface temperature which is periodic with the same frequency. However, the amplitude of the response dies out exponentially fast with increasing depth so at some (relatively small) finite depth the amplitude of the response is effectively zero. In addition, the subsurface temperature is out of phase with the forcing temperature at the surface by an amount that varies with the depth. Thus, a pipeline that is buried sufficiently deep beneath the surface of the earth is impervious to the daily (periodic) heating and cooling effects produced by the rising and setting of the sun or the periodic seasonal variations in surface temperature.

PROBLEM 9.13

Find $u = u(x, t)$ satisfying

$$\partial_{tt}u(x,t) = a^2\partial_{xx}u(x,t) \quad \text{for } x>0 \text{ and } t>0 \tag{1}$$

$$u(x,0) = f(x) \quad \text{and} \quad \partial_t u(x,0) = g(x) \quad \text{for } x>0 \tag{2}$$

$$u(0,t) = h(t) \quad \text{for } t>0. \tag{3}$$

This problem could arise in modeling acoustic waves in a long thin pipe, one end of which is subject to time dependent pressure disturbances described by the function $h(t)$. The other end is sufficiently distant that we can ignore the effect on the solution of conditions at that end.

SOLUTION 9.13

Once again it will prove to be convenient to divide this problem into two subproblems. The inhomogeneous initial value problem will be treated by the Fourier transform, while the subproblem that has the inhomogeneous boundary condition and homogeneous initial conditions will be solved using the Laplace transform.

Subproblem A. Let $v(x, t)$ solve

$$\partial_{tt}v(x,t) = a^2\partial_{xx}v(x,t) \quad \text{for } x>0 \text{ and } t>0 \tag{4}$$

$$v(x,0) = f(x) \quad \text{and} \quad \partial_t v(x,0) = g(x) \quad \text{for } x>0 \tag{5}$$

$$v(0,t) = 0 \quad \text{for } t>0 \tag{6}$$

Subproblem B. Let $w(x, t)$ satisfy

$$\partial_{tt}w(x,t) = a^2\partial_{xx}w(x,t) \quad \text{for } x>0 \text{ and } t>0 \tag{7}$$

$$w(x,0) = 0 \quad \text{and} \quad \partial_t w(x,0) = 0 \quad \text{for } x>0 \tag{8}$$

$$w(0,t) = h(t) \quad \text{for } t>0 \tag{9}$$

To solve subproblem A, we use the usual device of converting the semi-infinite problem to an equivalent problem on the whole real line. If we extend the data functions $f(x)$ and $g(x)$ to the whole line as odd functions, then the boundary condition $v(0, t) = 0$ will automatically be satisfied. Thus, we replace (4), (5), (6) by

$$\partial_{tt}v(x,t) = a^2\partial_{xx}v(x,t) \quad \text{for all } x \text{ and } t>0$$

$$v(x,0) = f_0(x) \quad \text{and} \quad \partial_t v(x,0) = g_0(x) \quad \text{for all } x,$$

The solution of this problem is given by

$$v(x,t) = \frac{1}{2}\left(f_0(x+at) + f_0(x-at)\right) + \frac{1}{2a}\int_{x-at}^{x+at} g_0(s)\,ds$$

and in view of the definition of the odd extension for functions this becomes

$$v(x,t) = \begin{cases} \dfrac{1}{2}\left(f(x+at) + f(x-at)\right) + \dfrac{1}{2a}\displaystyle\int_{x-at}^{x+at} g(s)\,ds & \text{if } 0 < t < x/a \\ \dfrac{1}{2}\left(f(x+at) - f(x-at)\right) + \dfrac{1}{2a}\displaystyle\int_{at-x}^{at+x} g(s)\,ds & \text{if } x/a < t \end{cases} \tag{10}$$

Note that (10) implies

$$v(0,t) = \frac{1}{2}\left(f(at) - f(at)\right) + \frac{1}{2a}\int_{at}^{at} g(s)\,ds = 0 \quad \text{for all } t > 0$$

To solve subproblem B, we let $W(x, s)$ denote the Laplace transform in t of the solution $w(x, t)$ for (7), (8), (9). Then

$$s^2 W(x,s) - s \cdot 0 - 0 = a^2 W''(x,s) \quad \text{for } x > 0 \text{ and } W(0,s) = H(s) = L\{h\}$$

The solution of this ordinary differential equation in x may be written in the form

$$W(x,s) = C_1 e^{-xs/a} + C_2 e^{xs/a} \quad \text{for } x > 0 \tag{11}$$

In order to choose the arbitrary constants C_1 and C_2, we must recall that the exponential functions in (11) act as "shift operators" on functions whose transforms they multiply. Since we are interested in a solution of the initial value problem (7), (8), (9) that corresponds to the propagation of the pressure disturbance $h(t)$ *into* the thin tube that occupies the region $x > 0$, we want to choose the wave that travels from left to right and not the wave that moves in the opposite direction. Thus, we choose $C_2 = 0$ in (11) and then the boundary condition $W(0, s) = H(s)$ implies $C_1 = H(s)$. Then

$$W(x,s) = H(s)\,e^{-xs/a} \quad \text{for } x > 0$$

and

$$w(x,t) = h(t - x/a)H(t - x/a) \quad \text{for } x > 0, t > 0 \tag{12}$$

Thus, for each fixed x, the pressure disturbance $w(x, t)$ at position $x > 0$ is equal to zero until t achieves the value x/a at which time it assumes the "delayed" value $h(t - x/a)$.

By the principle of superposition, the solution of the original initial-boundary value problem is the sum of the solutions to the two subproblems; i.e., $u = v + w$, where v represents the response to the initial state and w is the response to the input from the boundary.

Boundary Value Problems on Bounded Sets

PROBLEM 9.14 A DIRICHLET PROBLEM

For a given function $f = f(y)$, find $u = u(x, y)$ satisfying

$$\partial_{xx} vu(x, y) + \partial_{yy} u(x, y) = 0 \quad \text{for } 0 < x, y < 1 \tag{1}$$

$$u(0, y) = 0, \quad u(1, y) = f(y) \quad \text{for } 0 < y < 1 \tag{2}$$

$$u(x, 0) = 0, \quad u(x, 1) = 0 \quad \text{for } 0 < x < 1 \tag{3}$$

Here $u(x, y)$ could represent the steady-state temperature distribution in a heat conducting plate of sidelength 1 having specified temperature f along one edge.

SOLUTION 9.14

We begin with the assumption that $u(x, y) = X(x)Y(y)$ and substitute into equation (1) to obtain $X''(x)Y(y) + X(x)Y''(y) = 0$; i.e.,

$$\frac{X''(x)}{X(x)} = -\frac{Y''(y)}{Y(y)} \tag{4}$$

The left side of equation (4) depends only on x, whereas the right side depends only on y. The only way in which a function of x can equal a function of y as x and y range independently over the interval (0, 1) is for each function to be constant. If we denote this common constant value by λ, then (4) is equivalent to the coupled pair of equations:

$$-Y''(y) = \lambda Y(y) \tag{5a}$$

and

$$X''(x) = \lambda X(x) \tag{5b}$$

In addition, it follows from the boundary conditions (2) and (3) that

$$Y(0) = Y(1) = 0 \tag{6a}$$

and

$$X(0) = 0 \tag{6b}$$

The inhomogeneous boundary condition $u(1, y) = f(y)$ is set aside for now. The boundary value problem (5a), (6a) is seen to be a Sturm-Liouville problem; in fact (for $L = 1$), it is Problem 1 listed in the summary of chapter 6 with eigenvalues and eigenfunctions given by

$$\lambda_n = (n\pi)^2, \quad Y_n(y) = \sin n\pi y \quad \text{for } n = 1, 2, \ldots \tag{7}$$

Then equation (5b) becomes $X_n''(x) = (n\pi)^2 X_n(x)$. The general solution of this equation is a linear combination of any two of the following six functions:

$$e^{n\pi x},\ e^{-n\pi x},\ \sinh n\pi x,\ \cosh n\pi x,\ \sinh n\pi(1-x),\ \cosh n\pi(1-x).$$

If we choose to write $X_n(x) = A \sinh n\pi x + B \sinh n\pi(1-x)$ then the condition (6b) implies $X_n(0) = B \sinh n\pi = 0$ i.e., $B = 0$. Then for any choice of A, the function $X_n(x) = A \sinh n\pi x$ satisfies (5b) and (6b). Choosing a different A for each n, we see that $u_n(x, y) = A_n \sinh n\pi x \sinh n\pi y$ satisfies the differential equation (1) together with all the homogeneous boundary conditions in (2), (3). The principle of superposition implies that

$$u(x, y) = \sum_{n=1}^{\infty} A_n \sinh n\pi x \sinh n\pi y \tag{8}$$

also satisfies all of these conditions (if the series converges). The inhomogeneous boundary condition is satisfied if

$$u(1, y) = \sum_{n=1}^{\infty} A_n \sinh n\pi x \sinh n\pi y = f(y) \tag{9}$$

But if (9) holds then for any positive integer m,

$$\sum_{n=1}^{\infty} A_n \sinh n\pi \int_0^1 \sin n\pi y \sin m\pi y \, dy = \int_0^1 f(y) \sin m\pi y \, dy$$

and the orthogonality of the eigenfunctions causes all of the integrals on the left to vanish except the integral for $n = m$ which equals 1/2. Then

$$A_m = \frac{2}{\sinh m\pi} \int_0^1 f(y) \sin m\pi y \, dy \quad \text{for } m = 1, 2, \ldots \tag{10}$$

$$= \frac{f_m}{\sinh m\pi}$$

Using this result in (8), we get the solution of the Dirichlet boundary value problem (1), (2), (3)

$$u(x, y) = \sum_{m=1}^{\infty} f_m \frac{\sinh m\pi x}{\sinh m\pi} \sin m\pi y \tag{11}$$

The completeness of the family of eigenfunctions $\{\sin m\pi y\}$ guarantees that the series (11) converges to $f(y)$ when $x = 1$. It can also be shown that the series is absolutely and uniformly convergent on the subset $D_\varepsilon = \{\varepsilon \le x, y \le 1 - \varepsilon\}$ for any $\varepsilon > 0$. We can even show that we may differentiate the series (11) term by term any number of times with respect to x and y and the differentiated series is convergent on D_ε to the corresponding derivative of u. This statement is true independent of the square integrable function $f(y)$. Thus, the solution of this Dirichlet boundary value problem is infinitely smooth inside the region D, independent of the smoothness of the boundary data f. This smoothness of the solution is typical of solutions to elliptic boundary value problems.

PROBLEM 9.15

Let D denote a closed bounded domain in either $\Re^2$ or $\Re^3$ and let S denote the piecewise smooth boundary of D. Then show that the Neumann boundary value problem

$$\nabla^2 u = F \text{ in } D, \quad \mathbf{grad}\ u \cdot N = g \text{ on } S \tag{1}$$

has no solution unless F and g are such that

$$\int_D F = \int_S g \tag{2}$$

SOLUTION 9.15

For any smooth function u in D we have by part 3 of Theorem 3.9 that

$$\int_D \nabla^2 u = \int_S \mathbf{grad}\ u \cdot N \tag{3}$$

where N denotes the unit outward normal vector to S. If u satisfies (1), then (2) follows from (3). The condition (2) is a condition of compatibility on the data. This condition is mathematical but can be interpreted physically. For example, in the context of potential fluid flow, the value of F at each point of D is equal to the rate of fluid production at the point, whereas $g = \mathbf{grad}\ u \cdot N$ is the outward normal component of fluid velocity at points on the boundary. Then the integral of F over D equals the net rate of fluid production inside D and the integral of g over the boundary is the net flow of fluid across the boundary. These two quantities must be equal for the fluid system to remain in a state of equilibrium.

PROBLEM 9.16 A NEUMANN PROBLEM

For a given function $g = g(x)$, find $u = u(x,y)$ satisfying

$$\partial_{xx}u(x,y) + \partial_{yy}u(x,y) = 0 \quad \text{for } 0 < x,\ y < 1 \tag{1}$$

$$\partial_x u(0,y) = 0, \quad \partial_x u(1,y) = 0 \quad \text{for } 0 < y < 1 \tag{2}$$

$$\partial_y u(x,0) = g(x), \quad \partial_y u(x,1) = 0 \quad \text{for } 0 < x < 1. \tag{3}$$

Here $u(x, y)$ could represent the steady-state distribution of contaminant in a square diffusion cell of sidelength 1 having specified steady contaminant flux along one edge. This would be a continuous analogue of the discrete model described in Problem 1.2.

SOLUTION 9.16

Note that on the vertical sides of the square D, **grad** $u(x,y) \bullet N$ is equal to positive or negative $\partial_x u$, whereas, on the horizontal sides, it equals plus or minus $\partial_y u$. Thus, the boundary conditions (2), (3) are Neumann conditions on the piecewise smooth boundary of the square. By the result of the previous problem, we see that no solution exists unless the data function g is such that the integral of g over the interval (0, 1) is zero; i.e., g has average value zero on the interval (0, 1).

Assuming the compatibility condition is satisfied, we suppose that $u(x, y) = X(x)Y(y)$. Then equation (1) and the homogeneous boundary conditions imply that

$$-X''(x) = \lambda X(x), \qquad X'(0) = X'(1) = 0 \tag{4}$$

$$Y''(y) = \lambda Y(y), \qquad Y'(1) = 0 \tag{5}$$

As usual we put aside consideration of the one inhomogeneous boundary condition for now. One of these separated problems, the boundary value problem (4), is a Sturm-Liouville problem. In fact, for $L = 1$, it is the Sturm-Liouville problem 2 in the summary of chapter 6. The eigenvalues and eigenvectors are

$$\lambda_n = (n\pi)^2, \quad X_n(x) = \cos n\pi x \quad \text{for } n = 0, 1, \ldots \tag{6}$$

The general solution of the differential equation (5) may be written in terms of any two of the six functions mentioned previously in Problem 9.14. We will choose to write the solution in terms of the two functions from that list whose derivatives vanish at the endpoints of the interval (0, 1)

$$Y_n(y) = A_n \cosh n\pi y + B_n \cosh n\pi(1-y), \quad n = 1, 2, \ldots$$

$$Y_0(y) = A_0 y + B_0$$

Then $Y'(1) = 0$ implies $A_n = 0$ $n = 0, 1, \ldots$, and thus

$$y_n(y) = \begin{cases} B_0 & n = 0 \\ B_n \cosh n\pi(1-y) & n = 1, 2, \ldots \end{cases}$$

Aside from any consideration of convergence, the principle of superposition implies that for any choice of the constants B_n, the function

$$u(x, y) = B_0 + \sum_{n=1}^{\infty} B_n \cosh n\pi(1-y) \cos n\pi x \tag{7}$$

satisfies the equation (1) and all the homogeneous boundary conditions in the problem. It only remains to choose the constants B_n so that the inhomogeneous boundary condition is satisfied; i.e., we want the B_n to be such that

$$\partial_y u(x,0) = -\sum_{n=1}^{\infty} n\pi B_n \sinh n\pi \cos n\pi x = g(x) \tag{8}$$

For any positive integer m, we may multiply both sides of (8) by cos $m\pi x$ and integrate with respect to x from 0 to 1. The orthogonality of the eigenfunctions leads to

$$m\pi B_m \sinh m\pi (1/2) = \int_0^1 g(x) \cos m\pi x \, dx, \quad m = 1, 2, \ldots$$

We solve this equation for B_m and substitute into (7) to obtain the final form of the solution

$$u(x,y) = B_0 + \sum_{n=1}^{\infty} g_n \frac{\cosh n\pi (1-y)}{n\pi \sinh n\pi} \cos n\pi x$$

where

$$g_n = 2\int_0^1 g(x) \cos n\pi x \, dx, \quad n = 1, 2, \ldots$$

and B_0 is arbitrary. Note that if g does not satisfy the compatibility condition, then (8) produces a contradiction.

PROBLEM 9.17 THE DIRICHLET PROBLEM IN A DISC

Find the steady-state temperature distribution in a heat conducting disc of radius $a > 0$ if the temperature on the circumference of the disc is prescribed by a given function $f = f(\vartheta)$ for $-\pi < \vartheta < \pi$.

SOLUTION 9.17

After all the transient thermal behavior inside the disc has died out, the temperature distribution resulting from the given temperature distribution on the boundary satisfies Laplace's equation inside the disc. Because the domain is a disc, it is only natural to express Laplace's equation and the boundary condition in polar coordinates

$$r^{-1}\partial_r (r\partial_r u(r,\vartheta)) + r^{-2}\partial_{\vartheta\vartheta} u(r,\vartheta) = 0 \quad \text{for } 0 < r < a, \ -\pi < \vartheta < \pi \tag{1}$$

$$u(a,\vartheta) = f(\vartheta), \quad \text{for } -\pi < \vartheta < \pi \tag{2}$$

In addition, we require that $u(r, \vartheta)$ satisfy the periodicity conditions

$$u(r,\pi) = u(r,-\pi) \text{ and } \partial_\vartheta u(r,\pi) = \partial_\vartheta u(r,-\pi) \text{ for } 0 < r < a \tag{3}$$

Finally, since the unknown function $u = u(r, \vartheta)$ represents a temperature, we are looking for a bounded solution to the problem.

If we assume that $u(r, \vartheta) = R(r)\,T(\vartheta)$ then (1) together with (3) implies

$$-T''(\vartheta) = \lambda T(\vartheta), \quad T(-\pi) = T(-\pi), \quad T'(-\pi) = T'(\pi) \tag{4}$$

$$r(rR'(r))' = \lambda R(r), \quad R(r) \text{ bounded for } 0 < r < a \tag{5}$$

The boundary value problem (4) is recognizable as the Sturm-Liouville problem 5 (with $L = \pi$) from the summary at the end of chapter 6. Then the eigenvalues and eigenfunctions are

$$\lambda_n = n^2, \quad T_n(\vartheta) = A_n \cos n\vartheta + B \sin n\vartheta, \quad n = 0, 1, \ldots \tag{6}$$

Then equation (5) becomes

$$r^2 R''(r) + rR'(r) - n^2 R(r) = 0$$

for which the general solution can be written

$$R_n(r) = \begin{cases} c_o + d_0 \log r, & n = 0 \\ c_n r^n + d_n r^{-n}, & n = 1, 2, \ldots \end{cases}$$

The boundedness condition leads to the choice $d_n = 0$ for $n = 0, 1, \ldots$ Then it follows that for all choices of the constants a_n and b_n, the function

$$u(r, \vartheta) = a_0/2 + \sum_{n=1}^{\infty} (a_n \cos n\vartheta + b_n \sin n\vartheta)\, r^n \tag{7}$$

is bounded on the disc, satisfies Laplace's equation there, and is periodic in ϑ. It only remains to choose the constants a_n and b_n such that the temperature condition on the boundary is satisfied. This requires

$$u(a, \vartheta) = a_0/2 + \sum_{n=1}^{\infty} (a_n \cos n\vartheta + b_n \sin n\vartheta)\, a^n = f(\vartheta) \tag{8}$$

We multiply both sides of (8) by $\sin m\vartheta$ for a fixed but arbitrary positive integer m and both integrate with respect to ϑ from $-\pi$ to π. The orthogonality of the eigenfunctions leads to the result

$$\pi b_m a^m = \int_{-\pi}^{\pi} f(\vartheta) \sin m\vartheta \, d\vartheta, \quad m = 1, 2, \ldots$$

In a similar manner we find

$$\pi a_m a^m = \int_{-\pi}^{\pi} f(\vartheta) \cos m\vartheta \, d\vartheta, \quad m = 0, 1, 2, \ldots$$

We solve these equations for a_m and b_m and use the result in (7) to complete the solution of the problem. Note that

$$\lim_{r \to 0} u(r, \vartheta) = \frac{a_0}{2} = \frac{1}{2\pi} \int_{-\pi}^{\pi} f(\vartheta) \, d\vartheta$$

Thus, the value of $u(r, \vartheta)$ at the center of the disc is equal to the average of its values over the circumference of the disc.

Initial Boundary Value Problems

PROBLEM 9.18 THE HEAT EQUATION

Find $u = u(x, t)$ such that

$$\partial_t u(x, t) = D \partial_{xx} u(x, t) \quad \text{for } 0 < x < L, \ t > 0 \tag{1}$$

$$u(x, 0) = f(x) \quad \text{for } 0 < x < L \tag{2}$$

$$u(0, t) = 0, \quad \partial_x u(L, t) = 0 \quad \text{for } t > 0 \tag{3}$$

Then $u(x,t)$ describes the time dependent temperature at position x in a heat conducting bar of length L when the initial temperature distribution is given by $f(x)$. The end of the bar at $x = L$ is insulated whereas the end at $x = 0$ is maintained constantly at temperature zero. Here D denotes a constant, material dependent parameter and $L > 0$.

SOLUTION 9.18

If we begin with the usual assumption that $u(x, t) = X(x)T(t)$, then it follows from (1) and (3) that

$$-X''(x) = \lambda X(x), \quad X(0) = X'(L) = 0 \tag{4}$$

$$T'(t) = -\lambda T(t) \tag{5}$$

The boundary value problem (4) is the Sturm-Liouville problem 4 from the summary of chapter 6. The eigenvalues and eigenfunctions are

$$\lambda_n = ((n + 1/2)\pi/L)^2 \quad X_n(x) = \sin((n + 1/2)\pi x/L), \quad n = 0, 1, \ldots \tag{6}$$

and

$$T_n(t) = A_n e^{-\lambda_n t} \tag{7}$$

Then for all choices of the arbitrary constants A_n, the function

$$u(x,t) = \sum_{n=0}^{\infty} A_n e^{-\lambda_n t} \sin((n+1/2)\pi x/L) \tag{8}$$

satisfies equation (1) and the boundary conditions (3). The initial condition (2) also holds if

$$u(x,0) = \sum_{n=0}^{\infty} A_n 1 \sin((n+1/2)\pi x/L) = f(x)$$

Multiplying both sides of this expression by $X_m(x)$ and integrating from 0 to L leads, via the orthogonality for the eigenfunctions, to

$$A_m = (2/L)\int_0^L f(x)X_m(x)\,dx, \quad m = 0, 1, \ldots$$

Using these values for the constants in (8) produces a function which satisfies all of the conditions of the problem. Note that $0 < \lambda_1 < \lambda_2 < \cdots$ and, thus, $u(x,t)$ given by (8) tends to zero as t tends to infinity; i.e., the steady-state solution for this problem is the zero function. This is what we would expect physically because one end of the heat conducting bar is maintained at the temperature zero. In fact, because the boundary conditions are independent of time we can find the steady-state solution directly by solving the problem

$$0 = U''(x) \quad \text{for } 0 < x < L; \quad U(0) = 0, \quad U'(L) = 0$$

PROBLEM 9.19 INPUT FROM THE BOUNDARY

Find $u = u(x, t)$ such that

$$\partial_t u(x,t) = D\partial_{xx} u(x,t) \quad \text{for } 0 < x < L,\ t > 0 \tag{1}$$

$$u(x,0) = f(x) \quad \text{for } 0 < x < L \tag{2}$$

$$u(0,t) = p(t), \quad \partial_x u(L,t) = 0 \quad \text{for } t > 0 \tag{3}$$

Then $u(x, t)$ could represent the time dependent temperature in a heat conducting bar which is initially at temperature zero throughout, the end at $x = L$ is insulated, and the temperature at the end $x = 0$ is controlled according to the prescribed function $p(t)$. In particular, find $u(x, t)$ if $p(t) = 1 - \cos \Omega t$ and find the steady-state solution.

SOLUTION 9.19

Since the boundary conditions (3) are not homogeneous, we cannot proceed as we did in the last problem because the assumption $u = X(x)T(t)$ does not then lead to a Sturm-Liouville problem in x. However, if we let

$$v(x, t) = u(x, t) - p(t)h(x) \tag{4}$$

where $h(x)$ is any smooth function that satisfies $h(0) = 1$ and $h'(L) = 0$, then it is easy to check that $v(x, t)$ solves the problem

$$\partial_t v(x,t) = D\partial_{xx} v(x,t) - p'(t)h(x) + Dp(t)h''(x) \tag{5}$$

$$v(x, 0) = -p(0)h(x) \tag{6}$$

$$v(0, t) = 0 \quad \text{and} \quad \partial_x v(L,t) = 0. \tag{7}$$

Since the choice of $h(x)$ is free except for the conditions $h(0) = 1$ and $h'(L) = 0$, we may as well choose $h(x)$ so as to make (5) as simple as possible. For example, the choice $h(x) = 1$ reduces the problem for $v(x,t)$ to

$$\partial_t v(x,t) = D\partial_{xx} v(x,t) - p'(t) \tag{8}$$

$$v(x, 0) = -p(0) = 0 \tag{9}$$

$$v(0, t) = 0 \quad \text{and} \quad \partial_x u(L,t) = 0 \tag{10}$$

This is a problem in which the initial condition and boundary conditions are all homogeneous but the partial differential equation is inhomogeneous. However, in the previous problem we have seen that the heat equation with these boundary conditions leads to the eigenfunctions $X_n(x) = \sin((n+1/2)\pi x/L)$, $n = 0, 1, \ldots$. Therefore, we suppose that the solution to this problem also can be written in terms of these eigenfunctions; i.e., we assume a solution of the form

$$v(x,t) = \sum_{n=0}^{\infty} v_n(t) X_n(x) \tag{11}$$

where the unknown functions $v_n(t)$ are to be found. Then

$$\partial_t v(x,t) = \sum_{n=0}^{\infty} v_n'(t) X_n(x)$$

$$\text{and} \quad \partial_{xx} v(x,t) = \sum_{n=0}^{\infty} v_n(t) X_n''(x) = \sum_{n=0}^{\infty} -\lambda_n v_n(t) X_n(x)$$

Thus, substituting (11) into the initial boundary value problem (8), (9), (10) leads to

$$\sum_{n=0}^{\infty} v_n'(t) X_n(x) = -D \sum_{n=0}^{\infty} \lambda_n v_n(t) X_n(x) - p'(t)$$

$$\sum_{n=0}^{\infty} v_n(0) X_n(x) = 0$$

Multiplying these equations on both sides by $X_m(x)$ and integrating from 0 to L then implies, because of the orthogonality of the eigenfunctions

$$v_m'(t) = -\lambda_m D v_m(t) - p'(t) C_m \tag{12}$$

for $m = 0, 1, \ldots$

$$v_m(0) = 0 \tag{13}$$

where, for $m = 0, 1, \ldots,$

$$C_m = \int_0^L 1X_m(x)\,dx = \frac{2L}{\pi}\,\frac{1}{2m+1} \tag{14}$$

We can use the Laplace transform to solve (12), (13) for $v_m(t)$

$$v_m(t) = \frac{C_m\Omega}{\Omega^2 + \lambda_m^{\ 2} D^2}\left(\Omega\cos\Omega t - \lambda_m D\sin\Omega t - \Omega e^{-\lambda_m t}\right)$$

$$= C_m \cos\alpha_m\left(\cos(\Omega t + \alpha_m) - \cos\alpha_m e^{-\lambda_m t}\right) \tag{15}$$

where $\tan\alpha_m = \lambda_m D/\Omega$. Then substituting these results into (11) gives $v(x, t)$ and, finally,

$$u(x, t) = v(x, t) + p(t) = v(x, t) + 1 - \cos\Omega t$$

Note that as t tends to infinity, $v_m(t)$ tends to the limit

$$C_m \cos\alpha_m \cos(\Omega t + \alpha_m)$$

and then $u(x,t)$ tends to

$$1 - \cos\Omega t + \sum_{m=0}^{\infty} C_m \cos\alpha_m \cos(\Omega t + \alpha_m) X_m(x)$$

Thus, $u(x, t)$ tends to a periodic steady state having the same frequency as $p(t)$, the periodic forcing function on the boundary.

PROBLEM 9.20 LOWER ORDER TERMS IN THE HEAT EQUATION

Find $u = u(x, t)$ such that

$$\partial_t u(x, t) = D\partial_{xx} u(x, t) + B\partial_x u(x, t) \qquad \text{for } 0 < x < L,\ t > 0 \tag{1}$$

$$u(x, 0) = f(x) \qquad \text{for } 0 < x < L, \tag{2}$$

$$u(0, t) = 0 \quad \text{and} \quad \partial_x u(L, t) = 0 \qquad \text{for } t > 0 \tag{3}$$

This problem models heat conduction in a moving medium where the constant B is related to the convective velocity of the moving medium. B is negative if the medium moves from left to right and B is positive if the movement is in the opposite direction.

SOLUTION 9.20

Instead of applying the separation of variables procedure directly to this problem, let

$$u(x, t) = e^{bt - ax} v(x, t) \tag{4}$$

so that

$$\partial_t u(x,t) = e^{bt-ax}(\partial_t v + bv)$$

$$\partial_x u(x,t) = e^{bt-ax}(\partial_x v - av)$$

$$\partial_{xx} u(x,t) = e^{bt-ax}(\partial_{xx} v - 2a\partial_x v + a^2 v)$$

Substituting these results into (1), we find

$$\partial_t v(x,t) = D\partial_{xx} v(x,t) + (B - 2aD)\,\partial_x v(x,t) + (a^2D - aB - b)\,v(x,t)$$

Then if we choose the as yet unspecified constants a and b to have the values

$$a = B/2D \quad \text{and} \quad b = -B^2/4D \tag{5}$$

then it follows that $v(x,t)$ satisfies

$$\partial_t v(x,t) = D\partial_{xx} v(x,t) \tag{6}$$

$$v(x,0) = f(x)e^{ax} \tag{7}$$

$$v(0,t) = 0 \quad \text{and} \quad \partial_x v(L,t) - av(L,t) = 0 \tag{8}$$

Now we suppose $v(x,t) = X(x)T(t)$ which implies that $X(x)$ and $T(t)$ must satisfy the separated problems

$$-X''(x) = \lambda X(x), \quad X(0) = 0, \quad X(L) - cX'(L) = 0 \tag{9}$$

$$T'(t) = -\lambda D T(t) \tag{10}$$

where we have let $c = -1/a$ so that the problem (9) assumes the form of the Sturm-Liouville problem in Problem 6.5. In that problem we found that the Sturm-Liouville problem (9) has one negative eigenvalue if $0 < c < 1$. Here we have $c = -1/a = -2D/B$ and it follows that there is a negative eigenvalue if B satisfies $B < -2D < 0$. This corresponds to the medium moving left to right with sufficiently high speed. If this condition is satisfied, then (9) has one negative eigenvalue $\lambda_0 = -p_0{}^2$ where p_0 is a root of equation (2) in Problem 6.5. The corresponding eigenfunction is $X_0(x) = \sinh p_0 x$.

The Sturm-Liouville problem (9) also has an infinite number of positive eigenvalues $\lambda_n = p_n{}^2$, $n = 1, 2, \ldots$, where p_n is the nth positive root of equation (2) in Problem 6.6. The corresponding eigenfunctions are $X_n(x) = \sin p_n x$.

The corresponding solutions of (10) are then

$$T_0(t) = C_0 e^{p_0{}^2 Dt} \quad \text{and} \quad T_n(t) = C_n e^{-p_n^2 Dt}, \quad n = 1, 2, \ldots$$

and then

$$v(x,t) = C_0 e^{p_0{}^2 Dt} X_0(x) + \sum_{n=1}^{\infty} C_n e^{-p_n^2 Dt} X_n(x)$$

We use the orthogonality of the eigenfunctions to solve for the constants

C_m such that the initial condition (7) is satisfied. Once this is done, then

$$u(x,t) = e^{bt-ax} v(x,t)$$

$$= C_0 e^{(p_0{}^2 D - B^2/4D)t} e^{-ax} X_0(x) + e^{bt-ax} \sum_{n=1}^{\infty} C_n e^{-p_n{}^2 Dt} X_n(x)$$

Note that $u(x, t)$ grows exponentially with increasing t if the quantity

$$p_0{}^2 D - B^2/4D = D(p_0{}^2 - a^2)$$

is positive.

PROBLEM 9.21 THE WAVE EQUATION

Find $u = u(x, t)$ such that

$$\partial_{tt} u(x,t) = a^2 \partial_{xx} u(x,t) \quad \text{for } 0 < x < L,\ t > 0 \tag{1}$$

$$u(x,0) = f(x) \text{ and } \partial_t u(x,0) = g(x) \quad \text{for } 0 < x < L \tag{2}$$

$$u(0,y) = 0 \quad \text{and} \quad u(L,t) = 0 \quad \text{for } t > 0 \tag{3}$$

Find the solution value $u(.4L, 4.9L/a)$ if $f(x) = x(L - x)$ and $g(x) = 0$.

SOLUTION 9.21

This initial-boundary value problem could describe acoustic waves in a thin pipe of length L. Here $u(x,t)$ could represent the deviation at position x and time t from an equilibrium pressure condition, and the initial conditions (2) describe the initial pressure state in the tube. The boundary conditions (3) are then consistent with the ends of the pipe being open.

If we suppose $u(x, t) = X(x)T(t)$, then we arrive in the usual way at the two separated problems

$$-X''(x) = \lambda X(x), \quad X(0) = 0, \quad X(L) = 0 \tag{4}$$

$$-T''(t) = \lambda a^2 T(t) \tag{5}$$

Then (4) is a Sturm-Liouville problem with eigenvalues and eigenfunctions as listed in the summary of chapter 6

$$\lambda_n = (n\pi/L)^2, \quad X_n(x) = \sin n\pi x/L \quad \text{for } n = 1, 2, \ldots$$

The general solution of (5) can then be written in the form

$$T_n(t) = A_n \cos n\pi at/L + B_n \sin n\pi at/L \quad \text{for } n = 1, 2, \ldots$$

Superposition implies that for all choices of the arbitrary constants A_n and B_n, the function

$$u(x,t) = \sum_{n=1}^{\infty} (A_n \cos n\pi at/L + B_n \sin n\pi at/L)\, X_n(x) \tag{6}$$

satisfies (1) and (3). The initial conditions (2) are satisfied as well if the constants are such that

$$u(x,0) = \sum_{n=1}^{\infty} A_n X_n(x) = f(x) \quad \text{and} \quad \partial_t u(x,0) = \sum_{n=1}^{\infty} n\pi a B_n X_n(x) = g(x)$$

Using orthogonality we can show that these conditions are both satisfied if

$$A_n = \frac{2}{L}\int_0^L f(x)\, X_n(x)\, dx \quad \text{and}$$

$$B_n = \frac{1}{n\pi a}\,\frac{2}{L}\int_0^L g(x)\, X_n(x)\, dx \tag{7}$$

for $n = 1, 2, \ldots$. That is, $A_n = f_n$ and $B_n = g_n / n\pi a$ where f_n and g_n denote the Fourier sine series coefficients for f and g, respectively, as prescribed in (7.7). Then (6) becomes

$$u(x,t) = \sum_{n=1}^{\infty} f_n \cos\frac{n\pi at}{L}\sin\frac{n\pi x}{L} + \sum_{n=1}^{\infty}\frac{g_n}{n\pi a}\sin\frac{n\pi at}{L}\sin\frac{n\pi x}{L} \tag{8}$$

But simple identities can be used to show that

$$\cos\frac{n\pi at}{L}\sin\frac{n\pi x}{L} = \frac{1}{2}\sin\frac{n\pi(x+at)}{L} + \sin\frac{n\pi(x-at)}{L}$$

$$\sin\frac{n\pi at}{L}\sin\frac{n\pi x}{L} = \frac{1}{2}\left(\cos\frac{n\pi(x-at)}{L} - \cos\frac{n\pi(x+at)}{L}\right)$$

$$= \frac{n\pi}{2}\int_{x-at}^{x+at} \sin n\pi s\, ds$$

Using these results in (8) leads to

$$\begin{aligned} u(x,t) &= \frac{1}{2}\sum_{n=1}^{\infty} f_n\left(\sin\frac{n\pi(x+at)}{L} + \sin\frac{n\pi(x-at)}{L}\right) \\ &\quad + \frac{1}{2a}\sum_{n=1}^{\infty} g_n \int_{x-at}^{x+at} \sin\frac{n\pi s}{L}\, ds \end{aligned} \tag{9}$$

Since the Fourier sine series for a function converges to the odd $2L$-periodic extension of that function, it follows that

$$\sum_{n=1}^{\infty} f_n \sin\frac{n\pi(x \pm at)}{L} = f_0(x \pm at) \quad \text{and} \quad \sum_{n=1}^{\infty} g_n \sin\frac{n\pi s}{L} = g_0(s)$$

thus,

$$u(x,t) = \frac{1}{2}(f_0(x+at) + f_0(x-at)) + \frac{1}{2a}\int_{x-at}^{x+at} g_0(s)\,ds \qquad (10)$$

Here f_0 and g_0 denote the odd $2L$-periodic extensions of f and g. The solution (10) is the finite interval analogue of the D'Alembert solution of the initial value problem for the wave equation.
In the special case that $f(x) = x(L-x)$ and $g(x) = 0$, we can evaluate u at $x = .4L$, and $t = 4.9L/a$ using (10). That is,

$$u(0.4L, 4.9L/a) = \frac{1}{2}(f_0(5.3L) + f_0(-4.5L))$$

But f_0 is $2L$-periodic and odd and, thus,

$$f_0(5.3L) = f_0(5.3L - 6L) = f_0(-0.7L) = -f(0.7L) = -0.7L(0.3L) = -0.21L$$

$$f_0(-45L) = f_0(-4.5L + 4L) = f_0(-0.5L) = -f(0.5L) = -0.5L(0.5L) = -0.25L$$

i.e., $u(0.4L, 4.9L/a) = \frac{1}{2}(-0.21L - 0.25L) = -0.23L$

PROBLEM 9.22

Find $u = u(x,t)$ such that

$$\partial_{tt}u(x,t) = a^2\partial_{xx}u(x,t) \quad \text{for } 0 < x < L,\ t > 0 \qquad (1)$$

$$u(x,0) = f(x) \quad \text{and} \quad \partial_t u(x,0) = g(x) \quad \text{for } 0 < x < L \qquad (2)$$

$$u(0,t) = 0 \quad \text{and} \quad \partial_x u(L,t) = 0 \quad \text{for } t > 0 \qquad (3)$$

Find the solution value $u(.4L, 4.9L/a)$ if $f(x) = x(L-x)$ and $g(x) = 0$.

SOLUTION 9.22

We have seen in the previous problem that the solution of the initial-boundary value problem for the wave equation is given by

$$u(x,t) = \frac{1}{2}(f_b(x+at) + f_b(x-at)) + \frac{1}{2a}\int_{x-at}^{x+at} g_b(s)\,ds \qquad (4)$$

where f_b and g_b denote periodic extensions of the initial functions f and g. The nature of these extensions is determined by the symmetry properties of the eigenfunctions which are determined, in turn, by the boundary conditions. In fact, for simple boundary conditions, the extension can be discovered directly from the boundary conditions without finding the eigenfunctions. The data functions are extended with odd symmetry across a boundary where the boundary condition is $u = 0$ and the functions are extended with even symmetry across a boundary where the condition $\partial_x u = 0$ applies. Thus, when the same type of boundary condition applies at both ends of the interval $(0, L)$, the extension is periodic of period $2L$, whereas the extension must have period $4L$ if the conditions at the two ends are not of the same type. That is:

(a) For $u(0, t) = 0$ and $u(L,t) = 0$, the extension f_b is defined as

$$\begin{aligned} f_b(-x) &= -f(x) && \text{for } 0 < x < L \\ f_b(L+x) &= -f(L-x) && \text{for } 0 < x < L \\ &f_b \text{ is } 2L\text{-periodic} \end{aligned}$$

This is the odd $2L$-periodic extension of f.

(b) For $\partial_x u(0, t) = 0$ and $\partial_x u(L, t) = 0$, the extension f_b is defined as

$$\begin{aligned} f_b(-x) &= f(x) && \text{for } 0 < x < L \\ f_b(L+x) &= f(L-x) && \text{for } 0 < x < L \\ &f_b \text{ is } 2L\text{-periodic} \end{aligned}$$

This is the even $2L$-periodic extension of f.

(c) For $u(0, t) = 0$ and $\partial_x u(L, t) = 0$, the extension f_b is defined as

$$\begin{aligned} f_b(-x) &= -f(x) && \text{for } 0 < x < L \\ f_b(L+x) &= f(L-x) && \text{for } 0 < x < L \\ &f_b \text{ is } 4L\text{-periodic} \end{aligned}$$

This is the odd-even $4L$-periodic extension of f.

(d) For $\partial_x u(0, t) = 0$ and $u(L, t) = 0$, the extension f_b is defined as

$$\begin{aligned} f_b(-x) &= f(x) && \text{for } 0 < x < L \\ f_b(L+x) &= -f(L-x) && \text{for } 0 < x < L \\ &f_b \text{ is } 4L\text{-periodic} \end{aligned}$$

This is the even-odd $4L$-periodic extension of f.

The boundary conditions (3) are covered by case (c) and, thus, using this definition of f_b in (10) for $f(x) = x(L-x)$ and $g(x) = 0$, leads to

$$u(0.4L, 4.9L/a) = \frac{1}{2}(f_b(5.3L) + f_b(-4.5L))$$

Then since f_b is $4L$-periodic with odd symmetry at $x = 0$ and even symmetry at $x = L$, we have

$$f_b(5.3L) = f_b(5.3L-4L) = f_b(1.3L) = f(0.7L) = 0.7L(0.3L) = 0.21L$$

$$f_b(-4.5L) = f_b(-4.5L+4L) = f_b(-0.5L) = -0.5L(0.5L) = -0.25L$$

and

$$u(0.4L, 4.9L/a) = \frac{1}{2}(f_b(5.3L) + f_b(-4.5L)) = -0.02L$$

Problems in Several Dimensions

PROBLEM 9.23

Find $u = u(x, y, t)$ such that

$$\partial_{tt}u(x, y, t) = a^2(\partial_{xx}u(x, y, t) + \partial_{yy}u(x, y, t)) \quad \text{for } 0 < x, y < \pi, \ t > 0 \qquad (1)$$

$$u(x, y, 0) = f(x, y), \quad \partial_t u(x, y, 0) = 0 \quad \text{for } 0 < x, y < \pi \qquad (2)$$

$$\begin{aligned} &u(0, y, t) = 0, \quad u(\pi, y, t) = 0 \quad \text{for } 0 < y < \pi, \ t > 0 \\ &\partial_y u(x, 0, t) = 0, \quad \partial_y u(x, \pi, t) = 0 \quad \text{for } 0 < x < \pi, \ t > 0 \end{aligned} \qquad (3)$$

Equation (1) models the transverse vibrations of a thin membrane. Then $u(x, y, t)$ represents the out of plane deflection of a square membrane of sidelength π. The boundary conditions (3) are consistent with a membrane which is simply supported along opposing edges $x = 0$ and $x = \pi$, whereas the other two edges are free. The initial conditions (2) indicate the membrane has been given an initial displacement described by $f(x, y)$ and it released from rest.

SOLUTION 9.23

We begin with the assumption that $u(x, y, t) = U(x, y)T(t)$. Then it follows that

$$\begin{aligned} &\partial_{xx}U(x, y) + \partial_{yy}U(x, y) = \mu U \quad \text{for } 0 < x, \ y < \pi \\ &U(0, y) = 0, \quad U(\pi, y) = 0 \quad \text{and} \quad \partial_y U(x, 0) = 0, \ \partial_y U(x, \pi) = 0 \end{aligned} \qquad (4)$$

$$T''(t) = \mu a^2 T(t), \quad T'(0) = 0 \qquad (5)$$

Next, we proceed to find the nontrivial solutions for (4). We suppose, as usual, that $U(x, y) = X(x)Y(y)$ which implies

$$\frac{X''(x)}{X(x)} + \frac{Y''(y)}{Y(y)} = \mu$$

This leads to

$$-X''(x) = \lambda X(x), \quad X(0) = 0, \quad X(\pi) = 0 \tag{6}$$

$$-Y''(y) = aY(y), \quad Y'(0) = 0, \quad Y'(\pi) = 0 \quad \text{for } \alpha = -\lambda - \mu \tag{7}$$

Each of the problems (6) and (7) is a previously encountered Sturm-Liouville problem. The eigenvalues and eigenfunctions are as follows:

$$\lambda_n = n^2, \quad X_n(x) = \sin nx \quad \text{for } n = 1, 2, \ldots$$

$$a_m = m^2, \quad Y_m(y) = \cos my \quad \text{for } m = 0, 1, \ldots$$

Then

$$\mu_{mn} = -\lambda_n - \alpha_m = -n^2 - m^2 \quad \text{for } m = 0, 1, \ldots, n = 1, 2, \ldots$$

are the eigenvalues of (4). The corresponding two-parameter family of eigenfunctions are

$$U_{mn}(x, y) = \cos my \sin nx \quad \text{for } m = 0, 1, \ldots, \; n = 1, 2, \ldots$$

The solutions of (5) are then given by

$$T_{mn}(t) = A_{mn} \cos at\sqrt{m^2 + n^2}$$

and, thus, for all choices of the arbitrary constants A_{mn}, the function

$$u(x, y, t) = \sum_{m=0}^{\infty} \sum_{n=1}^{\infty} A_{mn} \cos\left(at\sqrt{m^2 + n^2}\right) X_n(x) Y_m(y)$$

satisfies all the homogeneous conditions of (1), (2), (3). The inhomogeneous initial condition is satisfied as well if the constants A_{mn} are chosen such that

$$u(x, y, 0) = \sum_{m=0}^{\infty} \sum_{n=1}^{\infty} A_{mn} X_n(x) Y_m(y) = f(x, y)$$

The orthogonality of the eigenfunctions can be used to show that this equality holds if

$$A_{mn} = (2/\pi)^2 \int_0^{\pi} \int_0^{\pi} f(x, y) X_n(x) Y_m(y)\, dx dy$$

Note that, in the context of a vibrating membrane, the eigenvalues μ_{mn} have the interpretation of being the fundamental frequencies of the membrane and the eigenfunctions $U_{mn}(x, y)$ are the corresponding mode shapes.

PROBLEM 9.24

Consider a cylindrical heat conducting solid of radius b and length π.

Find the equilibrium temperature distribution in the solid if the ends are maintained at temperature zero and the temperature of the lateral surface of the solid is prescribed by the function $f = f(\vartheta, z)$ where the z axis lies along the centerline of the solid cylinder.

SOLUTION 9.24

The equilibrium temperature distribution $u = u(r, \vartheta, z)$ satisfies Laplace's equation which we express in cylindrical coordinates together with the appropriate boundary conditions

$$\partial_{rr}u(r,\vartheta,z) + r^{-1}\partial_r u(r,\vartheta,z) + r^{-2}\partial_{\vartheta\vartheta}u(r,\vartheta,z) + \partial_{zz}u(r,\vartheta,z) = 0 \tag{1}$$

$$u(r,\vartheta,0) = 0 \text{ and } u(r,\vartheta,\pi) = 0 \quad \text{for } 0 < r < b,\ -\pi < \vartheta < \pi \tag{2}$$

$$u(b,\vartheta,z) = f(\vartheta,z) \qquad \text{for } -\pi < \vartheta < \pi,\ 0 < z < \pi \tag{3}$$

In addition, we have the implied conditions that u remains bounded inside the solid and $u(r, \vartheta, z)$ is periodic in ϑ.

If we assume $u(r,\vartheta,z) = R(r)H(\vartheta)Z(z)$, then

$$\frac{R''(r) + r^{-1}R'(r)}{R(r)} + r^{-2}\frac{H''(\vartheta)}{H(\vartheta)} = -\frac{Z''(z)}{Z(z)} = \lambda$$

which leads to

$$-Z''(z) = \lambda Z(z), \quad Z(0) = 0, \quad Z(\pi) = 0 \tag{4}$$

$$-H''(\vartheta) = \mu H(\vartheta), \quad H(-\pi) = H(\pi), \quad H'(\pi) = H'(-\pi) \tag{5}$$

$$r^2R''(r) + rR'(r) - (\mu + r^2\lambda)R(r) = 0 \tag{6}$$

Problems (4) and (5) are Sturm-Liouville problems with eigenvalues and eigenfunctions as follows:

$$\lambda_n = n^2, \quad Z_n(z) = \sin nz \quad \text{for } n = 1, 2, \ldots$$

$$\mu_m = m^2, \quad m = 0, 1, \ldots \text{ (double eigenvalues for } m > 0)$$

$$H_0(\vartheta) = 1, \quad H_m(\vartheta) = \cos m\vartheta, \sin m\vartheta \text{ or } H_m(\vartheta) = e^{im\vartheta}$$

Then (6) is the modified Bessel equation (see 6.19)

$$r^2R''(r) + rR'(r) - (m^2 + r^2n^2)R(r) = 0$$

with the general solution expressed in terms of modified Bessel functions

$$R_{mn}(r) = A_{mn}I_m(nr) + B_{mn}K_m(nr) \quad m = 0, 1, \ldots \quad n = 1, 2, \ldots$$

The modified Bessel function of the second kind, $K_m(nr)$ does not remain bounded as r tends to zero and thus we choose $B_{mn} = 0$.

For all choices of A_{mn} the function

$$u(r, \vartheta, z) = \sum_{m=-\infty}^{\infty} \sum_{n=1}^{\infty} A_{mn}I_m(nr)\, e^{im\vartheta} \sin nz$$

satisfies the homogeneous conditions of the problem. Here we chose to use the exponential form of the eigenfunctions in ϑ. The inhomogeneous boundary condition will be satisfied if

$$u(b, \vartheta, z) = \sum_{m=-\infty}^{\infty} \sum_{n=1}^{\infty} A_{mn}I_m(nb)\, e^{im\vartheta} \sin nz = f(\vartheta, z)$$

Using the orthogonality of the eigenfunctions, we find

$$A_{mn} = \frac{1}{I_m(nb)} \frac{2}{\pi^2} \int_0^{\pi} \int_{-\pi}^{\pi} f(\vartheta, z)\, e^{-im\vartheta} \sin nz \, d\vartheta dz$$

PROBLEM 9.25

Solve the preceding problem in the case that the boundary conditions are changed to model the situation where the lateral surface $r = b$ and the end $z = \pi$ of the solid are held at temperature zero while at the end $z = 0$ we have $u(r, \vartheta, 0) = g(r, \vartheta)$.

SOLUTION 9.25

The boundary conditions (2) and (3) of the previous problem are replaced by

$$u(r, \vartheta, 0) = g(r, \vartheta), \quad u(r, \vartheta, \pi) = 0 \quad \text{for } 0 < r < b, \ -\pi < \vartheta < \pi$$

$$u(b, \vartheta, z) = 0 \quad \text{for } -\pi < \vartheta < \pi, \ 0 < z < \pi$$

The governing equation is still Laplace's equation in cylindrical coordinates and the boundedness and periodicity conditions remain in effect. Then the separation assumption leads to the problems

$$Z''(z) = \lambda Z(z), \quad Z(\pi) = 0 \tag{1}$$

$$-H''(\vartheta) = \mu H(\vartheta), \quad H(-\pi) = H(\pi), \quad H'(\pi) = H'(-\pi) \tag{2}$$

$$r^2 R''(r) + rR'(r) + (r^2\lambda - \mu) R(r) = 0, \quad R(b) = 0 \tag{3}$$

Here the z problem is not a Sturm-Liouville problem as only one boundary condition is homogeneous. Problem (2) is a Sturm-Liouville problem with the eigenvalues and eigenfunctions noted previously as

$$\mu_m = m^2, \quad H_m(\vartheta) = e^{im\vartheta} \quad \text{for } m = 0, 1, -1, 2, -2, \ldots$$

Problem (3) is the other Sturm-Liouville problem. Note that

$$r^2 R''(r) + rR'(r) + (r^2\lambda - m^2) R(r) = 0, \quad R(b) = 0 \tag{3}$$

is the problem considered in Problem 6.8 with eigenfunctions

$$R_{mn}(r) = J_m(\lambda_{mn} r) \quad n = 1, 2, \ldots, \quad \text{and all integer values of } m$$

where λ_{mn} denotes the nth root of the eigenvalue equation $J_m(\lambda b) = 0$. Here the solution $Y_m(\lambda r)$ for (3) has been excluded by the condition that $R(r)$ is bounded. Now the general solution of (1) can be written as

$$Z_{mn}(z) = A_{mn} \sinh \lambda_{mn} z + B_{mn} \sinh \lambda_{mn} (\pi - z)$$

and the homogeneous boundary condition at $z = \pi$ implies $A_{mn} = 0$. Then for all values of B_{mn}, the function

$$u(r, \vartheta, z) = \sum_m \sum_{n=1}^{\infty} B_{mn} \sinh \lambda_{mn} (\pi - z)\, e^{im\vartheta} J_m(\lambda_{mn} r)$$

satisfies the homogeneous parts of the problem. The inhomogeneous boundary condition requires

$$u(r, \vartheta, 0) = \sum_m \sum_{n=1}^{\infty} B_{mn} \sinh \lambda_{mn} \pi e^{im\vartheta} J_m(\lambda_{mn} r) = g(r, \vartheta)$$

and, in the usual way, we find

$$B_{mn} = \frac{1}{\sinh \lambda_{mn} \pi} \frac{1}{\pi C_{mn}} \int_0^b \int_{-\pi}^{\pi} g(r, \vartheta)\, e^{-im\vartheta} J_m(\lambda_{mn} r)\, r \, dr d\vartheta$$

where

$$C_{mn} = \int_0^b J_m(\lambda_{mn} r)^2 r \, dr$$

PROBLEM 9.26

Find the equilibrium temperature distribution inside a solid heat conducting sphere or radius b if the temperature on the surface of the sphere is given by the prescribed function $f = f(\vartheta, \varphi)$.

SOLUTION 9.26

The equilibrium temperature inside the sphere conforms to Laplace's equation. To facilitate the statement of the boundary condition, we express Laplace's equation in spherical coordinates (r, ϑ, φ), where ϑ denotes the *polar azimuth* angle and ranges over the values $(-\pi, \pi)$, and φ denotes the *altitude* angle and ranges over $(0, \pi)$. Then Laplace's equation reads

$$r^{-2}(\partial_r(r^2\partial_r u) + (\sin\varphi)^{-1}\partial_\varphi(\sin\varphi\partial_\varphi u) + (\sin\varphi)^{-2}\partial_{\vartheta\vartheta}u) = 0 \tag{1}$$

The boundary condition becomes

$$u(b, \vartheta, \varphi) = f(\vartheta, \varphi) \quad \text{for } -\pi < \vartheta < \pi \text{ and } 0 < \varphi < \pi \tag{2}$$

In addition we have the implied conditions that u is bounded and periodic in ϑ on the interior of the sphere. If we assume that $u(r, \vartheta, \varphi) = R(r)H(\vartheta)G(\varphi)$ then (1) implies

$$\frac{\sin^2\varphi}{R(r)}\frac{d}{dr}(r^2R'(r)) + \frac{\sin\varphi}{G(\varphi)}\frac{d}{d\varphi}((\sin\varphi G'(\varphi))) = -\frac{1}{H(\vartheta)}H''(\vartheta)$$

Thus,

$$-H''(\vartheta) = \mu H(\vartheta), \quad H(-\pi) = H(\pi), \quad H'(-\pi) = H'(\pi) \tag{3}$$

$$(r^2R'(r))' = \lambda R(r) \tag{4}$$

$$\sin\varphi(\sin\varphi G'(\vartheta))' + (\lambda\sin^2\varphi - \mu)G(\varphi) = 0 \tag{5}$$

Problem (3) is a Sturm-Liouville problem with eigenvalues and eigenfunctions noted previously as

$$\mu_m = m^2, \quad H_m(\vartheta) = e^{im\vartheta} \quad \text{for } m = 0, 1, -1, 2, -2, \ldots$$

If we make the change of variable $x = \cos\varphi$ and $G(\varphi) = G(\varphi(x)) = g(x)$ in (5), this equation reduces to

$$((1-x^2)g'(x))' + (\lambda - \frac{\mu}{1-x^2})g(x) = 0 \quad \text{for } -1 < x < 1 \tag{6}$$

Equation (6) is called the *associated Legendre equation* and has the form of a singular Sturm-Liouville problem; its general solution can be constructed by power series methods. Note that (6) reduces to Legendre's equation (6.12) for $\mu = 0$. If the boundary function f does not depend on ϑ, then the solution u is independent of ϑ and the Laplace equation (1) simplifies to

$$r^{-2}(\partial_r(r^2\partial_r u)) + (\sin\varphi)^{-1}\partial_\varphi(\sin\varphi\partial_\varphi u) = 0 \tag{7}$$

The assumption that $u(r, \varphi) = R(r)G(\varphi)$ separates this equation into the problems (4) and (5) (with $\mu = 0$) and the aforementioned change of variable in (5) leads to

$$((1-x^2)g'(x))' + \lambda g(x) = 0 \quad \text{for } -1 < x < 1 \tag{8}$$

We saw in Chapter 6 that equation (8) has bounded nontrivial solutions only if $\lambda = n(n+1)$ for n a non-negative integer. In fact, we have

$$\lambda_n = n(n+1), \quad g_n(x) = P_n(x), \quad n = 0, 1, \ldots$$

where $P_n(x)$ denotes the Legendre polynomial of degree n. The general solutions of (4) corresponding to $\lambda = \lambda_n$ are given by

$$R_n(r) = A_n r^n + B_n r^{-n-1}, \quad 0 < r < b$$

and we choose $B_n = 0$ to preclude solutions that become unbounded at $r = 0$. Then for all choices of the constants A_n, the function

$$u(r, \vartheta) = \sum_{n=0}^{\infty} A_n r^n P_n(\cos\varphi) \tag{9}$$

satisfies the homogeneous conditions of the problem. The inhomogeneous condition on the boundary is satisfied if

$$u(b, \varphi) = \sum_{n=0}^{\infty} A_n b^n P_n(\cos\varphi) = f(\varphi)$$

which implies

$$A_n = b^{-n}(n + 1/2) \int_0^{\pi} f(\varphi) P_n(\cos\varphi) \sin\varphi \, d\varphi \tag{10}$$

The presence of $\sin\varphi$ in the integrand in (10) is due to the fact that the family $P_n(x)$ is an orthogonal family for $-1 < x < 1$ and for $x = \cos\varphi$ we have

$$\int_{-1}^{1} P_n(x) P_m(x)\, dx = \int_0^{\pi} P_n(\cos\varphi) P_m(\cos\varphi) \sin\varphi \, d\varphi$$

Using (10) in (9) leads to a solution of the boundary value problem.

Integral transforms, including the transforms of Laplace and Fourier, and the method of eigenfunction expansion are two useful solution methods for solving second order, linear, constant coefficient partial differential equations. Fortunately, many of the partial differential equations of mathematical physics are included in this category. Indications of when each method applies are as follows:

Fourier Transform. *The transform variable x ranges from minus infin-*

ity to plus infinity. In case the variable x ranges only from zero to infinity but boundary conditions require the solution u or its derivatives $\partial_x u$ *to vanish at* $x = 0$, *we can replace the problem by one where x ranges over the whole real line. We do this by extending the boundary data as an odd function (if* $u = 0$ *at* $x = 0$*) or as an even function (if* $\partial_x u = 0$ *at* $x = 0$*).*

Laplace Transform. *The transform variable t ranges from zero to infinity. If the equation contains t derivatives of order n, then u together with all its t derivatives of order less than n must be specified at* $t = 0$. *Thus the Laplace transform in y is not suitable for solving Laplace's equation on the half plane, but the Laplace transform in t is suitable for the initial value problems for the heat and wave equations.*

Eigenfunction Expansion. *At least one of the independent variables must be restricted to a bounded interval. In addition, the boundary conditions that apply at the ends of the bounded interval must be homogeneous so as to produce a Sturm-Liouville problem in the separated problems. If the boundary conditions are not homogeneous, it may be possible to make them homogeneous. In the case of Laplace's equation, this may be accomplished by splitting a problem into two or more subproblems. In the case of the heat or wave equations this may be accomplished by subtracting an appropriate function as in Problem 9.19.*

The Laplace, heat and wave equations are each typical of a whole class of partial differential equations: elliptic, parabolic and hyperbolic, respectively. This means that what is true for the Laplace equation is likely to be true about all elliptic partial differential equations, with similar statements true for the heat and wave equations. The solutions of each of these three equations exhibit characteristic mathematical behavior that is typical for it but not for either of the others. This mathematical behavior can be associated with certain physical behavior that is typical of the type of physical process that is modeled by the corresponding partial differential equation. For example:

Laplace's Equation. *The solutions to Laplace's equation are infinitely smooth inside the domain, independent of the smoothness of the data on the boundary. In addition, at each point in the domain the solution value at the point depends on all of the data values on the boundary. Such mathematical behavior generally describes a physical system that is in a state of equilibrium.*

Heat Equation. *Solutions to the heat equation are infinitely smooth for* $t > 0$ *everywhere inside the domain, independent of the smoothness of the initial state or the data on the boundary. The influence of the initial state can be said to propagate with infinite speed although it is more meaningful to view this as instantaneous smoothing of the data. This mathematical behavior is consistent with "diffusionlike" evolution, a term meant to suggest irreversible physical behavior like diffusion and conduction. In* *diffusionlike evolution, it is possible to distinguish between forward time and backward time because as the process evolves, information is lost. Thus, it is not generally possible to use information about a current state to recover a previous state of the system in diffusionlike evolution.*

Wave Equation. *Solutions to the wave equation are not smoother than the functions which prescribe the initial or boundary data. In addition, solutions to the wave equation propagate information with finite speed. Such mathematical behavior is associated with physical behavior that could be characterized as "wavelike" evolution. For example, the delay between a flash of distant lightning and the sound of thunder indicates that acoustic waves propagate at finite speed; of course, acoustic waves can be modeled by the wave equation.*

There are many other partial differential equations to which the methods of this chapter apply and there are many more to which they do not apply. Some of these latter will be considered in later chapters along with additional methods of solution.

10

Linear and Nonlinear Partial Differential Equations

The solution methods considered in the last chapter were limited to a special class of problems for partial differential equations — linear equations with constant coefficients. Here we are going to consider a different class of problems including equations with variable coefficients as well as nonlinear equations.

Linear partial differential equations of order one reduce to ordinary differential equations along certain families of curves called characteristics. This provides the basis for a solution method known as the method of characteristics. This method even extends to a class of mildly nonlinear (i.e., quasilinear) partial differential equations. We will see that solution behavior for nonlinear equations can differ markedly from the behavior of solutions for linear equations.

We will begin by considering equations of order one and later we will examine the interaction between linear and nonlinear behavior for some higher order equations.

FIRST ORDER EQUATIONS

Terminology

We will restrict our attention for the most part to first order partial differential equations in two independent variables. The most general such equation for the unknown function $u = u(x, t)$ has the form

$$F(x, t, u(x, t), \partial_x u(x, t), \partial_t u(x, t)) = 0 \tag{10.1}$$

Consideration of this fully nonlinear equation poses too broad a problem for this discussion. Therefore, we will consider a more special form of (10.1)

$$a(x, t, u)\,\partial_x u(x, t) + b(x, t, u)\,\partial_t u(x, t) = c(x, t, u) \tag{10.2}$$

This equation is said to be *quasilinear.* If $a(x, t, u)$ and $b(x, t, u)$ are independent of u, then the equation is said to be *almost linear*, and if, in addition, $c(x, t, u)$ has the form $C(x, t)u + D(x, t)$, then the equation is said to be *linear.* If $c(x, t, u) = 0$, the equation is said to be *homogeneous*; otherwise it is *inhomogeneous*. We will generally be interested in finding solutions u of (10.2) that satisfy an initial condition

$$u(x, 0) = f(x) \tag{10.3}$$

Initial conditions may be specified along curves in the xt plane other than the x axis, but (10.3) will be sufficiently general for our purposes.

The Method of Characteristics

THE LINEAR INITIAL VALUE PROBLEM

Consider the linear homogeneous initial value problem for the unknown function $u = u(x, t)$:

$$a(x, t)\,\partial_x u(x, t) + b(x, t)\,\partial_t u(x, t) = 0 \quad \text{for all } x \text{ and } t > 0 \tag{10.4}$$

$$u(x, 0) = f(x) \tag{10.5}$$

Let $\Gamma = \{(x, t)\colon\ x = x(s),\ t = t(s)\}$ denote a solution curve in the xt-plane for the system of ordinary differential equations

$$\frac{dx}{ds} = a(x, t), \qquad \frac{dt}{ds} = b(x, t) \tag{10.6}$$

Then Γ is said to be a *characteristic curve* for (10.4). A first order partial differential equation reduces to a first order ordinary differential equation along its characteristic curves. If the equation is linear, then the family of characteristic curves is *coherent*; i.e., distinct characteristics never cross. If the coefficients are constant, the characteristics are straight lines. The characteristic equations (10.6) are equivalent to

$$\frac{dx}{dt} = \frac{a(x, t)}{b(x, t)} \quad \text{at points where } b(x, t) \text{ is nonzero}$$

$$\text{or} \quad \frac{dt}{dx} = \frac{b(x, t)}{a(x, t)} \quad \text{at points where } a(x, t) \text{ is nonzero} \tag{10.7}$$

A point where $a(x, t)$ or $b(x, t)$ (or both) vanish is said to be a *singular* point for the equation.

Example 10.1

(a) For $f(x)$ a given function of x, consider the linear initial value problem

$$4\partial_x u(x,t) + 3\partial_t u(x,t) = u, \quad u(x,0) = f(x).$$

The characteristic equations are

$$\frac{dt}{ds} = 3 \quad \text{and} \quad \frac{dx}{ds} = 4$$

i.e.,

$$t = 3s + C_1 \quad \text{and} \quad x = 4s + C_2 \quad \text{or} \quad 3x - 4t = C_3$$

The characteristics are a family of straight lines; for any x_0, the line $3x - 4t = 3x_0$ crosses the x axis at $(x_0, 0)$ and as the value x_0 is varied we obtain various lines in the family. We say the family is *parameterized* by the parameter x_0.

Along the characteristic curves

$$\partial_x u \; x'(s) + \partial_t u \; t'(s) = \frac{du}{ds} = u$$

hence,

$$u = C(x_0)\, e^s \quad \text{or} \quad u(x,t) = F(3x-4t)\, e^{t/3} = F(3x-4t)\, e^{x/4}$$

for an arbitrary differentiable function of one variable F. This is the general solution of the equation. Note that as we vary x_0 in $C(x_0)\, e^s$, we obtain the value of u (for a fixed s) on various characteristic lines, and as we vary s, we see how u varies as we move along the characteristic which originates at $(x_0, 0)$.

The initial condition will be satisfied if $u(x, 0) = F(3x) = f(x)$; i.e., if $F(y) = f(y/3)$. Thus,

$$u(x,t) = f(x - 4t/3)\, e^{t/3}$$

solves the initial value problem. The solution is a travelling wave, moving from left to right with speed 4/3 and as the waveform propagates, the amplitude increases exponentially with t.

(b) Consider the linear initial value problem

$$t\partial_x u(x,t) + x\partial_t u(x,t) = tu(x,t) \quad \text{and} \quad u(x,0) = \sin x$$

The characteristic equations are $dt/ds = x$, $dx/ds = t$ and, thus, along solution curves of these equations, the partial differential equation reduces to

$$\partial_x u \; x'(s) + \partial_t u \; t'(s) = \frac{du}{ds} = tu(x(s), t(s))$$

The characteristic equations are equivalent to $dx/dt = t/x$ which leads to the solution curves $\Gamma = \{(x,t) : x^2 - t^2 = C = \text{constant}\}$. As the constant C varies, we obtain a family of hyperbolas. Note that the origin $x = t = 0$ is a singular point for this problem and, hence, there is no characteristic curve originating at this point. At every point $(x_0, 0)$ with x_0 different from zero, we have a characteristic curve

$$x^2 - t^2 = x_0^{\ 2}.$$

Along characteristics, we have $du/ds = tu$ or $du/u = t\,ds = dx$ and this can be solved to obtain $u = Ce^x$. Then the general solution of the partial differential equation may be written as

$$u(x, t) = F(x^2 - t^2)\, e^x$$

for an arbitrary differentiable function F. The initial condition is satisfied if

$$F(x^2)\, e^x = \sin x$$

i.e.,

$$F(y) = e^{-\sqrt{y}} \sin \sqrt{y}$$

Then $$u(x, t) = \exp\left[x - \sqrt{x^2 - t^2}\right] \sin \sqrt{x^2 - t^2}$$

solves the initial value problem. Note that the solution is defined only in that part of the xt plane where $x^2 - t^2$ is non-negative. This is a result of the singular point at the origin. Note that the variable coefficients produce characteristics which are curved, whereas the constant coefficient equation in part (a) led to straight line characteristics. Straight line characteristics are equivalent to constant speed of propagation, whereas curved characteristics are associated with variable speed. This is just the observation that the slope of the characteristic is equal to $dt/dx = b(x, t)/a(x, t)$.

THE QUASILINEAR INITIAL VALUE PROBLEM

We consider the quasilinear initial value problem (10.2), (10.3). Then it will be convenient to generalize the notion of a characteristic curve slightly. We refer to the space curve

$$\Gamma = \{(x, t, u) : x = x(s), t = t(s), u = u(s)\}$$

as a *space characteristic* for (10.2) if $x(s)$, $t(s)$, $u(s)$ satisfy

$$\frac{dx}{ds} = a(x, t, u), \quad \frac{dt}{ds} = b(x, t, u), \quad \frac{du}{ds} = c(x, t, u) \tag{10.8}$$

The solution to the initial value problem may be obtained by solving

$$\begin{aligned} \frac{dx}{ds} &= a(x, t, u), & x(0) &= x_0 \\ \frac{dt}{ds} &= b(x, t, u), & t(0) &= 0 \\ \frac{du}{ds} &= c(x, t, u) & u(x_0, 0) &= f(x_0) \end{aligned} \tag{10.9}$$

Example 10.2

Consider the quasilinear initial value problem

$$xu\partial_x u(x,t) + u\partial_t u(x,t) = -xt, \quad u(x,0) = x+1$$

The characteristic equations for this problem are

$$\frac{dx}{ds} = xu, \quad \frac{dt}{ds} = u, \quad \frac{du}{ds} = -xt$$

The first two equations imply $x^{-1}dx/ds = dt/ds$; i.e.,

$$\ln x = t + C_1(x_0) \quad \text{or} \quad x = C_2(x_0)e^t$$

where $C_1(x_0)$ and $C_2(x_0)$ denote arbitrary functions of one variable. The initial condition $x(0) = x_0$ from (10.9) implies that $C_2(x_0) = x_0$. Now the second two characteristic equations imply

$$u\, du = -te^t x_0 dt$$

i.e.,

$$u^2 = 2x_0(1-t)e^t + C_3(x_0)$$

The initial condition $u(x_0, 0) = x_0 + 1$ implies that $C_3(x_0) = x_0^2 + 1$. Therefore,

$$u(x,t) = \sqrt{2x(1-t)e^t + x^2 + 1}$$

solves the initial value problem.

Conservation Law Equations

Instead of considering nonlinear equations in general, we will focus on a special class of nonlinear equations known as *conservation law equations*. A conservation law equation asserts that the change in the amount $u(x, t)$ of some quantity contained in a region Ω is due to the flux of the quantity across the boundary of Ω plus the loss or gain due to internal sources or sinks. The 1- dimensional equation has the form

$$\partial_t Q(u) + \partial_x F(u) = G(u) \tag{10.10}$$

where $Q(u)$ denotes the density field for the material, whereas F and G denote, respectively, the flux and source density. We shall further restrict our attention to problems with $Q(u) = u$ and, at least initially, we will consider the homogeneous case where $G(u) = 0$.

THE INITIAL VALUE PROBLEM

Consider the initial value problem for a homogeneous conservation law equation

$$\partial_t u(x,t) + \partial_x F(u) = 0, \quad u(x,0) = f(x) \tag{10.11}$$

Note that $\partial_x F(u) = F'(u)\partial_x u$ and let $a(u) = F'(u)$. The characteristic curves for (10.10) are the solution curves of the ordinary differential equation

$$\frac{dx}{dt} = a(u) \tag{10.12}$$

Along characteristics, (10.11) reduces to

$$\partial_t u(x, t) + a(u)\partial_x u(x, t) = \frac{du}{dt} = 0$$

i.e., u is constant along characteristics. Then (10.12) implies the characteristics for (10.11) are straight lines and the characteristic that originates at $(x_0, 0)$ has the equation $x - a(u)t = x_0$. Along $x = a(u)t + x_0$, the solution for (10.11) is equal to the constant, $u(x, t) = f(x_0) = f(x - a_0 t)$, where $a_0 = a(f(x_0))$. In general, it is difficult to construct an explicit solution $u = u(x, t)$.

Theorem 10.1

Theorem 10.1. Suppose the functions $a(u) = F'(u)$ and $f(x)$ are continuously differentiable. Then the solution $u = u(x, t)$ for the initial value problem (10.11) is defined implicitly by $u(x, t) = f(x - a(u(x, t))t)$ for as long as the quantity $1 + t\,a'(u)f'(x - a(u)t)$ remains positive.

Example 10.3

(a) Consider the initial value problem

$$\partial_t u(x, t) + u\partial_x u(x, t) = 0, \quad u(x, 0) = f(x) = 1 + x/3$$

Then $a(u) = u$ and $x = a(u_0)t + x_0$ is the equation of the characteristic originating at $(x_0, 0)$. Along this characteristic line, $u(x, t)$ is equal to the constant u_0 where $u_0 = f(x_0) = 1 + x_0/3$. Since $a(u) = u$, Theorem 10.1 leads to the following implicit relation for the solution: $u(x, t) = f(x - ut) = 1 + (x - ut)/3$. In this case, we can solve to obtain an explicit formula for $u(x, t)$

$$u(x, t) = \frac{x + 1}{t + 3} \quad \text{for all } x \text{ and } t > 0$$

Note that the slope, dt/dx, of a straight line characteristic for (10.11) equals $1/a(u)$. In this example, the slope of the characteristic originating at $(x_0, 0)$ equals $1 + x_0/3$ and hence the characteristic lines are nonparallel, but there are no characteristics that intersect for $t > 0$.

(b) If the initial function $f(x)$ is changed to $f(x) = 1/(1 + x^2)$, then u is equal to the constant $u_0 = f(x_0) = 1/(1+x_0^2)$ along the straight line $x = a(u_0)t + x_0$; but in this case the implicit equation

$$u(x, t) = f(x - ut) = 1/(1 + (x - ut)^2)$$

is not explicitly solvable for u in terms of x and t. Note that along the characteristic originating at (0, 0), we have $u_0 = f(x_0) = 1$, and on the characteristic that originates at (1, 0), we have $u_0 = f(x_0) = 1/2$. Since $a(u_0) = u_0$, the equations of these two straight lines are $x = t$ and $x = 1 + t/2$, respectively. These lines meet at $(x, t) = (2, 2)$, leading to the impossible conclusion that $u(x, t)$ is equal to two different values at the same point. This situation is resolved by the formation of *shock curve*. Shocks are discussed in more detail below.

(c) If the initial data in the problem of part (a) is changed to

$$f(x) = \begin{cases} 0 & \text{if } x < 0 \\ x & \text{if } 0 < x < 1 \\ 1 & \text{if } x > 1 \end{cases}$$

then $f(x)$ is not continuously differentiable; so Theorem 10.1 does not apply, but we can still construct a piecewise description of the solution of the initial value problem. For $x_0 < 0$, we have $a_0 = u_0 = f(x_0) = 0$ and, thus, $u = 0$ along $x = x_0$. The characteristics $x = x_0$ for $x_0 \le 0$ cover the set $x \le 0$ in the xt half plane. For $x_0 > 1$, we have $a_0 = u_0 = f(x_0) = 1$; hence, $u = 1$ along the line $x - t = x_0$. The characteristics $x - t = x_0 \ge 1$ cover the portion $x - t \ge 1$ of the xt half plane. Finally, for $0 < x_0 < 1$, the characteristic originating at $(x_0, 0)$ is given by $x - u_0 t = x_0$, where $u_0 = f(x - u_0 t) = x - u_0 t$; i.e., $u_0 = x/(1 + t)$. Then $u = x/(1 + t)$ for $0 < x < 1 + t$ and for each $t > 0$.

$$u(x, t) = \begin{cases} 0 & \text{if } x < 0 \\ x/(1+t) & \text{if } 0 < x < 1 + t \\ 1 & \text{if } x > 1 + t \end{cases}$$

WEAK SOLUTIONS

The solution constructed in Example 10.3(c) is piecewise smooth but not smooth, and in Example 10.2(b), we see an initial value problem for which the solution is apparently only piecewise continuous. We refer to solutions for (10.11) which are only piecewise smooth or piecewise continuous as *weak solutions*. A formal definition of a weak solution can be stated in terms of a family of test functions.

TEST FUNCTIONS

Test Function. A function $g = g(x, t)$ is a test function if g is continuously differentiable with respect to both x and t in the half space $H = \{\text{all } x \text{ and } t \geq 0\}$ and if, in addition, the set of points where $g(x, t)$ is different from zero is contained in a closed, bounded subset of H.

Weak Solution. The bounded piecewise continuous function $u = u(x, t)$ is a weak solution of the initial value problem

$$\partial_t u(x, t) + \partial_x F(u) = S(u), \quad u(x, 0) = f(x) \tag{10.13}$$

if, for all test functions $g = g(x, t)$,

$$\int_0^\infty \int_{-\infty}^\infty \{u \partial_t g + F(u) \partial_x g + S(u) g\} \, dx dt + \int_{-\infty}^\infty f(x) g(x, 0) \, dx = 0 \tag{10.14}$$

Theorem 10.2

Theorem 10.2. Suppose $u = u(x, t)$ is a continuously differentiable function of x and t that satisfies (10.13). Then $u(x, t)$ satisfies (10.14) for all test functions $g(x, t)$. However, (10.14) may have piecewise continuous solutions which do not satisfy (10.13).

A weak solution need only satisfy the partial differential equation in an integral average sense; i.e., in the neighborhood of any point (x, t), we can choose a non-negative test function g that differs from zero only within the neighborhood. Then equation (10.14) must hold for this g which implies equation (10.13) is satisfied in this average sense near the point. The first type of weak solution we will consider is called an *expansion fan* or *expansion wave* solution.

EXPANSION FANS

Consider the initial value problem (10.11) with a discontinuous initial condition

$$u(x, 0) = \begin{cases} u_1 & \text{if } x < x_0 \\ u_2 & \text{if } x > x_0 \end{cases}$$

If $a(u_1) < a(u_2)$, then

$$u(x, t) = \begin{cases} u_1 & \text{if } x < (a(u_1)t + x_0) \quad \text{for } t > 0 \\ u_2 & \text{if } x > (a(u_2)t + x_0) \quad \text{for } t > 0 \end{cases}$$

Note that there are no characteristic lines and, hence, no solution values specified in the wedge shaped region $W = \{a(u_1)t < x - x_0 < a(u_2)t\}$. We can show that (10.14) is satisfied in this case by a solution of the form

$$u(x,t) = \begin{cases} u_1 & \text{if } x - x_0 < a(u_1)t \quad \text{for } t > 0 \\ b((x-x_0)/t) & \text{if } a(u_1)t < x - x_0 < a(u_2)t \\ u_2 & \text{if } x - x_0 > a(u_2)t \quad \text{for } t > 0 \end{cases} \tag{10.15}$$

where $b(a(u)) = u$ for all u; i.e., b is the inverse of the function $a(u)$. Then (10.15) is a weak solution of the initial value problem (10.13) resulting from a discontinuous initial state. We refer to the part of the solution defined by $b(x - x_0/t)$ as an *expansion fan*.

Example 10.4

The initial value problem

$$\partial_t u(x,t) + u\partial_x u(x,t) = 0, \quad u(x,0) = \begin{cases} 1+x & \text{if } x < 0 \\ 2+x & \text{if } x > 0 \end{cases}$$

has a discontinuity in the initial function occurring at $x = 0$. Although the initial function is not piecewise constant, we can solve this initial value problem by an expansion fan since $u_1 = f(0-) = 1$ is less than $u_2 = f(0+) = 2$ and, therefore, $a(u_1) = u_1$ is less than $a(u_2) = u_2$. The discontinuity generates characteristic lines $x = t$ and $x = 2t$ originating from the origin. Thus, the wedge shaped region where the expansion fan is defined is the set $W = \{t < x < 2t\}$.

The region $x < t$ is covered by characteristics originating on the half line $x < 0$ where $f(x) = 1 + x$. Then $u(x, t) = f(x - a(u)t) = 1 + x - ut$ in this region and we can solve explicitly for u in terms of x and t

$$u(x,t) = \frac{x+1}{t+1} \quad \text{for } x < t$$

Similarly, in the region $x > 2t$, $u(x, t) = 2 + x - ut$ and we find

$$u(x,t) = \frac{x+2}{t+1} \quad \text{for } x > 2t$$

Then since $a(u) = u$ we have $b(u) = u$ and the solution of the initial value problem is

$$u(x,t) = \begin{cases} \dfrac{x+1}{t+1} & \text{for } x < t \\ x/t, & t < x < 2t \\ \dfrac{x+2}{t+1} & \text{for } x > 2t \end{cases}$$

SHOCKS

A discontinuity in the initial condition where $f(x)$ jumps from the value u_1 to the value u_2 produces an expansion fan solution if $a(u_1) < a(u_2)$. If $a(u_1) > a(u_2)$, then the discontinuity persists and is propagated along a curve $x = s(t)$ in the xt plane originating at the initial discontinuity $(x_0, 0)$. The curve is called a *shock curve* and the discontinuous weak solution is called a *shock wave*. A shock discontinuity in the solution of a conservation law equation may occur even with a smooth initial condition if conflicting characteristics intersect.

Consider the characteristic lines originating at distinct points $(x_1, 0)$, $(x_2, 0)$ with $x_1 < x_2$ and $u_j = f(x_j)$, $a_j = a(u_j), j = 1, 2$. If $a_1 > a_2$, then the two characteristics meet at $t = T$ such that $x_1 + a_1 T = x_2 + a_2 T$; i.e.,

$$T = \frac{x_2 - x_1}{a_1 - a_2} = \text{break time}, \qquad X = x_1 + a_1 T = x_2 + a_2 T$$

When an intersection of characteristics occurs at (X, T) and no characteristics intersect before $t = T$ then we have a solution for (10.11) of the form

$$u(x, t) = \begin{cases} u_c(x, t) & \text{for } 0 < t < T \\ u_1(x, t) & \text{for } x < s(t) \text{ and } t > T \\ u_2(x, t) & \text{for } x > s(t) \text{ and } t > T \end{cases} \tag{10.16}$$

where, in this case, the shock curve $x = s(t)$ originates at (X, T). A shock curve originating from the intersection of two characteristics is not a characteristic curve.

Theorem 10.3

Theorem 10.3. Let the weak solution $u = u(x, t)$ for the conservation law equation $\partial_t u(x, t) + \partial_x F(u) = 0$ have a jump discontinuity across the curve $S = \{x = s(t)\}$. That is, $u(x,t)$ is of the form (10.16) where $u_1(x, t)$ and $u_2(x, t)$ are smooth in a neighborhood of S and $u_1 = u_1(s(t), t)$ does not equal $u_2 = u_2(s(t), t)$. Then $s(t)$ must satisfy the *jump condition*

$$\frac{ds}{dt} = \frac{F(u_1) - F(u_2)}{u_1 - u_2} \tag{10.17}$$

If, in addition to (10.17), the *entropy condition*

$$a(u_1) > \frac{ds}{dt} > a(u_2) \tag{10.18}$$

is satisfied, then this weak solution is unique.

Note that if the partial differential equation is linear [i.e., if $F(u) = a(x, t)u$], then the jump condition (10.17) reduces to $s'(t) = a(x, t)$ and the shock curve is a characteristic curve. For nonlinear equations, shock curves are not characteristic curves.

Travelling Wave Solutions

In the previous chapter we saw that the linear equations $\partial_t u(x, t) - a\partial_x u(x, t) = 0$ and $\partial_{tt} u(x, t) - a^2 \partial_{xx} u(x, t) = 0$ each admit a so-called travelling wave solution to the initial value problem; that is, a solution of the form $f(x - ct)$. We have also seen that a nonlinear initial value problem of the form (10.11) generally propagates initial signals in such a way that distortion occurs and spontaneous discontinuities in the form of shocks may even develop. It is often of interest to know whether a partial differential equation admits a travelling wave solution, and in the solved problems, we consider the existence of such solutions for the nonlinear equations

$$\partial_t u(x, t) + u\partial_x u(x, t) - D\partial_{xx} u(x, t) = 0 \tag{10.19}$$

$$\partial_t u(x, t) + u\partial_x u(x, t) - k\partial_{xxx} u(x, t) = 0 \tag{10.20}$$

These equations, known as Burger's equation and the Korteweg-deVries equation, respectively, will be used to illustrate that interaction between nonlinear and higher order terms in a partial differential equation can produce behavior that is not possible when only one term or the other is present. For example, higher order terms can serve to modify the shock inducing effect of the nonlinear terms so as to permit a travelling wave solution to the equation. Travelling wave solutions, waveforms that propagate at constant speed without distortion, are also called *solitary waves*. Some nonlinear equations admit solitary wave solutions that retain their identity after interacting with other solitary waves (e.g., waves moving in opposite directions pass one through the other and retain their precollision waveforms). Such solutions are called *soliton* wave solutions.

SOLVED PROBLEMS

Method of Characteristics

PROBLEM 10.1

Solve the initial value problem: $\partial_t u(x, t) + \sin 2\pi t \partial_x u(x, t) = 0$, $u(x, 0) = x^2$.

SOLUTION 10.1

The characteristic equation $dx/dt = \sin 2\pi t$ leads to the result $2\pi x + \cos 2\pi t = C$, a constant. Then for an arbitrary smooth function F,

$$u(x, t) = F(2\pi x + \cos 2\pi t)$$

is a solution of the differential equation. Then $u(x, 0) = F(2\pi x + 1) = x^2$ if we choose $F(z) = ((z - 1)/2\pi)^2$, and the solution of the initial value problem is

$$u(x, t) = ((2\pi x + \cos 2\pi t - 1)/2\pi)^2$$

PROBLEM 10.2

Solve the initial value problem: $\partial_t u(x, t) + u\partial_x u(x, t) = -x$, $u(x, 0) = 0$.

SOLUTION 10.2

The characteristic equations for this quasilinear problem are, as prescribed by (10.8),

$$\frac{dx}{ds} = u, \quad \frac{dt}{ds} = 1, \quad \frac{du}{ds} = -x \tag{1}$$

Then $x\,dx = -u\,du$ and it follows that on the characteristic originating at $(x_0, 0)$ we have $x^2 + u^2 = C_1(x_0)$. Then the initial condition implies $x_0{}^2 + 0 = C_1(x_0)$; thus,

$$u(x, t) = \sqrt{x_0{}^2 - x^2} \tag{2}$$

The first two equations in (1) imply $dx/dt = u$ and using (2) in this equation leads to

$$t = \arccos(x/x_0) \quad \text{or} \quad x(t) = x_0 \cos t \tag{3}$$

Then $$u^2 = x_0{}^2(1 - \cos^2 t) = x_0 \sin t \tag{4}$$

For each $x_0 > 0$, the solution $u(x, t)$ starts from zero at $(x_0, 0)$ and varies along the characteristic given by (3) as specified by either (2) or (4). At $t = \pi/2$, the characteristics converge at $(0, \pi/2)$ and the solution ceases to exist.

PROBLEM 10.3

Solve the initial value problem: $\partial_t u(x, t) + u\partial_x u(x, t) = x$, $u(x, 0) = 0$.

SOLUTION 10.3

Proceeding as in the previous problem, noting the slight difference in the equation, we arrive at

$$u(x, t) = \sqrt{x^2 - x_0{}^2} \tag{1}$$

This time $dx/dt = u$ leads to

$$t = \log\left(x/x_0 + \sqrt{(x/x_0)^2 - 1}\right) \quad \text{or} \quad x(t) = x_0 \cosh t \tag{2}$$

Then

$$u(x, t) = \sqrt{x^2(1 - \operatorname{sech}^2 t)} = x \tanh t \tag{3}$$

In this case for any real x_0, the solution starts from zero and varies along the characteristic $x(t) = x_0 \cosh t$ as prescribed by (3). The solution exists for all $t > 0$.

PROBLEM 10.4

Solve the initial value problem: $\partial_t u(x, t) + (1/u)\partial_x u(x, t) = 0$, $u(x, 0) = f(x)$, where

$$f(x) = \begin{cases} 2 & \text{if } x < 0 \\ 2 - x & \text{if } 0 < x < 1 \\ 1 & \text{if } x > 1 \end{cases}$$

SOLUTION 10.4

Since the equation is homogeneous, the solution is constant along characteristics. Thus, the characteristics are straight lines of the form $x - a(u_0)t = x - t/u_0 = x_0$, where $u_0 = f(x_0)$.

If $x_0 < 0$, then $f(x_0) = 2$ and we have $u = 2$ along the lines $x - t/2 = x_0$. These characteristics cover the set $x < t/2$.

If $x_0 > 1$, then $f(x_0) = 1$ and we have $u = 1$ along $x - t = x_0$. These characteristics cover the set $x > 1 + t$.

Finally, for x_0 between zero and one, we have $u(x_0, 0) = f(x_0) = 2 - x_0$. Then, in the set $t/2 < x < 1 + t$, the solution $u(x, t)$ is given implicitly by $u(x, t) = f(x - t/u) = 2 - (x - t/u)$. We can solve this explicitly for u in terms of x and t to find

$$u(x, t) = \frac{1}{2}\left(2 - x + \sqrt{(2 - x)^2 + 4t}\right) \quad \text{for } t/2 < x < t + 1$$

This solution is piecewise smooth and exists for all x and $t > 0$.

PROBLEM 10.5

Solve the initial value problem: $\partial_t u(x, t) + u^2 \partial_x u(x, t) = 0$ and $u(x, 0) = f(x)$ where $f(x) = 0$ for $x < 0$ and $f(x) = \sqrt{x}$ for $x > 0$.

SOLUTION 10.5

Since $f(x_0) = 0$ for $x_0 < 0$, all characteristics that originate at $x_0 < 0$ are of the form $x = x_0$ and we have $u(x,t) = 0$ for $x < 0$ for all $t > 0$.

For $x_0 > 0$, the characteristics are the lines $x - u_0{}^2 t = x_0$ and, in the set $x > 0$, the solution is given implicitly by

$$u = f(x - u^2 t) = \sqrt{x - u^2 t}$$

i.e.,

$$u(x,t) = \sqrt{x/(1+t)} \qquad \text{for } x > 0 \text{ and } t > 0.$$

This solution is piecewise smooth and exists for all x and $t > 0$.

Expansion Fans

PROBLEM 10.6

Solve the initial value problem: $\partial_t u(x,t) + u^2 \partial_x u(x,t) = 0$ and $u(x, 0) = f(x)$, where $f(x) = 1$ for $x < 1$ and $f(x) = 2$ for $x > 1$.

SOLUTION 10.6

The initial condition here is discontinuous at $x = 1$ with $u_1 = f(1-) = 1$ and $u_2 = f(1+) = 2$. Then since $a(u) = u^2$, $a(u_1) < a(u_2)$, and it follows that an expansion fan solution applies. We have $b(u) = \sqrt{u}$ in this case and, thus, the solution has the form

$$u(x,t) = \begin{cases} 1 & \text{if } x - 1 < t \\ \sqrt{(x-1)/t} & \text{if } t < x - 1 < 4t \\ 2 & \text{if } x - 1 > 4t \end{cases}$$

This solution is piecewise smooth and exists for all x and all $t > 0$.

PROBLEM 10.7

Consider the initial value problem for the nonlinear differential equation $\partial_t u(x,t) + a(u)\partial_x u(x,t) = 0$ with a discontinuous initial condition

$$u(x,0) = \begin{cases} u_1 & \text{if } x < x_0 \\ u_2 & \text{if } x > x_0 \end{cases}$$

Show that if $b(s)$ is a smooth, nonconstant function such that $u(x, t) = b(x - x_0/t)$ solves the equation in the wedge shaped region $W = \{a(u_1)t < x - x_0 < a(u_2)t\}$ then $b(a(u)) = u$; i.e., b is the inverse of the function a.

SOLUTION 10.7

Suppose $u(x, t) = b(x - x_0/t)$ solves $\partial_t u(x,t) + a(u)\partial_x u(x,t) = 0$ in the region W. Then

$\partial_x u(x,t) = b'(x-x_0/t)\,t^{-1}$ and $\partial_t u(x,t) = b'(x-x_0/t)\,(-(x-x_0)/t^2)$
and
$\partial_t u(x,t) + a(u)\,\partial_x u(x,t) = b'(x-x_0/t)\,(-(x-x_0)/t + a(b(x-x_0/t))) = 0$

We have assumed that b is not a constant function and, thus, $b'(x-x_0/t)$ is not the zero function. Then the differential equation requires that

$$a(b(x-x_0/t)) = x-x_0/t \quad \text{for } (x, t) \text{ in } W$$

Note that on the edges of the wedge W, we have $x - x_0 = a(u_j)\,t$, $j = 1, 2$. Then on the edges of W, $u = b(x-x_0/t)$ reduces to $u_j = b(a(u_j))$ $j = 1, 2$.

Shocks

PROBLEM 10.8

Solve the initial value problem for $\partial_t u(x,t) + u\partial_x u(x,t) = 0$, $u(x, 0) = f(x)$, where

$$f(x) = \begin{cases} 1 & \text{if } x < 0 \\ 1-x & \text{if } 0 < x < 1 \\ 0 & \text{if } x > 0 \end{cases}$$

SOLUTION 10.8

The characteristics for this homogeneous differential equation are the straight lines $x - u_0 t = x_0$, where $u_0 = f(x_0)$. If $x_0 < 0$, then $f(x_0) = 1$ and $u(x, t) = 1$ along all characteristics of the form $x - t = x_0 < 0$. These characteristics cover the set $\{x < t\}$ in the xt plane. If $x_0 > 1$, then $f(x_0) = 0$ and $u(x, t) = 0$ along the characteristics of the form $x = x_0 > 1$. These characteristics cover the set $\{x > 1\}$ in the xt plane.

If $x_0 > 1$, then $f(x_0) = 0$ and $u(x,t) = 0$ along the characteristics of the form $x = x_0 > 1$. These characteristics cover the set $\{x > 1\}$ in the xt plane.

In the set $\{t < x < 1\}$, we have characteristics originating at $(x_0, 0)$ for $0 < x_0 < 1$, where we have $f(x_0) = 1 - x_0$. In this region, $u(x, t)$ is given implicitly by

$$u(x, t) = f(x - ut) = 1 - (x - ut)$$

hence

$$u(x,t) = \frac{1-x}{1-t} \quad \text{for } t < x < 1$$

Note that this solution is defined only for $0 < t < 1$. At $t = 1$, the characteristic $x - t = 0$, originating at the origin, intersects the characteristic $x = 1$, which originates at $(1, 0)$. Since $u = 1$ along the characteristic from the origin and $u = 0$ along the other characteristic, we expect a shock to form at (1, 1).

Since $F(u) = u^2/2$, we note that

$$\frac{F(u_1) - F(u_2)}{u_1 - u_2} = \frac{1}{2}(u_1 + u_2)$$

and, therefore, the jump relation for this problem reduces to

$$s'(t) = \frac{1}{2}(1 + 0) = 1/2 \quad \text{and} \quad s(1) = 1$$

Then $s(t) = t/2 + C$ and the condition that the shock start from (1, 1) forces the result $s(t) = (t + 1)/2$. Finally,
$a(u_1) = u_1 = 1 > 1/2 = s'(t) > 0 = a(u_2)$, so the entropy condition is satisfied. Then Theorem 10.3 implies this is the only weak solution of the initial value problem.

The full solution is:

$$\text{for } 0 < t < 1: \quad u(x, t) = \begin{cases} 1 & \text{if } x < t \\ (1 - x)/(1 - t) & \text{if } t < x < 1 \\ 0 & \text{if } x > 1 \end{cases}$$

$$\text{for } t > 1: \quad u(x, t) = \begin{cases} 1 & \text{if } x < 1 + (t - 1)/2 \\ 0 & \text{if } x > 1 + (t - 1)/2 \end{cases}$$

The solution for $0 < t < 1$ is piecewise smooth, whereas the solution for $t > 1$ is piecewise continuous, but not continuous across the shock.

PROBLEM 10.9

Solve the previous initial value problem if the initial function is changed to $f(x) = |x|^{1/2}$ if $x < 0$ and $f(x) = 0$ for $x > 0$.

SOLUTION 10.9

As in the previous problem, the characteristic originating at $(x_0, 0)$ has the form $x - u_0 t = x_0$, where $u_0 = f(x_0)$.

If $x_0 > 0$, then $f(x_0) = 0$ and then $u(x, t) = 0$ along the characteristics of the form $x = x_0 > 0$.

If $x_0 < 0$, then along the characteristics $x - u_0 t = x_0$, u is given implicitly by

$$u = |x - ut|^{1/2}$$

i.e.,

$$u^2 = |x - ut| = -(x - ut)$$

We choose the negative value for $|x - ut|$ in order to have $x_0 < 0$ for $t = 0$. Now solving the quadratic equation for u, we find

$$u(x, t) = \frac{1}{2}(t + \sqrt{t^2 - 4x}) \tag{1}$$

Here we choose the positive square root in the quadratic formula in order to have $u(x_0, 0) = |x_0|^{1/2}$.

Note that the characteristics $x = x_0 > 0$ are vertical, whereas the characteristics $x = u_0 t + x_0$ have positive slope. Then the two types of characteristics intersect and a shock must form. The shock condition in this example reads as

$$s'(t) = \frac{F(u_1) - F(u_2)}{u_1 - u_2} = \frac{1}{2}(u_1 + u_2) = \frac{1}{4}(t + \sqrt{t^2 - 4s(t)}) \quad (2)$$

If we assume a solution of the form $s(t) = kt^2$ and substitute this into (2) we find that the equation is satisfied if $k = 3/16$. Then the shock curve is given by $s(t) = 3t^2/16$ and the solution to the initial value problem is given by (1) for $x < 3t^2/16$ and is equal to zero for $x > 3t^2/16$.

Note that the entropy condition is satisfied since

$$a(u_1) = u_1 > s'(t) = \frac{1}{2}u_1 > u_2 = 0$$

Note also that the shock in this example is a curve and not a straight line.

PROBLEM 10.10

Solve the initial value problem from Problem 10.8 if $f(x)$ is given by

$$f(x) = \begin{cases} a & \text{if } x < 0 \\ b & \text{if } 0 < x < 1 \\ c & \text{if } x > 1 \end{cases} \quad \text{for } a > b > c$$

SOLUTION 10.10

As usual, the characteristic originating at $(x_0, 0)$ has the equation $x - u_0 t = x_0$ for $u_0 = f(x_0)$. Then

for $x_0 < 0$: $u(x, t) = f(x_0) = a$ along $x - at = x_0$

for $0 < x_0 < 1$: $u(x, t) = b$ along $x - bt = x_0$

for $x_0 > 1$: $u(x, t) = c$ along $x - ct = x_0$

Since $a > b$, a shock forms at (0, 0). The shock curve there is given by $x = s(t)$ where $s(t)$ satisfies

$$s'(t) = \frac{1}{2}(u_1 + u_2) = \frac{1}{2}(a + b), \quad s(0) = 0 \quad (1)$$

Then $s(t) = (a + b)t/2$ is the equation of this shock curve.

Since $b > c$, a shock also forms at (1, 0) where f has another discontinuity. This shock curve is given by $x = s(t)$, where $s(t)$ satisfies

$$s'(t) = \frac{1}{2}(u_1 + u_2) = \frac{1}{2}(b + c) \qquad s(0) = 0 \tag{2}$$

Then this shock curve has the equation $s(t) = 1 + (b + c)t/2$ and the solution of the initial value problem is given by

$$u(x, t) = \begin{cases} a & \text{if } x < (a+b)\,t/2 \\ b & \text{if } (a+b)\,t/2 < x < 1 + (b+c)\,t/2 \\ c & \text{if } x > 1 + (b+c)\,t/2 \end{cases} \tag{3}$$

But the two shock curves intersect at $t = T$, satisfying $(a + b)T/2 = 1 + (b + c)T/2$ and we find $T = 2/(a - c)$ and $X = (a + b)/(a - c)$. At this intersection point, since the shock on the left carries the value a which is greater than c, the value that is carried by the shock on the right, a new shock forms and, for $t > T$, we have the solution

$$u(x, t) = \begin{cases} a & \text{if } x < (a+c)\,(t-T)/2 + X \\ c & \text{if } x > (a+c)\,(t-T)/2 + X \end{cases} \qquad \text{for } t > T \tag{4}$$

Thus, the solution (3) is in force only up to the time T and then the solution (4) is in effect. Note that the solution (4) could also arise from other initial states. Therefore, shock solutions are *irreversible* in the sense that we cannot recover the initial state from a subsequent state by reversing the time. The implication is that a physical process modeled by a conservation law equation is irreversible.

PROBLEM 10.11

Let the weak solution $u = u(x, t)$ for the conservation law equation

$$\partial_t u(x, t) + \partial_x F(u) = 0$$

have a jump discontinuity across the curve $S = \{x = s(t)\}$. That is, $u(x, t)$ is of the form (10.16) where $u_1(x, t)$ and $u_2(x, t)$ are smooth in a neighborhood of S and $u_1 = u_1(s(t), t)$ does not equal $u_2 = u_2(s(t), t)$. Then show that $s(t)$ must satisfy the *jump condition* (10.17)

SOLUTION 10.11

Let $x_1, x_2, t_1 > 0$ and $t_2 > t_1$ be chosen such that $x_1 < s(t) < x_2$ for $t_1 < t < t_2$, and for t in (t_1, t_2), let

$$Q(t) = \int_{x_1}^{x_2} u(x, t)\,dx = \int_{x_1}^{s(t)} u_1(x, t)\,dx + \int_{s(t)}^{x_2} u_2(x, t)\,dx$$

Then Leibniz's rule implies

$$Q'(t) = u_1(s(t),t)s'(t) - u_2(s(t),t)s'(t)$$

But $u(x, t)$ can be interpreted as the linear density of some substance and then $Q(t)$ represents the amount of that substance in the interval (x_1, x_2) at time t. It follows that $Q'(t)$ must equal the inflow minus the outflow of substance to the interval, that is,

$$Q'(t) = F(u_1(x_1,t)) - F(u_2(x_2,t)).$$

We equate the two expressions for $Q'(t)$ and note that since x_1 and x_2 are arbitrary as long as $s(t)$ lies between them, we can let x_1 and x_2 approach $s(t)$ from the left and right, respectively. This leads to the jump relation

$$[u_1(s(t),t) - u_2(s(t),t)]s'(t) = F(u_1(x_1,t)) - F(u_2(x_2,t)).$$

The meaning of Theorem 10.3 is that if a conservation law equation has a shock type solution, then the shock curve must satisfy the jump relation. However, a conservation law equation may have other weak solutions in addition to a shock type solution. In this case, we apply the entropy condition. If the shock does not satisfy the entropy condition, then it is physically inadmissible. See the following problem for examples.

PROBLEM 10.2

Show that the initial value problem $\partial_t u(x,t) + u\partial_x u(x,t) = 0$, $u(x, 0) = H(x)$, where $H(x)$ denotes the Heaviside step function, has a shock type solution and an expansion wave solution. Show also that the shock solution does not satisfy the entropy condition.

SOLUTION 10.12

Since $a(u) = u$, it is easy to find the expansion wave solution for this problem

$$u(x,t) = \begin{cases} 0 & \text{if } x < 0 \\ x/t & \text{if } 0 < x < t \\ 1 & \text{if } x > t \end{cases}$$

It is also easy to see that $u(x, t) = H(x - t/2)$ is a shock type solution for the problem. That is, $s(t) = t/2$ and $F(u) = u^2/2$, so

$$s'(t) = \frac{1}{2} = \frac{F(u_1) - F(u_2)}{u_1 - u_2} = \frac{1}{2}(u_1 + u_2) = \frac{1}{2}(0+1)$$

Then $u(x, t) = H(x - t/2)$ is mathematically admissible in that it satisfies the differential equation in the weak sense and it satisfies the jump condition across the shock curve at $x = t/2$. However, $a(u_1) = u_1 = 0 < s'(t) = 1/2 < 1 = a(u_2)$ and so the entropy condition is not satisfied. Then the shock solution is not physically admissible and the expansion fan solution is the weak solution which is both mathematically and physically admissible.

Weak Solutions

PROBLEM 10.13

Show that if $u = u(x, t)$ is a smooth solution for (10.13), then it must also be a weak solution; i.e., $u(x, t)$ must satisfy (10.14) for all test functions g.

SOLUTION 10.13

Suppose $u(x, t)$ is a continuously differentiable solution for (10.13). We sometimes refer to a smooth solution as a *classical solution* for the initial value problem. Let $g(x, t)$ denote an arbitrary test function, and suppose the set where g is different from zero is contained in the bounded set $M = \{(x, t)\colon\ x_0 < x < x_1,\ 0 \le t \le t_1\}$, but $g(x, 0)$ need not be zero. Then

$$\int_0^{\infty}\int_{-\infty}^{\infty} (\partial_t u(x,t) + \partial_x F(u(x,t)) - S(u(x,t)))\, g(x,t)\, dxdt = 0 \quad (1)$$

But

$$\int_0^{\infty}\int_{-\infty}^{\infty} \partial_t u(x,t)\, g(x,t)\, dxdt = \int_0^{t_1}\int_{x_0}^{x_1} \partial_t u(x,t)\, g(x,t)\, dxdt$$

$$= \int_{x_0}^{x_1} ug|_0^{t_1} dx - \int_0^{t_1}\int_{x_0}^{x_1} \partial_t g(x,t)\, u(x,t)\, dxdt$$

$$= -\int_{-\infty}^{\infty} u(x,0)\, g(x,0)\, dx - \int_0^{\infty}\int_{-\infty}^{\infty} \partial_t g(x,t)\, u(x,t)\, dxdt \quad (2)$$

Similarly,

$$\int_0^{\infty}\int_{-\infty}^{\infty} \partial_x F(u(x,t))\, g(x,t)\, dxdt = -\int_0^{\infty}\int_{-\infty}^{\infty} F(u(x,t))\, \partial_x g(x,t)\, dxdt \quad (3)$$

Then combining (2) and (3) leads to the result

$$\int_0^{\infty}\int_{-\infty}^{\infty} (u\partial_t g(x,t) + F(u)\, \partial_x g(x,t) + S(u)\, g(x,t))\, dxdt + \int_{-\infty}^{\infty} f(x)\, g(x,0)\, dx = 0 \quad (4)$$

Since the test function $g(x, t)$ is arbitrary, then (4) must hold for all test functions. This shows that (10.14) must hold for every smooth solution of

(10.13) but it is apparent that a function $u(x, t)$ satisfying (10.14) need not even be continuous, much less differentiable.

Travelling Wave Solutions

PROBLEM 10.14

Show that the linear equations

(a) convection-diffusion equation

$$\partial_t u(x,t) + a\partial_x u(x,t) - D\partial_{xx} u(x,t) = 0$$

(b) convection-dispersion equation

$$\partial_t u(x,t) + a\partial_x u(x,t) + k\partial_{xxx} u(x,t) = 0$$

have no interesting travelling wave solutions.

SOLUTION 10.14

Substituting $u(x, t) = f(x - at)$ into the convection-diffusion equation leads to

$$-af'(z) + af'(z) + kf''(z) = 0$$

Then $f(z) = C_1 z + C_2$ and $u(x,t) = C_1(x - at) + C_2$. This is a travelling wave with $\partial_{xx} u(x,t) = 0$; i.e., there is convection but no diffusion. The only travelling wave solution to the convection-diffusion equation is, in reality, just a solution of the simple advection equation $\partial_t u(x,t) + a\partial_x u(x,t) = 0$.

Similarly, substituting $f(x - at)$ into the convection-dispersion equation leads to

$$-af'(z) + af'(z) + kf'''(z) = 0.$$

Then $f(z) = C_1 z^2 + C_2 z + C_3$ and $u(x, t) = C_1(x - at) + C_2(x - at) + C_3$. Note that $\partial_{xxx} u(x,t) = 0$; i.e., there is no dispersion.

This problem illustrates that the effect of the higher order terms in these linear equations is to suppress all but the simplest travelling wave solutions to the equations. As we shall see, for nonlinear equations, this is not the case.

PROBLEM 10.15

Show that for $D > 0$, Burger's equation

$$\partial_t u(x,t) + u\partial_x u - D\partial_{xx} u(x,t) = 0 \qquad (1)$$

has a travelling wave solution.

SOLUTION 10.15

Note first that substituting $u(x, t) = f(x - at)$ into Burger's equation with $D = 0$ leads to $(f(z) - a)f'(z) = 0$. This implies $f(z)$ is constant and the corresponding solution $u(x, t)$ = constant is not a travelling wave except in the most trivial sense. In fact, we have seen in several previous examples that Burger's equation with $D = 0$ may propagate even a smooth initial waveform with an ever-steepening front that eventually breaks to form a shock.

For $D > 0$, on the other hand, assuming $u(x, t) = f(x - at)$ leads to

$$-af'(z) + f(z)f'(z) - Df''(z) = 0$$

i.e.,

$$Df''(z) = d/dz\left(\frac{1}{2}f^2(z) - af(z)\right).$$

Then

$$Df'(z) = \left(\frac{1}{2}f^2(z) - af(z)\right) + C \qquad (1)$$

For each positive constant of integration C, we have two positive roots for the quadratic equation

$$f^2(z) - 2af(z) + 2C = (f - f_1)(f - f_2) = 0$$

It is not hard to show that $f'(z) < 0$ for $f_1 < f(z) < f_2$ and that $f'(z) > 0$ for $f(z) > f_2$ and for $f(z) < f_1$. Then (1) can be seen to have a solution $f(z)$ whose orbit tends from the unstable critical point at f_2 toward the stable critical point at f_1. In fact, we can separate variables and integrate (1) to find

$$\frac{2D}{f_2 - f_1}\log\frac{f_2 - f(z)}{f(z) - f_1} = z$$

Then

$$f(z) = \frac{f_2 + f_1 e^{pz}}{1 + e^{pz}} \quad \text{for} \quad p = \frac{f_2 - f_1}{2D}$$

and

$$u(x, t) = \frac{f_2 + f_1 e^{p(x - at)}}{1 + e^{p(x - at)}}$$

Note that for each $t > 0$, $u(x, t)$ tends to f_2 as x tends to minus infinity and tends to f_1 as x tends to plus infinity. Note also that the roots f_1, f_2 of the quadratic equation satisfy $(f_1 + f_2)/2 = a$, where a is the speed of the travelling wave.

This example illustrates that without the diffusion term, solutions to equation (1) would shock up, whereas if there is diffusion without the nonlinear term, all travelling wave solutions are suppressed. However, when both terms are present, they compete in such a way as to allow travelling wave solutions to exist.

PROBLEM 10.16

Show that the Korteweg-deVries equation

$$\partial_t u(x, t) + u\partial_x u(x, t) + k\partial_{xxx} u(x, t) = 0 \qquad (1)$$

has a travelling wave solution.

SOLUTION 10.16

We have already seen that no travelling wave solution is possible if $k = 0$ and no interesting solution is possible when the nonlinear term is absent. However, substituting $u(x, t) = f(x - at)$ into (1) leads to the following ordinary differential equation for $f(z)$:

$$-af'(z) + \left(\frac{1}{2}f(z)^2\right)' + kf'''(z) = 0 \tag{2}$$

i.e.,

$$2kf''(z) + f(z)^2 - 2af(z) = C_1 \tag{3}$$

For a travelling wave, we require that $f(z)$ tend to zero for large $|z|$ together with $f'(z)$ and $f''(z)$. Thus, $C_1 = 0$. Now we multiply (3) by $f'(z)$ and integrate once more to get

$$D\,f'(z)^2 = a\,f(z)^2 - \frac{1}{3}f(z)^3 + C_2 \tag{4}$$

The boundary conditions at infinity imply $C_2 = 0$ and thus

$$\sqrt{3D}\,f'(z) = f(z)\sqrt{3a - f(z)} \tag{5}$$

The ordinary differential equation (5) is separable and can be integrated to give

$$f(z) = 3a\ \text{sech}^2(cz), \quad \text{where} \quad 2c = \sqrt{a/D} \tag{6}$$

Thus the competition between the nonlinear term and the dispersive term in equation (1) leads to the travelling wave solution

$$u(x, t) = 3a\ \text{sech}^2(c(x - at)).$$

Other interesting solutions to this equation result from different choices for the constants C_1 and C_2.

In this chapter we have used the equation $\partial_t u + u\partial_x u = 0$, the simplest example of a quasilinear equation of order one to illustrate the following differences between linear and nonlinear partial differential equations:

(a) ***Existence***. *For sufficiently smooth initial data, the linear initial value problem has a unique smooth solution that exists for all $t > 0$. For the nonlinear problem, existence of a solution for all $t > 0$ cannot be guaranteed, no matter how smooth the initial data (see Example 10.3(b)) Global solutions to the nonlinear problem exist if we allow weak solutions.*

(b) ***Uniqueness***. *The solution of the linear initial value problem is uniquely determined by the data. Weak solutions for nonlinear problems are not uniquely determined by the data (see Problem 10.12). The entropy condition is a criterion for selecting a unique (physically relevant) solution for a nonlinear initial value problem.*

(c) ***Smoothness.*** *Discontinuities in solutions of linear partial differential equations do not arise spontaneously; they must originate with the data and then propagate along characteristic curves. Nonlinear equations may exhibit spontaneous discontinuities (shocks) even with smooth data. These discontinuities then propagate along noncharacteristic curves that depend on the equation and on the data. Discontinuities in the initial data for a nonlinear problem may be resolved as shocks or as expansion fans.*

(d) ***Reversibility.*** *Linear equations are generally reversible in the sense that the initial state can be computed from a subsequent state simply by reversing the time variable. In nonlinear problems, it is generally not possible to recover the initial state from a subsequent state (see Problem 10.10).*

As higher order terms are added to quasilinear equation of order one, more complicated solution behavior may develop as a result of competition between the nonlinear and higher order terms.

11

Functional Optimization and Variational Methods

The calculus of variations, functional optimization, is an old and very extensive branch of mathematics. In this chapter we will very briefly survey some of the classical aspects of the subject before discussing more modern developments which are related to weak formulations of boundary value problems and provide the theoretical basis for the finite element method.

In elementary calculus, we learn that the extreme points of a function defined over a set of real numbers can often be found by finding the zeroes of the function's first derivative. Classical calculus of variations extends this concept to problems where the function to be optimized is real valued but is defined over a set of functions rather than a set of numbers. Such functions are called functionals and the types of problems that can be formulated and solved by the methodology of functional optimization are many and varied. We survey a few of them in the solved problems.

An application of functional optimization that provides a link between the classical and modern treatments of the subject involves so-called minimum principles of physics. Many physical systems conform to such principles – the principle of least action, the principle of minimum potential energy, Fermat's principle of least time are the most notable examples. When these postulates apply, they take the place of force balances and conservation statements for deriving the mathematical model for the system in question and lead to the variational form of the governing equations.

Finally, the variational form of the governing equations for a system provides a natural framework in which to consider the relaxation of the smoothness requirements for solutions to differential equations. The notion of a weak solution was introduced in the last chapter and here we see it again in a slightly different context. We will see that certain boundary value problems are equivalent to related variational problems and in other cases, there is no such correspondence. In either case, however, we can give a weak formulation for the problem and try to approximate the solution. It must be emphasized that weak or generalized solutions for differential equations are no less physically relevant than smooth solutions. On the contrary, many physical problems have no smooth solution and this theory provides a setting in which such problems can be studied and understood.

FUNCTIONAL OPTIMIZATION

Let $f(\boldsymbol{x})$ denote a real valued function on $\Re^n$ and consider the problem of finding the extreme values of f; i.e., finding the maximum or minimum value for f. The following results are useful for this purpose:

(a) If f is continuous on a closed bounded set Ω, then f assumes its maximum and its minimum at points of Ω.

(b) If f is continuously differentiable on Ω and if x_0 in Ω is an interior extreme point for f, then **grad** $f(x_0) = \mathbf{0}$.

(c) If f is a positive, quadratic function, then f has a unique minimum at the point x_0 where **grad** $f(x_0) = \mathbf{0}$.

FUNCTIONALS

In this chapter we will consider the problem of finding extreme values for real valued functions whose domain is a set of functions rather than a set in $\Re^n$. Such functions are called *functionals*.

Example 11.1 Functionals

(a) Let $D(J)$ denote the set of all functions $y = y(x)$ defined and continuously differentiable on $[a, b]$ that satisfy the conditions $y(a) = p$ and $y(b) = q$ for given constants p and q. Then consider

$$J[y] = \int_a^b \sqrt{1 + y'(x)^2}\, dx$$

Then J is a functional and $D(J)$ is referred to as the *domain* of J.

(b) The functional J in part (a) is a special case of the more general functional

$$J[y] = \int_a^b F(x, y(x), y'(x))\ dx$$

where F denotes a smooth function of three variables. The domain of this functional is some subset of the set of all continuously differentiable functions.

(c) Let $u = u(x_1, \ldots, x_n)$ denote a smooth real valued function of n variables and consider the functional

$$J[u] = \int_\Omega F(x, u, \textbf{grad } u)\, dx$$

Then J is a functional whose domain could be a subset of the functions that are continuously differentiable on Ω. For example,

$$J[u] = \int_\Omega (\partial_x u(x, y)^2 + \partial_y u(x, y)^2 - g(x, y)\, u(x, y))\, dxdy$$

with $D(J)$ equal to the smooth functions that vanish on the boundary S of Ω is a functional of this type with $n = 2$.

FUNCTION SPACES

Functionals are defined on sets of functions rather than on sets of numbers. In order to provide these sets of functions with some of the structure that is available in $\Re^n$, we define a function space to be a vector space in which the vectors are functions. We have already introduced the function space of square integrable functions in chapter 6. Here, we define several other function spaces as well.

Example 11.2 Function Spaces

(a) Let $[a, b]$ denote a closed bounded interval and, for non-negative integer m, let $C^m[a, b]$ denote the set of all functions $f(x)$ such that f and its derivatives of order less than or equal to m are continuous on $[a, b]$. Then C^m is a vector space of functions for each m. For $m = 0$, $C^0[a, b]$ is just the space of continuous functions on $[a,b]$. More generally for Ω a closed bounded set in $\Re^n$, we can define $C^m(\Omega)$ to be the space of functions that are continuous on Ω together with all partial derivatives up to order m.

(b) We may define $H^m[a, b]$ to be the space of all functions f such that f and all its derivatives up to order m are square integrable on $[a,b]$. For

f and g in $H^m[a, b]$, we can define an inner product by

$$(f, g)_m = \int_a^b \sum_{j=0}^{m} f^{(j)}(x)\, g^{(j)}(x)\, dx \tag{11.1}$$

The norm associated with this inner product is then

$$\|f\|_m = (f, f)_m^{\ 1/2} = \left(\int_a^b \sum_{j=0}^{m} f^{(j)}(x)^2 dx \right)^{1/2} \tag{11.2}$$

When $m = 0$, this is just the space of square integrable functions introduced in chapter 6. We can generalize this definition to higher dimensions. For example, when $m = 1$ and $n = 2$. $H^1(\Omega)$ is the set of square integrable functions with square integrable first derivatives. For $u = u(x, y)$, $v = v(x, y)$ in $H^1(\Omega)$

$$(u, v)_1 = \int_\Omega (uv + \partial_x u \partial_x v + \partial_y u \partial_y v)\, dxdy \tag{11.3}$$

$$= \int_\Omega (uv + \mathbf{grad}\ u \cdot \mathbf{grad}\ v)\, dxdy$$

and

$$\|u\|_1 = \left(\int_\Omega (u^2 + \partial_x u^2 + \partial_y u^2)\, dxdy \right)^{1/2} \tag{11.4}$$

HILBERT SPACES

For each non-negative integer m the space $H^m[a, b]$ has the property that if $\{f_n\}$ is a sequence of functions in $H^m[a, b]$ such that $\|f_p - f_q\|_m$ tends to zero as p and q tend to infinity, then there exists a function f that belongs to $H^m[a, b]$ such that $\|f - f_n\|_m$ tends to zero as n tends to infinity. A space with this property is said to be *complete*. If the norm on the space is associated with an inner product (as the norm on $H^m[a, b]$ is), then the complete space is called a *Hilbert space*. Thus, $H^m[a, b]$ is a Hilbert space for each non-negative integer m. More generally, for Ω a closed bounded set in $\Re^n$, the spaces $H^m(\Omega)$ are Hilbert spaces when the inner product and norm are defined in a manner analogous to (11.1) and (11.2).

SUBSPACES

Recall that in chapter 1 we introduced the notion of a subspace of a vector space. In chapter 1, we were referring to the vector space of n-tuples of real numbers, but the notion of a subspace is a general vector space concept.

A subset M of a function space is a *subspace* if M is closed under the operation of forming linear combinations; i.e., for arbitrary f and g in M and arbitrary scalars A and B, the function $Af + Bg$ belongs to M. Then for Ω a closed bounded set in $\Re^n$, we have:

1. $C^p(\Omega)$ is a subspace of $C^q(\Omega)$ if $p > q$.

2. $H^p(\Omega)$ is a subspace of $H^q(\Omega)$ if $p > q$. (11.5)

3. $C^p(\Omega)$ is a subspace of $H^p(\Omega)$ for $p \geq 0$.

It can be shown that for $[a, b]$ a closed bounded interval

4. $H^1[a, b]$ is a subspace of $C^0[a, b]$. (11.6)

DENSE SUBSPACE

Let M be a subspace of a function space X in which the norm is denoted by $\|f\|$. Then M is said to be *dense in X* if for each f in X and every $\varepsilon > 0$ there is a function g in M such that $\|g - f\| < \varepsilon$.

Theorem 11.1

Theorem 11.1. Let Ω denote a closed bounded subset in $\Re^n$. For each non-negative integer m, the subspace $M = C^m(\Omega)$ is dense in the function space $X = H^0(\Omega)$ for the norm

$$\|f\|_0 = \left(\int_\Omega f(x)^2 dx\right)^{1/2} \tag{11.7}$$

The conclusion is also true if M is the subspace $\{f \text{ in } C^m(\Omega) : f = 0 \text{ on } S\}$, where S denotes the boundary of Ω.

Theorem 11.2

Theorem 11.2. Let M be a dense subspace of Hilbert space H and let the inner product on H be denoted by (f, g). If f in H is such that $(f,g) = 0$ for all g in M, then it follows that $f = 0$.

TRANSLATE OF A SUBSPACE

A set D in a vector space is said to be a *translate* of the subspace M if there is a subspace M in the vector space such that for every u and v in D, the difference $u - v$ belongs to M.

Example 11.3

(a) Let D denote the set of all y in $C^1[a, b]$ such that $y(a) = p$ and $y(b) =$

q and let M denote the set D when $p = q = 0$. Then M is a subspace of C^1 and D is a translate of the subspace M, but D is not a subspace.

(b) Let Ω denote a bounded set in $\Re^3$ having smooth boundary S. Let D denote the set of all functions $u(x,y,z)$ in $C^1(\Omega)$ such that $u = g$ on S, where g is a given function defined on S. Let M denote the set D in the case that g is identically zero. Then M is a subspace of C^1 and D is a translate of the subspace M.

(c) Let D denote the set of pairs (x, y) such that $2x + 3y = 6$ and let M denote the set of pairs (x, y) such that $2x + 3y = 0$. Then M is a line through the origin in $\Re^2$, whereas D is a line parallel to M but not passing through the origin. M is a subspace of $\Re^2$ and D is a translate of M.

Variation of a Functional

Let $J[u]$ be a functional whose domain $D(J)$ is a translate of a subspace M in a function space. Then, for u in $D(J)$ and v in M, we define the *variation* of J at u in the direction v to be

$$\delta J[u;v] = \lim_{h \to 0} \frac{J(u+hv) - J(u)}{h}$$

if the limit exists. Note that $\delta J[u;v]$ is just $f'(0)$ for the function $f(h) = J[u + hv]$.

Example 11.4

(a) Consider the functional J of Example 11.1(a) with domain equal to the set D of Example 11.3(a). Then, for y in D and v in M,

$$f(h) = J[y+hv] = \int_a^b \sqrt{1+(y'(x)+hv'(x))^2}\,dx$$

and

$$f'(h) = \int_a^b \frac{(y'(x)+hv'(x))v'(x)}{\sqrt{1+(y'(x)+hv'(x))^2}}\,dx$$

Thus,

$$\delta J[y;v] = \int_a^b \frac{y'(x)v'(x)}{\sqrt{1+y'(x)^2}}\,dx$$

(b) Consider the functional

$$J[u] = \int_\Omega (\partial_x u(x,y)^2 + \partial_y u(x,y)^2 - 2g(x,y)\,u(x,y))\,dxdy$$

with domain equal to the set D of Example 11.3(b). Then, for u in D and v in M,

$$f(h) = \int_{\Omega} ((u_x + hv_x)^2 + (u_y + hv_y)^2 - 2g(u + hv))\,dxdy$$

and

$$f'(h) = 2\int_{\Omega} ((u_x + hv_x)v_x + (u_y + hv_y)v_y - gv)\,dxdy$$

Then

$$\delta J[u;v] = 2\int_{\Omega} (u_x v_x + u_y v_y - gv)\,dxdy$$

(c) Consider the functional

$$J[y] = \int_a^b F(x, y(x), y'(x))\,dx$$

with domain D as in part (a). Then proceeding as in part (a), we find

$$\delta J[u;v] = \int_a^b (\partial_y F(x, y(x), y'(x))\,v(x) + \partial_{y'} F(x, y(x), y'(x))\,v'(x))\,dx$$

LOCAL EXTREME POINT

Let J be a functional whose domain D is a translate of subspace M. We say u in D is a *local maximum* for J if $J(u) \geq J(u + hv)$ for all v in M and all h sufficiently small [if $J(u) \leq J(u + hv)$ u is a *local minimum*]. We say u is a *local extreme point* if it is either a local maximum or a local minimum.

Gradient of a Functional

We shall see that if J is a functional whose domain D is a translate of a dense subspace M, and if u is a local extreme point for J, then $\delta J[u;v] = 0$ for all v in M. We can translate this into a condition bearing only on u.

Example 11.5

(a) Consider the functional J of Example 11.4(a) with domain equal to the set D of Example 11.3(a). Suppose, for y in D, $\delta J[y;v] = 0$ for all v in M. Then integrating by parts,

$$\delta J[y;v] = \int_a^b \frac{y'(x)\,v'(x)}{\sqrt{1 + y(x)^2}}\,dx = \left.\frac{y'(x)\,v(x)}{\sqrt{1 + y(x)^2}}\right|_a^b - \int_a^b v(x)\frac{d}{dx}\frac{y'(x)}{\sqrt{1 + y(x)^2}}\,dx$$

$$= 0 - (f, v)_0 \quad \text{where } f(x) = \frac{d}{dx} \frac{y'(x)}{\sqrt{1 + y'(x)^2}}$$

If $y(x)$ is sufficiently regular that $f(x)$ is in $H^0[a, b]$, then $(f, v)_0 = 0$ for all v in M implies $f = 0$ since $M = \{y \text{ in } C^1[a, b] : y(a) = y(b) = 0\}$ is dense in $H^0[a, b]$. If y is twice continuously differentiable, then f is continuous and, therefore, in $H^0[a, b]$. In this case, we say y is in the *domain of the gradient of J* and we write

$$\nabla J[y](x) = \frac{d}{dx} \frac{y'(x)}{\sqrt{1 + y'(x)^2}} = \text{gradient of } J \text{ at } y$$

Note that $\nabla J[u] = 0$ is a differential equation for the function $y(x)$.

(b) Consider the functional J of Example 11.4(b) with domain equal to the set D of Example 11.3(b). Suppose for u in D, $\delta J[u;v] = 0$ for all v in M. But

$$\delta J[u;v] = 2\int_\Omega (u_x v_x + u_y v_y - gv)\, dxdy$$

and Green's identity implies

$$\int_\Omega (u_x v_x + u_y v_y) = \int_S v \,\mathbf{grad}\, u \bullet N - \int_\Omega v \operatorname{div} \mathbf{grad}\, u.$$

The integral over S vanishes for every v in $M = \{v \text{ in } C^1(\Omega) : v = 0 \text{ on } S\}$ and, thus,

$$\delta J[u, v] = 2(w, v)_0 \quad \text{where} \quad w = -\operatorname{div} \mathbf{grad}\, u - g.$$

Since M is dense in $H^0(\Omega)$, $\delta J[u, v] = 2(w, v)_0 = 0$ for all v in M implies that $w = 0$, provided u is sufficiently smooth that w belongs to $H^0(\Omega)$. If u is twice continuously differentiable on Ω, then w is continuous and, therefore, is square integrable on Ω. Then u is in the domain of the gradient of J and

$$\nabla J[u](x, y) = -2(\operatorname{div} \mathbf{grad}\, u(x, y) + g(x, y)) = \text{gradient of } J \text{ at } u.$$

In this case $\nabla J[u] = 0$ is a partial differential equation for $u = u(x, y)$.

Theorem 11.3

Theorem 11.3. Let D be a translate of dense subspace M. For functional J on domain D, suppose that u in D is a local extreme point for J. Then $\delta J[u;v] = 0$ for all v in M. In addition, if u is in the domain of the gradient of J, then u satisfies the associated *Euler's equation*, $\nabla J[u] = 0$.

INDIRECT METHOD OF CALCULUS OF VARIATIONS

Theorem 11.3 is the basis for the so-called *indirect method* for the fundamental problem of the calculus of variations. Finding the extreme values of a functional on a given domain D is the fundamental problem of the calculus of variations, and Theorem 11.3 allows us to replace this problem with the more familiar problem of finding a solution in D to a differential equation. If the differential equation can be solved, then the solution is at least a stationary point for the functional and is a candidate for an extreme point. Often there is additional information about the functional that guarantees that any stationary point is an extreme point. In such cases, the solution of the differential equation is the solution of the variational problem.

DIRECT METHODS FOR VARIATIONAL PROBLEMS

Frequently, the indirect approach to a variational problem leads to a very difficult nonlinear differential equation. In this case it may be necessary to attack the variational problem directly. One such approach is to approximate the problem by a simpler problem in which we seek the extreme values of the functional over a suitable (finite) n-dimensional subset D_n of the original domain D. This reduces the problem to one of finding the extreme values of a function of n variables, a problem that can be treated by the methods of multivariable calculus. The success of this method rests on finding a finite dimensional subset D_n that is a really "suitable" approximation to D. The finite element method provides a systematic way for selecting a suitable approximating subspace D_n. The rudiments of the finite element method are discussed in chapter 13.

VARIATIONAL METHODS IN MECHANICS

One way of obtaining a mathematical model for a physical system is to state governing principles of material balance or force balance in mathematical terms. Often this leads to a differential equation whose solution describes the behavior of the system. An alternative approach is to postulate a "least action principle" which states that the system behaves in such a way as to optimize a so-called action functional that depends on the state variables characterizing the system. Optimizing the functional leads to the associated Euler equations which are then the mathematical model for the physical system. This approach can only be applied to systems that are "conservative" in some physical sense. The mathematical property associated with physically conservative systems is self-adjointness. A discussion of self-adjointness is not within the scope of this text.

Lagrangian Mechanics

PRINCIPLE OF LEAST ACTION

Consider a system with one degree of freedom represented by the state variable $x = x(t)$. Suppose that the potential energy V of the system is a function of the position (only) $V = V(x)$ and that kinetic energy of the system $T = T(x')$ depends only the velocity $x'(t)$. Then define the *Lagrangian function* of the system by $L(x, x') = T(x') - V(x)$ and define the *action functional* to be

$$A[x] = \int_0^T L(x(t), x'(t))\, dt \tag{11.8}$$

The domain of this functional is taken to be the set $D = \{x \text{ in } C^1[0, T] : x(0) = p, x(T) = q\}$ for given p and q. Then D is a translate of the subspace $M = \{x \text{ in } C^1[0, T] : x(0) = x(T) = 0\}$. We have then the following postulate.

Principle of Least Action. The physical system whose action functional is given by (11.8) will behave in such a way that A is minimized over D.

The principle of least action implies that the Euler's equation $\nabla A[x] = 0$ is the governing differential equation for the physical system. For a functional of the form (11.8) this is going to be a single ordinary differential equation of order two.

SYSTEMS WITH SEVERAL DEPENDENT VARIABLES

More generally, for a system with n degrees of freedom we have $X(t) = [x_1(t), \ldots, x_n(t)]$ and

$$A[X] = \int_0^T L(x_1, \ldots, x_n; x_1', \ldots, x_n')\, dt \tag{11.9}$$

We show in the solved problems how to compute the variation and gradient for a functional of the form (11.9) with n dependent variables. In this case the Euler's equations turn out to be a system of n second order equations.

SYSTEMS WITH SEVERAL INDEPENDENT VARIABLES

We can adapt the action integral approach to systems with one or more state variables each depending on more than one independent variable. For example, for a system with a single state variable $u = u(x, y, t)$, depending on independent variables (x, y, t) we have the kinetic energy $T = T(\partial_t u)$ and the potential energy $V = V(u, \partial_x u, \partial_y u, \ldots)$ which we assume depends on u and its spatial derivatives but not on the time derivatives. Then the action functional has the form

$$A[u] = \int_0^T \int_\Omega L(u, u_t, u_x, u_y, \ldots)\, dxdydt \tag{11.10}$$

and we seek to minimize A over a domain D composed of functions $u(x, y, t)$ in a suitable smoothness class defined by the order of the derivatives appearing in the Lagrangian function and conforming to certain boundary conditions that apply on the boundary of some bounded region Ω in the xy plane. We show in the solved problems how to compute the variation and gradient for a functional of this form. The Euler's equation in this case will be a single partial differential equation for the unknown function $u = u(x, y, t)$.

PRINCIPLE OF MINIMUM POTENTIAL ENERGY

A principle analogous to the principle of least action asserts that a system that is in a state of equilibrium will assume the state of minimum potential energy. Thus, if the potential energy of the system can be expressed as functional depending on certain state variables (which are usually functions of two or more spatial variables, but are independent of time), then the functional can be minimized by the methods we have described. In such cases, the Euler's equation is generally an elliptic partial differential equation.

VARIATIONAL FORMULATION OF DIFFERENTIAL EQUATIONS

The indirect method of the calculus of variations seeks to replace a variational problem with a differential equation with the hope that a solution of the differential equation will produce a solution for the variational problem. In the variational approach to differential equations, we do the reverse; we seek to identify a differential equation problem as the Euler equation associated with some variational problem. Then a solution of the differential equation can be obtained by solving the variational problem. This approach is attractive for two reasons. First, there are abstract existence results available for certain variational problems, whereas results applying directly to the differential equation are more difficult to obtain. Second, the direct methods for approximating the solutions to variational problems can be effectively implemented on a computer and provide an alternative to the usual finite difference techniques for approximating the solution to a differential equation.

A CLASSICAL BOUNDARY VALUE PROBLEM

Let Ω denote a bounded region in $\Re^2$ having smooth boundary S composed of nonoverlapping arcs S_1 and S_2 whose union is S. Define a partial differential operator L on functions $u = u(x, y)$ by

$$\begin{aligned} L[u](x,y) &= -a_{11}(x,y)\,\partial_{xx}u - 2a_{12}(x,y)\,\partial_{xy}u - a_{22}(x,y)\,\partial_{yy}u \\ &\quad + b_1(x,y)\,\partial_x u + b_2(x,y)\,\partial_y u + c(x,y)\,u \end{aligned} \tag{11.11}$$

where $a_{ij}(x, y)$, $b_j(x, y)$, and $c(x, y)$ denote known functions that are defined and smooth on Ω. Note that $L[u]$ can be expressed using vector notation as follows:

$$L[u](x, y) = -\operatorname{div}(A \,\mathbf{grad}\, u) + \boldsymbol{B} \bullet \mathbf{grad}\, u + cu \tag{11.12}$$

where

$$A = \begin{bmatrix} a_{11} & a_{12} \\ a_{12} & a_{22} \end{bmatrix}, \quad \boldsymbol{B} = [b_1 + \partial_x a_{11} + \partial_y a_{12}, b_2 + \partial_x a_{12} + \partial_y a_{22}]^{\mathrm{T}}$$

Now consider the boundary value problem

$$L[u] = f(x, y) \text{ in } \Omega, \quad u = g \text{ on } S_1, \quad N \bullet A \,\mathbf{grad}\, u = h \text{ on } S_2 \tag{11.13}$$

where f, defined on Ω, g defined on S_1, and h defined on S_2 are all known functions and N denotes the outward unit normal to S_2. Since all of the equalities in (11.13) are satisfied in the classical pointwise sense, we refer to (11.13) as a *classical boundary value problem.*

WEAK FORMULATION OF THE BOUNDARY VALUE PROBLEM

Let M denote the set of functions v in $H^1(\Omega)$ such that $v = 0$ on S_1. Then M is a subspace that can be shown to be dense in the Hilbert space $H^0(\Omega)$. The set $D = \{u \text{ in } H^1(\Omega) : u = g \text{ on } S_1\}$ is a translate of this subspace. For arbitrary v in M

$$\int_\Omega vL[u]\, dxdy = \int_\Omega v(-\operatorname{div}(A \,\mathbf{grad}\, u) + \boldsymbol{B} \bullet \mathbf{grad}\, u + cu)$$

$$= \int_{S_2} vN \bullet A \,\mathbf{grad}\, u + \int_\Omega ((\mathbf{grad}\, v) A \,\mathbf{grad}\, u + v\boldsymbol{B} \bullet \mathbf{grad}\, u + cuv)\, dxdy$$

where we have used the divergence theorem together with the identity

$$\operatorname{div}(vA \,\mathbf{grad}\, u) = v \operatorname{div} A \,\mathbf{grad}\, u + (\mathbf{grad}\, v) A \,\mathbf{grad}\, u$$

If u in D solves (11.13), then u also satisfies

$$B[u, v] = F(v) \quad \text{for all } v \text{ in } M \tag{11.14}$$

where

$$B[u, v] = \int_\Omega ((\mathbf{grad}\, v) A \,\mathbf{grad}\, u + v\boldsymbol{B} \bullet \mathbf{grad}\, u + cuv)\, dxdy \tag{11.15}$$

and

$$F(v) = (f, v)_0 + \int_{S_2} vh\, dv \tag{11.16}$$

WEAK SOLUTIONS

A function u in $C^2(\Omega)$ that is continuously differentiable on S and satisfies (11.13) is said to be a *classical solution* of the boundary value problem. If u in D solves (11.14), then u is said to be a *weak solution* of the boundary value problem. Every classical solution is a weak solution, but not conversely.

Theorem 11.4

Theorem 11.4. Suppose u is a classical solution of (11.13). Then u belongs to $D = \{u \text{ in } H(\Omega) : u = g \text{ on } S_1\}$ and satisfies (11.14). A function u in D that satisfies (11.14) is not necessarily in $C^2(\Omega)$ and, thus, is not necessarily a classical solution for (11.13).

Linear Functionals and Bilinear Forms

The functional F in (11.16) is an example of a *linear functional* on the subspace M in $H^0(\Omega)$. That is, F is a functional and, for every u, v in M and all constants C_1, C_2

$$F(C_1 u + C_2 v) = C_1 F(u) + C_2 F(v)$$

The expression $B[u, v]$ in (11.15) is an example of a *bilinear form* on M in $H^0(\Omega)$. That is, for all constants C_1, C_2 and all u, v, w in M,

$$B[C_1 u + C_2 v, w] = C_1 B[u, w] + C_2 B[v, w]$$

$$B[u, C_1 v + C_2 w] = C_1 B[u, v] + C_2 B[u, w]$$

If B satisfies $B[u, v] = B[v, u]$ for all u, v in M, then we say B is *symmetric*. If $B[u, u] \geq 0$ for all u in M with $B[u, u] = 0$ if and only if $u = 0$, then we say B is *positive definite*.

Note that the bilinear form $B[u, v]$ in (11.15) is symmetric if

$$\mathbf{B} = [b_1 + \partial_x a_{11} + \partial_y a_{12}, b_2 + \partial_x a_{12} + \partial_y a_{22}]^{\mathsf{T}} = \mathbf{0}$$

and $B[u, v]$ is positive definite if $\mathbf{B} = \mathbf{0}$ and the matrix $A = A(x, y)$ in (11.12) is positive definite at each point (x, y) of the closed bounded set Ω. In this case, the operator L of (11.11) is said to be *elliptic*.

QUADRATIC FUNCTIONALS

For u in D, and B as defined in (11.15), define the functional

$$Q[u] = B[u, u] - 2(f, u)_0 + \int_{S_2} uh \, du = B[u, u] - 2F[u] \quad (11.16)$$

Then $Q[u]$ is said to be a *quadratic functional*.

Theorem 11.5

Theorem 11.5. Let the quadratic functional Q be defined by (11.16) on the domain D, where D is a translate of a subspace M dense in $H^0(\Omega)$. If the bilinear form B is symmetric, then $\delta Q[u;v] = 2(B[u,v] - F(v))$ for all u in D and v in M. If B is positive definite, then Q has a unique minimum u_0 in D which satisfies (11.14); i.e., u_0 is the unique weak solution of (11.13).

If $B[u, v]$ is not symmetric, then $B[u, v] - F(v)$ is not the variation of any quadratic functional Q, but we can still consider the weak formulation of the boundary value problem (11.14). If B is symmetric but not positive definite, then B is the variation of a quadratic functional Q. But when positive definiteness is lacking, even though (11.14) implies the variation of the functional Q is equal to zero, the existence of a unique minimum u is not guaranteed. Compare these results with those of Problems 1.16 through 1.18.

SOLVED PROBLEMS

Functional Optimization

PROBLEM 11.1

Compute $\delta J[u;v]$ for the functional

$$J[u] = \int_a^b 2\pi u(x)\sqrt{1+u'(x)^2}\,dx \tag{1}$$

on the domain $D = \{u(x) \text{ in } C^1[a,b] : u(a) = p \text{ and } u(b) = q\}$.

SOLUTION 11.1

Note that in this example, D is a translate of the subspace $M = \{u(x)$ in $C^1[a,b] : u(a) = u(b) = 0\}$ in $H^0[a,b]$. Then, for u in D and v in M,

$$f(h) = \int_a^b 2\pi(u(x) + hv(x))\sqrt{1 + (u'(x) + hv'(x))^2}\,dx$$

and

$$'(h) = \int_a^b v(x)\sqrt{1+(u'(x)+hv'(x))^2} + \frac{(u+hv)(u'+hv')v'}{\sqrt{1+(u'(x)+hv'(x))^2}}\,dx$$

Thus,

$$\delta J[u;v] = f'(0) = 2\pi\int_a^b\left(v(x)\sqrt{1+u'(x)^2} + \frac{uu'v'}{\sqrt{1+u'(x)^2}}\right)dx$$

Note that this is of the form

$$\delta J[u;v] = \int_a^b (\partial_u F(u,u')v + \partial_{u'} F(u,u')v')\,dx$$

where

$$F(u,u') = 2\pi u(x)\sqrt{1+u'(x)^2}$$

PROBLEM 11.2

Find the Euler's equation associated with the functional in the previous problem.

SOLUTION 11.2

We integrate by parts to show that

$$\int_a^b \frac{u(x)u'(x)}{\sqrt{1+u'(x)^2}} v'(x)\,dx = \frac{u(x)u'(x)}{\sqrt{1+u'(x)^2}} v(x)\Big|_a^b - \int_a^b v(x)\frac{d}{dx}\left(\frac{u(x)u'(x)}{\sqrt{1+u'(x)^2}}\right)dx$$

The boundary term vanishes for v in M and thus

$$\int_a^b \left(\partial_u F(u,u') - \frac{d}{dx}\partial_{u'} F(u,u')\right) v(x)\,dx = (G,v)_0 \quad \text{for all } v \text{ in } M$$

for

$$G(x) = \partial_u F(u,u') - \frac{d}{dx}(\partial_{u'} F(u,u'))$$

If u in D is twice continuously differentiable, then $\delta J[u,v] = (G,v)_0$ for all v in M. Then u is in the domain of the gradient of J and $G(x) = \nabla J[u](x)$. Now M is dense in $H^0[a,b]$ and, thus, Theorem 11.2 implies that if u is such that $\delta J[u,v] = 0$ for all v in M then $G(x) = \nabla J[u](x) = 0$. Then the Euler's equation for this functional J is

$$\nabla J[u](x) = \sqrt{1+u'(x)^2} - \frac{d}{dx}\left(\frac{u(x)u'(x)}{\sqrt{1+u'(x)^2}}\right) = 0$$

PROBLEM 11.3

Show that the Euler's equation for the functional

$$J[u] = \int_a^b F(u(x),u'(x))\,dx \tag{1}$$

(note that F does not depend explicitly on x) reduces to

$$F(u,u') = u'(x)\partial_{u'} F(u,u') = \text{constant} \tag{2}$$

SOLUTION 11.3

Let

$$Z(u,u') = F(u,u') - u'(x)\partial_{u'} F(u,u')$$

Then

$$\frac{d\, Z((u(x)), u'(x))}{dx} = \partial_u F(u, u')u'(x) + \partial_{u'} F(u, u')u''(x)$$
$$- u''(x)\partial_{u'} F(u, u') - u'(x)\partial_{u'} F(u, u')$$
$$= u'(x)\left(\partial_u F(u, u') - \frac{d}{dx}\partial_{u'} F(u, u')\right)$$

If $u(x)$ solves the Euler's equation for J, then $dZ/dx = 0$ and $Z =$ constant. If Z is constant, then u must solve the Euler's equation for J. Thus, Euler's equation reduces to equation (2) when F does not depend explicitly on x. The quantity Z is called a *first integral* of the Euler's equation.

PROBLEM 11.4

Suppose $y = u(x)$ is a smooth curve joining the points (a, p) and (b, q) in the xy plane. If p and q are positive constants, find the function $u(x)$ such that when this curve is rotated about the x axis, the minimum amount of surface area is swept out.

SOLUTION 11.4

The surface area generated when the curve $y = u(x)$ for $a < x < b$ is rotated about the x axis is equal to $J[u]$, the functional in Problem 11.1. Then the problem of finding the curve that minimizes the surface area is just the problem of minimizing the functional J over the domain D.
If $u = u(x)$ is a smooth function that produces the minimal area, it follows that u satisfies $\nabla J[u](x) = 0$. But according to the result of the previous problem, this reduces to

$$Z[u, u'] = u(x)\sqrt{1 + u'(x)^2} - u'(x)\left(\frac{u(x)u'(x)}{\sqrt{1 + u'(x)^2}}\right) = C_1$$

i.e.,

$$u(x) = C_1\sqrt{1 + u'(x)^2}$$

This differential equation for $u(x)$ separates and can be integrated to find

$$u(x) = C_1 \cosh\left(\frac{x + C_2}{C_1}\right) \tag{1}$$

The two constants of integration C_1, C_2 are then determined by the conditions $u(a) = p$ and $u(b) = q$. With these choices for the constants, $y = u(x)$ for u in (1) is the curve that minimizes the surface area of revolution.

PROBLEM 11.5

Suppose $u_0 = u_0(x)$ minimizes the functional

$$J(u) = \int_a^b F(x, u(x), u'(x))\,dx$$

over the domain $D = \{u \text{ in } C^0[a, b]\}$. Then show that u_0 must satisfy the *natural boundary conditions*

$$\partial_{u'} F(u, u')(a) = \partial_{u'} F(u, u')(b) = 0$$

SOLUTION 11.5

Note that the domain D in this example is itself a subspace and thus for u and v in D,

$$\delta J[u;v] = \int_a^b (\partial_u F(u, u')v + \partial_{u'} F(u, u')v')\,dx \tag{1}$$

If u in D is twice continuously differentiable then

$$\delta J[u;v] = \int_a^b ((\partial_u F(u, u') - \frac{d}{dx}\partial_{u'} F(u, u'))v(x)\,dx + \partial_{u'} F(u, u'))v(x)\Big|_a^b \tag{2}$$

If u_0 minimizes J over D, then $\delta J[u_0, v] = 0$ for every v in D. In particular, for v in $M = \{v \text{ in } D\colon v(a) = v(b) = 0\}$, equation (2) reduces to

$$\delta J[u_0, v] = (\nabla J[u_0], v)_0 = 0 \quad \text{for all } v \text{ in } M \tag{3}$$

Since M is dense in $H^0[a, b]$, it follows that $\nabla J[u_0] = 0$. Then (2) and (3) together imply

$$\delta J[u_0, v] = \partial_{u'} F(u_0, u_0')v(x)|_a^b = 0 \quad \text{for all } v \text{ in } D \tag{4}$$

But (4) must hold for all v in D, including those v which do not vanish at $x = a$ and $x = b$. Then (4) implies

$$\partial_{u'} F(u_0, u_0')(a) = \partial_{u'} F(u_0, u_0')(b) = 0 \tag{5}$$

The conditions (5) are satisfied *automatically* by the minimizing function in D. These conditions are called *natural boundary conditions*.

PROBLEM 11.6

Let p, q, and r denote given functions in $C^1[a, b]$ with $p(x)$ strictly positive on $[a, b]$ and consider the functional

$$J(u) = \int_a^b (p(x)u'(x)^2 + q(x)u(x)^2 + 2r(x)u(x))\,dx \tag{1}$$

Let $D_1 = \{u \text{ in } C^1[a, b]\}$, $D_2 = \{u \text{ in } D_1\colon u(a) = A\}$, and $D_3 = \{u \text{ in } D_1\colon u(a) = A,\ u(b) = B\}$. Find functions u_1, u_2, and u_3 that minimize J over the domains D_1, D_2, and D_3, respectively. Here A and B denote given constants.

SOLUTION 11.6

Let $M_3 = \{v \text{ in } C^1[a,b] : v(a) = v(b) = 0\}$ and $M_2 = \{v \text{ in } C^1[a,b] : v(a) = 0\}$ and note that D_3 is a translate of the dense subspace M_3, D_2 is a translate of the dense subspace M_2, and D_1 is itself a subspace. All of the subspaces are dense in $H^0[a,b]$ and D_1 contains D_2, which contains D_3.

For u and v in D_1

$$\delta J[u,v] = \int_a^b (2pu'v' + 2quv + 2rv)\,dx \tag{2}$$

and if u is twice differentiable, then

$$\delta J[u,v] = 2p(x)u'(x)v(x)\big|_a^b + 2\int_a^b (-(pu')' + qu + r)\,v(x)\,dx \tag{3}$$

In particular, if the twice differentiable function u is in D_3, then (3) implies

$$\delta J[u,v] = (\nabla J[u], v)_0 \quad \text{for all } v \text{ in } M_3 \tag{4}$$

where $\nabla J[u](x) = 2(-(pu')'(x) + qu(x) + r(x))$. If u_3 minimizes J over D_3, then (4) implies $\nabla J[u_3] = 0$. Thus, u_3, the function which minimizes J over D_3, is the solution of

$$-(p(x)u'(x))' + q(x)u(x) + r(x) = 0, \quad u(a) = u(b) = 0 \tag{5}$$

When u is in D_2, equation (3) reduces to

$$\delta J[u,v] = 2p(b)u'(b)v(b) + 2(\nabla J[u], v)_0 \quad \text{for all } v \text{ in } M_2 \tag{6}$$

If u_2 minimizes J over D_2, then

$$2p(b)u_2'(b)v(b) + 2(\nabla J[u_2], v)_0 = 0 \quad \text{for all } v \text{ in } M_2 \tag{7}$$

In particular, for those v in M_2 which are also in M_3, (7) reduces to (4). Then it follows as before that $\nabla J[u_2] = 0$. Substituting this into (7) and choosing a v in M_2 for which $v(b)$ is not zero, we conclude that $u_2'(b) = 0$ (recall $p > 0$ on $[a,b]$). Thus, u_2, the function that minimizes J over D_2, is the solution of

$$-(p(x)u'(x))' + q(x)u(x) + r(x) = 0, \quad u(a) = 0, \ u'(b) = 0 \tag{8}$$

Finally, if u_1 minimizes J over D_1, then the usual argument leads to the result that $\nabla J[u_1] = 0$. Using this result with (3) implies that

$$2p(x)u'(x)v(x)\big|_a^b = 0 \quad \text{for all } v \text{ in } D_1 \tag{9}$$

from which it follows that $u'(a) = u'(b) = 0$. Then u_1, the function that minimizes J over D_1, is the solution of the boundary value problem

$$-(p(x)u'(x))' + q(x)u(x) + r(x) = 0, \quad u'(a) = 0, \quad u'(b) = 0 \tag{10}$$

PROBLEM 11.7

Let $y = u(x)$ be the equation of a smooth curve joining the points (a, p) and (b, q) in the xy plane. Choose the function $u(x)$ so that the area under the curve is minimized but the arclength of the curve is equal to the positive constant L, where $L > b - a$.

SOLUTION 11.7

This problem amounts to minimizing the functional

$$J[u] = \int_a^b u(x)\,dx$$

over the domain $D = \{u \text{ in } C^1[a, b] : u(a) = p, u(b) = q\}$ subject to the constraint

$$K[u] = \int_a^b \sqrt{1 + u'(x)^2}\,dx = L$$

We can show that if u_0 is the minimizing element in D, then as long as $\nabla K[u_0]$ is not zero, there is a real number λ such that $\nabla(J + \lambda K)[u_0] = 0$. In this case, since

$$\delta J[u, v] = \int_a^b v(x)\,dx \quad \text{and} \quad \delta K[u, v] = \int_a^b \frac{u'(x)v'(x)}{\sqrt{1 + u'(x)^2}}\,dx$$

we find

$$\nabla(J + \lambda K)[u] = 1 - \lambda \frac{d}{dx}\left(\frac{u'(x)}{\sqrt{1 + u'(x)^2}}\right)$$

This gradient vanishes if $u(x)$ in D satisfies

$$\frac{u'(x)}{\sqrt{1 + u'(x)^2}} = \frac{x - C_1}{\lambda} \qquad C_1 = \text{constant}$$

This implies

$$u(x) = C_2 - \sqrt{\lambda^2 - (x - C_1)^2}, \qquad C_2 = \text{constant}$$

Choosing C_1, C_2 such that $u(a) = p$ and $u(b) = q$ produces the function u_0 that minimizes J over D subject to the constraint $K[u] = L$.

PROBLEM 11.8

Compute the variation and the gradient of the functional

$$J[X] = \int_a^b L(X(t), X'(t))\, dt \tag{1}$$

on the domain $D = \{X(t) = [x_1(t), \ldots x_n(t)] : \; x_j(t)$ in $C^1[a, b]$ with $x_j(a) = p_j$ and $x_j(b) = q_j$, $j = 1, \ldots, n\}$.

SOLUTION 11.8

Let M denote the subspace consisting of all n-tuples $Z(t)$ whose components are all in $C^1[a, b]$ and $Z(a) = Z(b) = \mathbf{0}$. Then M is a subspace that is dense in the space $(H^0[a, b])^n$ of n-tuples whose components are all in $H^0[a, b]$. Now let $f(h) = J[X + hZ]$ for X in D and Z in M and h a real number. Then

$$f'(0) = \int_a^b \sum_{j=1}^{n} (\partial_{x_j} L(X, X')\, z_j(t) + \partial_{x_j'} L(X, X')\, z_{j'}(t))\, dt \tag{1}$$

$$= \delta J[X, Z] \qquad \text{for } X \text{ in } D \text{ and } Z \text{ in } M$$

Integration by parts leads to

$$\delta J[X, Z] = \sum_{j=1}^{n} (G_j, z_j)_0 \qquad \text{for all } Z \text{ in } M \tag{2}$$

where

$$G_j(t) = \partial_{x_j} L(X, X') - \frac{d}{dt}\partial_{x_j'} L(X, X'), \quad j = 1, \ldots, n$$

Thus,

$$\nabla J[X](t) = [G_1(t), \ldots, G_n(t)]$$

for $X(t)$ in D such that each $x_j(t)$ is twice differentiable. The gradient of the action functional vanishes when the n equations

$$\partial_{x_j} L(X, X') = \frac{d}{dt}(\partial_{x_j'} L(X, X')), \quad j = 1, \ldots, n \tag{3}$$

are satisfied.

PROBLEM 11.9

A pendulum consisting of a ball of mass m suspended by a weightless rod of length L pivots freely about the pivot point P. The pendulum swings back and forth in a fixed plane and the angle between the rod and the vertical is denoted by $\vartheta(t)$. If the pivot point P moves periodically in the vertical direction according to the prescribed function $h(t)$, then use the principle of least action to derive the equations of motion for this system (see Figure 11.1)

SOLUTION 11.9

When the rod makes an angle $\vartheta = \vartheta(t)$ with the vertical, the ball is at position (x, y) where $x = L\sin\vartheta(t)$ and $y = -L\cos\vartheta(t) - h(t)$. Then the kinetic energy of the system is given by

$$T = \frac{m}{2}(x'(t)^2 + y'(t)^2) = \frac{m}{2}(L^2\vartheta'(t)^2 - 2L\vartheta'(t)h'(t)\sin\vartheta(t) + h'(t)^2)$$

The potential energy equals $V = mgy = -mg(h(t) + L\cos\vartheta(t))$; hence, the Lagrangian function

$$L(\vartheta, \vartheta') = \frac{m}{2}(L^2\vartheta'(t)^2 - 2L\vartheta'(t)h'(t)\sin\vartheta(t) + h'(t)^2)$$
$$+ mg(h(t) + L\cos\vartheta(t))$$

The principle of least action asserts that the motion of the pendulum system [described here by the state variable $\vartheta = \vartheta(t)$] must be such as to render the action integral (11.8) stationary; i.e., $\delta A[\vartheta, \varphi] = 0$ for all φ in M. It follows that the motion $\vartheta(t)$ must satisfy the associated Euler's equation $\partial_\vartheta L - (d/dt)(\partial_{\vartheta'} L) = 0$. Since

$$\partial_\vartheta L = -mL\vartheta'(t)h'(t)\cos\vartheta(t) - mgL\sin\vartheta(t)$$

$$\partial_{\vartheta'} L = mL^2\vartheta'(t) - mLh'(t)\sin\vartheta(t)$$

the equations of motion for this system are found to be

$$L\vartheta''(t) = (h''(t) - g)\sin\vartheta(t). \tag{1}$$

All motions of this forced pendulum must conform to the governing equation (1). Particular motions are selected by imposing initial conditions on $\vartheta(t)$ and $\vartheta'(t)$.

PROBLEM 11.10

Two pendulum systems of the type considered in the previous problem hang side by side, joined by an elastic spring of stiffness K as shown in Figure 11.2. The lengths and masses of the two systems are denoted L_1, L_2 and m_1, m_2, respectively, and $\vartheta_1(t), \vartheta_2(t)$ denote the angle each rod makes with the vertical. If the $L_1 > L_2$ and the spring is attached to the longer pendulum at a point that is a distance L_2 from the pivot point, and if the two pivot points are separated by a distance L_2, then use the principle of least action to derive the equations of motion for this two degree of freedom system.

SOLUTION 11.10

The kinetic energy of this system, expressed in terms of the state variables ϑ_1 and ϑ_2, is

$$T = \frac{1}{2}m_1L_1{}^2\vartheta_1{}'(t)^2 + \frac{1}{2}m_2L_2{}^2\vartheta_2{}'(t)^2$$

and the potential energy equals

$$V = -m_1 g L_1 \cos\vartheta_1(t) - m_2 g L_2 \cos\vartheta_2(t) + \frac{1}{2} K S(\vartheta_1, \vartheta_2)^2$$

where $S(\vartheta_1, \vartheta_2)$ denotes the amount by which the spring is stretched when the angular displacements of the two pendulums are ϑ_1 and ϑ_2. The Pythagorean theorem leads to the expression

$$S(\vartheta_1, \vartheta_2) = L_2\left(\sqrt{(\sin\vartheta_1 - \sin\vartheta_2 - 1)^2 + (\cos\vartheta_1 - \cos\vartheta_2)^2} - 1\right)$$

The action functional is stationary if $\nabla A(\vartheta_1, \vartheta_2) = (0, 0)$ or, recalling equations (3) in Problem 11.8,

$$\partial_{\vartheta_1} V = \frac{d}{dt}(\partial_{\vartheta_1'} T) \quad \text{and} \quad \partial_{\vartheta_2} V = \frac{d}{dt}(\partial_{\vartheta_2'} T)$$

Since

$$\partial_{\vartheta_1} S(\vartheta_1, \vartheta_2)^2 = 2L_2 S(\vartheta_1, \vartheta_2) \frac{\sin(\vartheta_1 - \vartheta_2) - \cos\vartheta_1}{\sqrt{(\sin\vartheta_1 - \sin\vartheta_2 - 1)^2 + (\cos\vartheta_1 - \cos\vartheta_2)^2}}$$

$$\partial_{\vartheta_2} S(\vartheta_1, \vartheta_2)^2 = 2L_2 S(\vartheta_1, \vartheta_2) \frac{\cos\vartheta_2 - \sin(\vartheta_1 - \vartheta_2)}{\sqrt{(\sin\vartheta_1 - \sin\vartheta_2 - 1)^2 + (\cos\vartheta_1 - \cos\vartheta_2)^2}}$$

Therefore, the principle of least action implies that the equations of motion for the coupled pendulum system are

$$L_1\vartheta_1''(t) = -g\sin\vartheta_1(t) + \frac{KL_2}{m_1 L_1} S(\vartheta_1, \vartheta_2) \frac{\sin(\vartheta_1 - \vartheta_2) - \cos\vartheta_1}{\sqrt{(\sin\vartheta_1 - \sin\vartheta_2 - 1)^2 + (\cos\vartheta_1 - \cos\vartheta_2)^2}}$$

$$L_2\vartheta_2''(t) = -g\sin\vartheta_2(t) + \frac{K}{m_2} S(\vartheta_1, \vartheta_2) \frac{\cos\vartheta_2 - \sin(\vartheta_1 - \vartheta_2)}{\sqrt{(\sin\vartheta_1 - \sin\vartheta_2 - 1)^2 + (\cos\vartheta_1 - \cos\vartheta_2)^2}}$$

PROBLEM 11.11 THE VIBRATING STRING

A 1-dimensional elastic string of linear density p and length L lies along the x axis when in its equilibrium state. Applying a time dependent transverse load of force density $f(x, t)$ produces a deflection of $u(x, t)$ in the string. Use the principle of least action to find an initial value problem for determining $u(x, t)$.

SOLUTION 11.11

If the string in equilibrium occupies the interval $[0, L]$ on the x axis, then the kinetic energy of the string in motion is given by

$$T = \frac{p}{2}\int_0^L \partial_t u(x,t)^2\,dx \tag{1}$$

If we suppose the string supports only tension (no shear or moments), then the string stores potential energy in an amount that is proportional to the "stretch" of the string. We have

$$\text{stretch} = \int_0^L \sqrt{1+\partial_x u(x,t)^2}\,dx - L$$

For $\partial_x u$ small, $(1+\partial_x u^2)^{1/2} \approx 1 + (1/2)\,\partial_x u^2$, hence we have the approximation

$$\text{stretch} = \frac{1}{2}\int_0^L \partial_x u(x,t)^2 dx$$

Then for k a material dependent constant, the total potential energy is equal to

$$V = \int_0^L \left(\frac{k}{2}\partial_x u(x,t)^2 + f(x,t)\,u(x,t)\right)dx \tag{2}$$

and

$$A[u] = \int_0^T\int_0^L \left(\frac{p}{2}\partial_t u^2 - \frac{k}{2}\partial_x u(x,t)^2 - f(x,t)\,u(x,t)\right)dx\,dt \tag{3}$$

We can choose the domain of the action functional to be the set D of functions that are continuously differentiable with respect to x and t and satisfy the conditions $u(0, t) = u(L, t) = 0$ and $u(x, 0) = P(x)$, $u(x, T) = Q(x)$. Then D describes a string with fixed ends. D is a translate of the subspace M of smooth functions $v(x, t)$ that vanish for $x = 0$ and L and also vanish for $t = 0$ and T. Then it is easy to show that, for u in D and v in M,

$$\delta A[u,v] = \int_0^T\int_0^L (p\partial_t u\partial_t v - k\partial_x u\partial_x v - fv)\,dx\,dt \tag{4}$$

and

$$\nabla A[u](x,t) = p\partial_{tt}u(x,t) - k\partial_{xx}u(x,t) - f(x,t) \tag{5}$$

Then $u(x, t)$ joins a $t = 0$ state $P(x)$ to a $t = T$ state $Q(x)$ if $u(x, t)$ satisfies

$$p\partial_{tt}u(x,t) - k\partial_{xx}u(x,t) - f(x,t) = 0, \quad 0 < x < L, \quad 0 < t < T$$

$$u(0,t) = u(L,t) = 0 \quad \text{for } 0 < t < T$$

A particular path is selected by imposing initial conditions on u and $\partial_t u$.

PROBLEM 11.12 THE VIBRATING BEAM

Repeat the previous problem when the string is replaced by an elastic beam.

SOLUTION 11.12

A beam is assumed not to stretch but it does bend and the bending moment is proportional to the curvature of the center line of the beam. Since this curvature is proportional to $\partial_{xx}u$, this leads to the result that the potential energy stored in a deformed beam is equal to

$$V = \int_0^L \left(\frac{k}{2}\partial_{xx}u(x,t)^2 + f(x,t)\,u(x,t)\right)dx \qquad (k = \text{constant})$$

Then

$$\nabla A[u] = \int_0^T\int_0^L \left(\frac{p}{2}\partial_t u^2 - \frac{k}{2}\partial_{xx}u(x,t)^2 - f(x,t)\,u(x,t)\right)dxdt$$

and we can take the domain D of this functional to be the set of functions $u(x, t)$ that are continuously differentiable with respect to t and are two times continuously differentiable with respect to x and satisfy the conditions $u(0, t) = u(L, t) = 0$, $\partial_x u(0,t) = \partial_x u(L,t) = 0$ and $u(x, 0) = P(x)$, $u(x, T) = Q(x)$. The boundary conditions at $x = 0$ and L here represent a beam that is clamped at both ends. Let M denote the subspace of similarly smooth functions satisfying the same boundary conditions and vanishing at $t = 0$ and T. Then D is a translate of the subspace M, and M is dense in the space of functions that are square integrable with respect to x and t. Then we can show that for each u in D and v in M

$$\delta A[u, v] = \int_0^T\int_0^L (p\partial_t u\partial_t v - k\partial_{xx}u\partial_{xx}v - fv)\,dxdt$$

and

$$\nabla A[u](x,t) = p\partial_{tt}u(x,t) - k\partial_{xxxx}u(x,t) - f(x,t)$$

where the domain of the gradient contains functions u in D that are C^2 in t and are C^4 in x. Any function u in D that minimizes the action functional must solve the following problem:

$$p\partial_{tt}u(x,t) - k\partial_{xxxx}u(x,t) - f(x,t) \qquad 0 < x < L,\ 0 < t < T$$

$$u(0,t) = u(L,t) = 0 \quad \text{and} \quad \partial_x u(0,t) = \partial_x u(L,t) = 0$$

To select a particular function, we must impose initial conditions on u and $\partial_t u$.

Principle of Minimum Potential Energy

PROBLEM 11.13 BENDING OF AN ELASTIC BEAM

Consider an elastic beam of length L occupying the interval $[0, L]$ of the x axis in its rest state. If the beam is subject to a distributed load $p(x)$ (in units of force per unit length), then the beam assumes a new equilibrium configuration. Derive the boundary value problem for this loaded configuration $u = u(x)$ under various conditions at the ends of the beam.

SOLUTION 11.13

The potential energy functional for the beam is

$$E[u] = \int_0^L \left(\frac{k}{2}u''(x)^2 - p(x)u(x)\right)dx \tag{1}$$

where k is a constant depending on the beam material and on the cross-sectional shape of the beam. We now choose the domain D for E in different ways we can model the following conditions at the ends of the beam:

1. End $x = 0$ clamped, end at $x = L$ free: $D_1 = \{u \text{ smooth: } u(0) = u'(0) = 0\}$, D_1 is a dense subspace of $H^0[0, L]$.

2. Both ends simply supported: $D_2 = \{u \text{ smooth: } u(0) = u(L) = 0\}$ D_2 is a dense subspace of $H^0[0, L]$.

For u and v in $C^2[0, L]$ but not otherwise specified, we have no difficulty in showing that

$$\delta E[u, v] = \int_0^L (ku''(x)v''(x) - p(x)u(x)v(x))\,dx \tag{2}$$

Then we integrate by parts twice to move all of the derivatives off of v onto u

$$\delta E[u, v] = ku''(x)v'(x)\Big|_0^L - ku'''(x)v(x)\Big|_0^L + \int_0^L (ku^{(4)} - p(x)u(x))v(x)\,dx \tag{3}$$

If u and v belong to the subspace D_1; then this reduces to

$$\delta E[u, v] = ku''(L)v'(L) - ku'''(L)v(L) + (G, v)_0 \quad \text{for all } v \text{ in } D_1 \tag{4}$$

for

$$G(x) = ku^{(4)}(x) - p(x)u(x), \quad 0 < x < L \tag{5}$$

If u_0 minimizes $E[u]$ over D_1, then (4) holds for all v in D_1, including those v such that $v(L) = v'(L) = 0$. Such functions v form a subspace M_1 of D_1, and M_1 can also be shown to be dense in $H^0[0, L]$. Then $\delta E[u_0, v] = (G, v)_0 = 0$ for all v in M_1 implies that $G = 0$. But if $G = 0$, then (4) implies that

$$ku_0''(L)v'(L) - ku_0'''(L)v(L) = 0 \quad \text{for all } v \text{ in } D_1$$

This, in turn, implies $u_o''(L) = u_0'''(L) = 0$. Then $\nabla E[u](x) = G(x)$ and the domain of the gradient is the set $DG_1 = \{u$ in D_1: $u''(L) = u'''(L) = 0\}$. The boundary value problem to find the equilibrium deflection configuration $u(x)$ for one end clamped and the other end free is then to find u in DG_1 such that $G(x) = 0$. We can show that the shearing force and moment at position x in the beam are proportional to $u'''(x)$ and $u''(x)$, respectively. Then these natural boundary conditions just express the fact that the $x = L$ end of the beam is free of shear and moment; i.e., it is free.

If u and v belong to D_2 then (3) reduces to

$$\delta E[u, v] = ku''(l)v'(L) - ku''(0)v(0) + (G, v)_0 \quad \text{for all } v \text{ in } D_2$$

and by the same style argument we conclude that if u_1 minimizes $E[u]$ over D_2 then $u_1''(L) = u_1''(0) = 0$. Then $\nabla E[u](x) = G(x)$ and the domain of the gradient for E on D_2 is $DG_2 = \{u \text{ in } D_2\text{: } u''(L) = u''(0) = 0\}$. Thus, there is no moment at an end that is simply supported.

PROBLEM 11.14

Consider the functional

$$J[u] = \int_\Omega F(x_1, ..., x_n; u, u_1, ..., u_n)\, d\Omega \tag{1}$$

where $u = u(x_1, \ldots, x_n)$ is in $C^1(\Omega)$ for Ω a bounded region in R^n and u_j denotes the derivative of u with respect to x_j. Let the boundary S of Ω be composed of complementary parts S_1 and S_2, and let the domain of the functional J be the set $D = \{u \text{ in } C^1(\Omega) : u = g \text{ on } S_1\}$ for some given function g defined on S_1. Then compute the variation and the gradient for J.

SOLUTION 11.14

For $M = \{v \text{ in } C^1(\Omega) : v = 0 \text{ on } S_1\}$, let $f(h) = J[u + hv]$ for u in D and v in M. Then

$$f(h) = \int_\Omega F(X; u + hv, u_1 + hv_1, ..., u_n + hv_n)\, d\Omega$$

and

$$f(0) = \int_\Omega \left(\partial_u F v + \sum_{j=1}^{n} \partial_{u_j} F v_j\right) d\Omega.$$

Hence,

$$\delta J[u, v] = \int_\Omega \left(\partial_u F v + \sum_{j=1}^{n} \partial_{u_j} F v_j\right) d\Omega \quad \text{for } u \text{ in } D \text{ and } v \text{ in } M \tag{2}$$

The identity div $(vH) = H \bullet$ **grad** $v + v$ div H for any smooth vector field H implies that

$$\sum_{j=1}^{n} \partial_{u_j} F v_j = \text{div } (vH) - v \sum_{j=1}^{n} \frac{\partial}{\partial x_j}(\partial_{u_j} F) \tag{3}$$

for $H = [\partial_{u_1} F, ..., \partial_{u_n} F]^\mathsf{T}$. Then the divergence theorem implies

$$\int_{\Omega} \sum_{j=1}^{n} \partial_{u_j} F \, v_j d\Omega = \int_{S} v \, \boldsymbol{H} \bullet \boldsymbol{N} ds - \int_{\Omega} v \operatorname{div} \boldsymbol{H} \, d\Omega$$

where N denotes the outward unit normal to S. Each v in M vanishes on S_1 and, thus,

$$\delta J[u, v] = \int_{\Omega} (\partial_u F - \operatorname{div} \boldsymbol{H}) v \, d\Omega + \int_{S_2} v \, \boldsymbol{H} \bullet \boldsymbol{N} \, ds \tag{4}$$

It follows from (4) that if u in D satisfies the natural boundary condition

$\boldsymbol{H}[u] = [\partial_{u_1} F, \ldots, \partial_{u_n} F]^{\mathrm{T}} = \boldsymbol{0}$ on S_2, then $\delta J[u, v] = (G, v)_0$ for all v in M. We conclude that

$$\nabla J[u](X) = \partial_u F - \operatorname{div} \boldsymbol{H} \tag{5}$$

and

domain of the gradient $= \{u \text{ in } D\colon \boldsymbol{H}[u] = \boldsymbol{0} \text{ on } S_2\}$

PROBLEM 11.15

Consider an elastic membrane stretched over a rigid frame lying in the plane and suppose the frame forms a simple nonclosed plane curve which we denote by S_1. Suppose also that the unsupported edge of the membrane forms a curve S_2 such that S_1 together with S_2 form a simple closed curve containing a region Ω in its interior. Then the membrane covers Ω and in the absence of any applied load, the membrane will lie flat on the plane. Suppose that the membrane is now subjected to a transverse load having force density equal to $F = F(x, y)$. If we let $u = u(x, y)$ denote the out-of-plane deflection of the membrane, then use the result of the previous problem in conjunction with the principle of minimum potential energy to derive a boundary value problem for determining the equilibrium deflection $u(x, y)$.

SOLUTION 11.15

We can show that as long as the membrane is capable of supporting a constant tension but no shears or moments, then the potential energy stored in the stretched membrane with equilibrium deflection mode shape $u = u(x, y)$ is

$$E[u] = \int_{\Omega} (T \, \textbf{grad} \, u \bullet \textbf{grad} \, u - uF) \, d\Omega$$

$$= \int_{\Omega} (T(u_x^{\,2} + u_y^{\,2}) - uF) \, d\Omega$$

where $F = F(x,y)$ denotes the applied load and T is a scalar related to the constant tension in the membrane. The domain of the functional E is the set $D = \{u \text{ in } C^1(\Omega) : u = 0 \text{ on } S_1\}$ which is a subspace of $C^1(\Omega)$ and is dense in $H^0(\Omega)$. By the results of the previous problem

$$\delta E[u,v] = \int_\Omega (2T(u_x v_x + u_y v_y) - vF)\, d\Omega \quad \text{for } u,v \text{ in } D$$

$$= \int_\Omega (-2T \text{ div } \mathbf{grad}\ u - F)\, v\, d\Omega + \int_S v\, \mathbf{grad}\ u \bullet N ds$$

$$= (G,v)_0 + \int_{S_2} v\, \mathbf{grad}\ u \bullet N\, ds$$

Then $E[u]$ is minimized by any function u in D such that $\mathbf{grad}\ u \bullet N = \mathbf{0}$ on S_2 and $G(x, y) = -2T$ div $\mathbf{grad}\ u - F = 0$ in Ω. That is, $u = u(x, y)$ solves the boundary value problem

$$\partial_{xx} u(x,y) + \partial_{yy} u(x,y) = \frac{-1}{2T} F(x,y) \quad \text{in } \Omega$$

$$u = 0 \text{ on } S_1 \quad \text{and} \quad \mathbf{grad}\ u \bullet N = \mathbf{0} \text{ on } S_2$$

The weak formulation of this boundary value problem is:

Find u in D such that $\delta E[u, v] = 0$ for all v in D.

PROBLEM 11.16

Let Ω denote a bounded region in the plane with smooth boundary S consisting of complementary arcs S_1 and S_2. For F and c in $C^0(\Omega)$ and p and g defined and continuous on S_2, consider the functional

$$Q[u] = \int_\Omega (\partial_x u^2 + \partial_y u^2 + cu^2 - 2Fu)\, dxdy + \int_{S_2} (pu^2 - g)\, ds \quad (1)$$

on the domain $D = \{u \text{ in } H^1(\Omega):\ u = 0 \text{ on } S_1\}$. Then compute $\delta Q[u, v]$ and $\nabla Q[u]$.

SOLUTION 11.16

For u and v in D and real number h, it is easy to see that for $f(h) = Q[u + hv]$,

$$f(h) = Q[u] + 2h \int_\Omega (\partial_x u\ \partial_x v + \partial_y u\ \partial_y v + cuv - Fv)$$

$$+ 2h \int_{S_2} (pu - g)\, v\, ds + h^2 \int_\Omega (\partial_x v^2 + \partial_y v^2 + qv^2)\, dxdy + h^2 \int_{S_2} pv^2 dx$$

hence,

$$f'(0) = \delta Q[u,v] = 2\int_\Omega (\partial_x u\ \partial_x v + \partial_y u\ \partial_y v + cuv - Fv) + 2\int_{S_2} (pu - g)\, v\, ds$$

Note that when $p = 0$, the bilinear form

$$B[u,v] = \int_{\Omega} (\partial_x u\, \partial_x v + \partial_y u\, \partial_y v + cuv)\, dxdy + \int_{S_2} puv\, ds \qquad (3)$$

corresponds to (11.15) in the case $\boldsymbol{B} = \mathbf{0}$ and A equals the identity matrix. Then since the identity is a symmetric positive definite matrix, it follows that the bilinear form B is symmetric positive definite if $c > 0$ on Ω and $p > 0$ on S_2.

We have now

$$\delta Q[u,v] = 2\int_{\Omega} \mathbf{grad}\ u \bullet \mathbf{grad}\ v + cuv - Fv + 2\int_{S_2} (pu-g)\, v\, ds$$

$$= 2\int_{S} vN \bullet \mathbf{grad}\ u\ ds - 2\int_{S_2} (v \operatorname{div} \mathbf{grad}\ u - cu + F) + 2\int_{S_2} (pu-g)\, v\, ds$$

Then

$$\delta Q[u,v] = 2\int_{S_2} (N \bullet \mathbf{grad}\ u + pu - g)\, v\, ds - 2\int_{\Omega} v(\operatorname{div} \mathbf{grad}\ u - cu + F) \qquad (4)$$

for all v in D. If u_0 in D minimizes Q over D, then $\delta Q[u_0, v] = 0$ for all v in D, including those that vanish on S_2 as well as on S_1. The set of all such v forms a subspace M of the subspace D and M can be shown to be dense in $H^0(\Omega)$. Then for v in M, (4) reduces to

$$\delta Q[u,v] = \int_{\Omega} v(\operatorname{div} \mathbf{grad}\ u_0 - cu_0 + F) = (\nabla Q[u_0], v)_0 = 0 \text{ for all } v \text{ in } M$$

Since M is dense in $H^0(\Omega)$, this implies

$$\nabla Q[u_0](x,y) = \operatorname{div} \mathbf{grad}\ u_0 - cu_0 + F = 0 \quad \text{in } \Omega \qquad (5)$$

But (4) must hold for all v in D not just those in M and thus (5) and (4) together imply

$$\int_{S_2} (N \bullet \mathbf{grad}\ u_0 + pu_0 - g)\, v\, ds = 0 \quad \text{for all } v \text{ in } D$$

This equation leads to the result $N \bullet \mathbf{grad}\ u_0 + pu_0 - g = 0$ on S_2. Then $\nabla Q[u]$ is given by (5) and the domain of the gradient is the set of functions u in D which satisfy the additional condition $N \bullet \mathbf{grad}\ u + pu - g = 0$ on S_2. Note that the problem of minimizing Q over D is equivalent to solving the problem:

$$\text{Find } u \text{ in } D \text{ such that } B[u,v] = (F,v)_0 - \int_{S_2} gv \text{ for all } v \text{ in } D \qquad (6)$$

If $c > 0$ on Ω and $p > 0$ on S_2, then Q is positive definite and both problems have the same unique solution. The problem (6) is the weak form of the following boundary value problem:

$$\text{div}\,\mathbf{grad}\,u - cu + F = 0 \quad \text{in } \Omega$$
$$u = 0 \text{ on } S_1 \quad \text{and} \quad N \bullet \mathbf{grad}\; u + pu - g \;=\; 0 \text{ on } S_2 \tag{7}$$

If the solution of (6) is in $C^2(\Omega)$, then that solution also solves (7). However, if the solution to (6) is in $H^1(\Omega)$ but not in $C^2(\Omega)$, then (7) has no smooth solution, only the weak solution (6).

PROBLEM 11.17

Let Ω be a bounded region in the plane having smooth boundary S composed of complementary arcs S_1 and S_2. For known functions $a > 0$, b_1, b_2, c, and F defined and smooth on Ω, and for f, g, and h defined and smooth on S, give the weak formulation of the boundary value problem

$$-\nabla(a(x,y)\nabla u(x,y)) + b_1(x,y)\,\partial_x u + b_2(x,y)\,\partial_y u + c(x,y)\,u \;=\; F(x,y) \;\text{ in } \Omega \tag{1}$$

$$u = f \quad \text{on } S_1 \tag{2}$$

and

$$aN \bullet \mathbf{grad}\,u + hu = g \quad \text{on } S_2 \tag{3}$$

where N denotes the unit outward normal vector to S. Is the weak form of this problem equivalent to a variational problem for a quadratic functional?

SOLUTION 11.17

Note that when the matrix A equals $a(x, y)$ times the identity, the operator L in (11.12) reduces to

$$L[u]\,(x,y) \;=\; -\nabla(a(x,y)\nabla u(x,y)) + b_1(x,y)\,\partial_x u + b_2(x,y)\,\partial_y u + c(x,y)\,u$$

which is the operator in (1). The boundary condition (3) is going to turn out to be a *natural boundary condition*, automatically satisfied by the solution of the weak problem. The boundary condition (2) is an example of what is often called a *stable boundary condition*. We force the solution to satisfy this condition by building it into the solution space as follows. Let $M = \{v \text{ in } H^1(\Omega) :\; v = 0 \text{ on } S_1\}$. Then M is a subspace dense in $H^0(\Omega)$ and the set $D = \{u \text{ in } H^1(\Omega) :\; u = f \text{ on } S_1\}$ is a translate of this subspace. Now multiply both sides of (1) by an arbitrary v in M and use the divergence theorem together with the identity

$$\text{div}\,(v\, a\,\mathbf{grad}\,u) = v\,\text{div}\,(a\,\mathbf{grad}\,u) + (\mathbf{grad}\,v)(a\,\mathbf{grad}\,u)$$

to write

$$\int_\Omega vL[u]\,dxdy \;=\; \int_\Omega v(-\text{div}\;(a\,\mathbf{grad}\;u) + \boldsymbol{B} \bullet \mathbf{grad}\;u + cu)\,dxdy \tag{4}$$

$$= \int_S vN \bullet (-a \text{ grad } u) + \int_\Omega (a \text{ grad } v \bullet \text{ grad } u + vB \bullet \text{grad } u + cuv)\, dxdy$$

Condition (3) implies that for all v in M

$$\int_S vN \bullet (a \text{ grad } u) = \int_{S_2} vN \bullet (a \text{ grad } u) = \int_{S_2} v(g - hu)$$

Then it follows that

$$\int_\Omega vL[u]\, dxdy = \int_\Omega (a \text{ grad } v \cdot \text{ grad } u + vB \cdot \text{grad } u + cuv)\, dxdy$$

$$- \int_{S_2} v(g - hu) = \int_\Omega Fv$$

i.e.,

$$B[u, v] = \int_\Omega Fv + \int_{S_2} vg \quad \text{for all } v \text{ in } M$$

where

$$B[u, v] = \int_\Omega (a \text{ grad } v \bullet \text{ grad } u + vB \bullet \text{grad } u + cuv)\, dxdy + \int_{S_2} huv$$

$$B = [b_1(x, y) + \partial_x a(x, y), b_2(x, y) + \partial_y a(x, y)]^{\mathrm{T}}$$

Then the boundary value problem (1), (2), (3) has the following weak formulation:

$$\text{Find } u \text{ in } D \text{ such that } B[u, v] = \int_\Omega Fv + \int_{S_2} vg \quad \text{for all } v \text{ in } M \tag{5}$$

Note that $B[u, v]$ does not, in general, equal $B[v, u]$. It is not difficult to show that for any quadratic functional $Q[u]$, the variation $\delta Q[u, v]$ is symmetric; i.e., $\delta Q[u, v] = \delta Q[v, u]$. Thus, $B[u, v]$ is not the variation of any quadratic functional and the weak problem (5) is not equivalent to the problem of minimizing some Q over D.

Suppose (5) has a solution u_0 in D. If u_0 is sufficiently smooth that $L[u_0]$ belongs to $H^0(\Omega)$, then we can write

$$(L[u_0], v)_0 = (F, v)_0 \quad \text{for all } v \text{ in } M$$

But (4) implies

$$(L[u_0], v)_0 = B_0[u_0, v] - \int_{S_2} vaN \bullet \nabla u_0 \quad \text{for all } v \text{ in } M$$

and (5) implies

$$(F, v)_0 = B_0 [u_0, v] + \int_{S_2} (hu_0 - g)\, v \qquad \text{for all } v \text{ in } M$$

where

$$B_0 [u_0, v] = \int_{\Omega} (a \,\textbf{grad}\, v \bullet \textbf{grad}\, u + v\boldsymbol{B} \bullet \textbf{grad}\, u + cuv)\, dxdy$$

Then it follows that

$$\int_{S_2} v\,(aN \bullet \nabla u_0 + hu_0 - g) = 0 \qquad \text{for all } v \text{ in } M$$

and, thus, in a generalized sense, u_0 must satisfy the natural boundary condition (3).

Let J[u] be a functional defined over domain D which is a translate of subspace M. If u_0 in D is an extreme point for J, then $\delta J [u_0, v] = 0$ for all v in M. If u_0 is sufficiently smooth, then u_0 solves the associated Euler's equation, $\nabla J [u_0] = 0$. We have:

$$\text{If } J[y] = \int_a^b F(x, y(x), y'(x))\, dx \qquad \text{for y in D}$$

then for all v in M

$$\delta Jy;v = \int_a^b (\ \partial_y F(x, y(x), y'(x))\, v(x) + \ \partial_{y'} F(x, yx, y'(x)\, v'(x))\,)\, dx$$

and

$$\nabla J[u](x) = \partial_y F(x, y(x), y'(x)) - \frac{d}{dx} \partial_{y'} F(x, y(x), y'(x))$$

Similar results are derived in the solved problems for functionals J that depend on several dependent variables or involve more than a single independent variable.

For many physical systems it is possible to postulate minimum principles that can be used to replace force balances or conservation laws in deriving the governing mathematical model for the physical system. For example, conservative systems conform to the principle of least action, conservative systems that are in an equilibrium state satisfy the principle of minimum energy, and light travels in space so as to minimize the travel time (Fermat's principle of least time). Each of these principles can be stated as a problem to find the extreme values of a functional over a suitable domain and then the governing equations for the physical system are the associated Euler's equations.

Certain boundary value problems in differential equations are equivalent to variational problems; i.e., u is a solution of the boundary value problem if and only if u is an extreme value for an associated functional. If u is an extreme point for J, then u solves the problem:

$$\text{Find u in D such that } \delta J[u, v] = 0 \text{ for all v in M}$$

This problem is referred to as the weak form of the associated boundary value problem. If the associated functional J is a positive definite quadratic functional, then the variational problem and the weak boundary value problem each have the same unique solution. Certain other boundary value problems are not equivalent to any variational problem, but have a weak form nonetheless. Whether associated with a variational problem or not, approximate solutions for the weak problems may be obtained by so-called direct methods of calculus of variations. These will be considered in chapter 13.

12

Finite Difference Methods for Differential Equations

In previous chapters we have discussed ways of constructing solutions to various ordinary and partial differential equations. The problems where it is possible to construct a solution in terms of elementary functions are a very small, special subclass of all problems that occur in applications. For problems not in this small class, we may wish to resort to numerical solutions. One of the two most common methods for obtaining numerical solutions is based on replacing the derivatives in the differential equation by suitable difference quotients. In this chapter, we provide an introductory survey of how this may be done.

ORDINARY DIFFERENTIAL EQUATIONS

PARTITIONS

The set of points $\{x_0, x_1, \ldots, x_n\}$ is a partition or mesh for the closed bounded interval (a,b) if $a = x_0 < x_1 < \cdots < x_n = b$. The largest difference $h_j = x_j - x_{j-1}$ for $j = 1, \ldots, n$ is referred to as the *mesh size*. Here we will always use a uniform mesh with

$$x_j = a + jh, \quad j = 0, 1, \ldots, n \quad \text{with} \quad h = (b-a)/n. \tag{12.1}$$

APPROXIMATE SOLUTION

Let $y = y(x)$ denote a solution for the initial value problem

$$y'(x) = F(x, y(x)), \qquad y(a) = y_0 \tag{12.2}$$

We say that $y(x)$ as an *exact solution* and the set of numbers $\{y_0, y_1, \ldots, y_n\}$ is a convergent *approximate solution* for the initial value problem if, for each j, $|y(x_j) - y_j|$ tends to zero as h tends to zero. For each j, we define the *local truncation error* E_j to be the difference $|y(x_j) - y_j|$, assuming that $y_{j-1} = y(x_{j-1})$. Note that the initial value y_0 of the solution is given.

Single Step Methods

Taylor's theorem in 1 dimension states that, for $y = y(x)$ smooth,

$$y(x+h) = y'(x) + y(x)h + \frac{1}{2}y''(x)h^2 + \frac{1}{6}y^{(3)}(x)h^3 + \cdots$$

Then $y(x_{j+1}) \approx y(x_j) + hy'(x_j)$ for each h and since (12.2) implies $y'(x_j) = F(x_j, y_j)$ we have the following examples of so-called *single step algorithms* for generating approximate solutions:

Euler's Method

$$y_{j+1} = y_j + hF(x_j, y_j) \quad \text{for } j = 0, 1, \ldots, n-1 \tag{12.3}$$

Runge-Kutta (Order 2)

$$\begin{aligned} y_{j+1} &= y_j + (K_1 + K_2)/2 \quad \text{for } j = 0, 1, \ldots, n-1 \\ K_1 &= hF(x_j, y_j) \quad \text{and} \quad K_2 = hF(x_j + h, y_j + K_1) \end{aligned} \tag{12.4}$$

Runge-Kutta (Order 4)

$$\begin{aligned} y_{j+1} &= y_j + (K_1 + 2K_2 + 2K_3 + K_4)/6 \quad \text{for } j = 0, 1, \ldots, n-1 \\ K_1 &= hF(x_j, y_j), \quad K_2 = hF(x_j + h/2, y_j + K_1/2) \\ K_3 &= hF(x_j + h/2, y_j + K_2/2), \quad K_4 = hF(x_j + h, y_j + K_3) \end{aligned} \tag{12.5}$$

Theorem 12.1

Theorem 12.1. Each of the approximate solutions generated by (12.3), (12.4), and (12.5) is convergent. There exist positive constants C_1, C_2, and C_3 such that for each j

(a) Euler's method: $E_j \le C_1 h^2$,

(b) Runge-Kutta 2: $E_j \le C_2 h^3$,

(c) Runge-Kutta 4: $E_j \le C_3 h^5$.

Each of these algorithms can be modified to apply to first order systems.

PARTIAL DIFFERENTIAL EQUATIONS

DIFFERENCE QUOTIENTS FOR DERIVATIVES

Theorem 3.3, Taylor's theorem in two variables, asserts that for $f = f(x, y)$ smooth

$$\begin{aligned} f(x+h, y+k) &= f(x, y) + \partial_x f(x, y)\, h + \partial_y f(x, y)\, k \\ &+ \frac{1}{2}\left(h^2 \partial_{xx} f(x, y) + 2hk\partial_{xy} f(x, y) + k^2 \partial_{yy} f(x, y)\right) + \cdots \end{aligned} \tag{12.6}$$

This result can be used to derive difference quotient approximations to the partial derivatives of f.

Theorem 12.2

Theorem 12.2. For $f = f(x, y)$ sufficiently smooth, there exist constants C_p that depend on the higher order derivatives of f such that

(a) $\left|\partial_y f(x, y) - \dfrac{f(x, y+k) - f(x, y)}{k}\right| \le C_1 k$ (forward difference for $\partial_y f$)

(b) $\left|\partial_x f(x, y) - \dfrac{f(x+h, k) - f(x-h, k)}{2h}\right| \le C_2 h^2$ (centered difference for $\partial_x f$)

(c) $\left|\partial_{xx} f(x, y) - \dfrac{f(x+h, y) - 2f(x, y) + f(x-h, y)}{h^2}\right| \le C_3 h^2$

MESH NOTATION

Let $\{x_0, x_1, \ldots, x_n\}$ and $\{y_0, y_1, \ldots y_m\}$ denote partitions of closed bounded intervals (a, b) and (c, d), respectively. Then the set of points (x_i, y_j) is said to form a *grid* or *mesh* in the xy plane and the individual points of the mesh are called *grid points* or *nodes*. If the x and y partitions are both uniform, then we refer to the mesh as uniform also. If $h = k$, then we refer to the mesh as *square*, otherwise it is *rectangular*. We use the notation $f_{ij} = f(x_i, y_j)$ to denote the values of the function $f(x, y)$ at the nodes.

Parabolic Equations

Let $u = u(x, t)$ denote a solution for the initial boundary value problem

$$\begin{aligned} L_1[u(x, t)] &= \partial_t u(x, t) - \partial_x (D\partial_x u(x, t)) \\ &= S(x, t) \qquad 0 < x < 1,\ 0 < t < T \end{aligned}$$

$$u(x,0) = f(x), \quad 0 < x < 1 \qquad (12.7)$$
$$u(0,t) = p(t), \quad D\partial_x u(1,t) = q(t), \quad 0 < t < T$$

where $D = D(x,t) > 0$ and S, f, p, and q are given functions. Then $L_1[u]$ is an example of a *parabolic* partial differential operator. If D is constant, then L_1 is the heat equation operator discussed in chapter 9.

CONVERGENT APPROXIMATE SOLUTIONS

Let (x_i, t_j) denote a rectangular mesh on $(0,1) \times (0,T)$ with $x_i = ih$ and $t_j = jk$ for $h = 1/n$ and $k = T/m$, and i, j running from zero to n and m, respectively. The set of numbers U_{ij} is a *convergent approximate solution* for (12.7) if for each i and j, $|U_{ij} - u_{ij}|$ tends to zero as h and k tend to zero. We have the following algorithms for generating approximate solutions for the differential equation in (12.7). For convenience, we let $r = k/h^2$.

Forward Difference (Explicit) Method

$$U_{ij+1} = rD_{i+1/2,j}U_{i+1,j} + (1 - r(D_{i+1/2,j} + D_{i-1/2,j}))U_{ij}$$
$$+ rD_{i-1/2,j}U_{i-1,j} + S_{ij}k$$

Backward Difference (Implicit) Method

$$-rD_{i+1/2,j+1}U_{i+1,j+1} + (1 + r(D_{i+1/2,j+1} + D_{i-1/2,j+1}))U_{ij+1}$$
$$-rD_{i-1/2,j+1}U_{i-1,j+1}$$
$$= U_{ij} + kS_{ij}$$

In either case, we have the initial condition: $U_{i0} = f_i$ for $i = 0, 1, \ldots, n$ and the boundary conditions:

$U_{0,j} = p_j$ and $U_{n+1,j} = U_{n-1,j} + 2hq_j/D_{nj}$ for $j = 1, 2, \ldots, m$

Here we have used the notation

$$D_{i+1/2,j} = (D_{i+1,j} + D_{i,j})/2 \quad \text{(arithmetic mean)}$$

or, in some applications, we use

$$D_{i+1/2,j} = (2D_{i+1,j}D_{i,j})/(D_{i+1,j} + D_{i,j}) \quad \text{(geometric mean)}$$

Note that the forward difference method is based on the difference equation

$$\Delta_t U_{ij} - \Delta_x(D_{ij}\Delta_x U_{ij}) = S_{ij} \qquad (12.8)$$

for approximating the equation (12.7), whereas the backward difference method uses

$$\Delta_t U_{ij} - \Delta_x(D_{i,j+1}\Delta_x U_{i,j+1}) = S_{ij} \qquad (12.9)$$

Thus, the forward difference method leads to an explicit formula for the unknown $U_{i,j+1}$ in terms of known solution values U_{ij}, whereas the backward difference method gives the unknown values of U_{ij+1} only implicitly since we have to solve a linear system of algebraic equations for the unknowns $U_{1j+1}, \ldots, U_{nj+1}$.

CONSISTENT APPROXIMATION SCHEMES

Let $D[u_{ij}] = S_{ij}$ denote a difference equation approximation to the differential equation $L_1[u](x, t)$ in (12.7). Then we say the difference equation is *consistent* with the differential equation if $|D[U_{ij}] - L_1[u]_{ij}|$ tends to zero as h and k tend to zero independently. Replacing derivatives in a differential equation with difference quotients does not always lead to a difference equation consistent with the differential equation. If the difference equation is not consistent with the differential equation yet the approximate solution is convergent, then it is convergent to the solution of some other differential equation.

STABLE APPROXIMATIONS

Letting h and k tend to zero is equivalent to letting n and m tend to infinity. Thus, as k tends to zero more time steps are required to "march forward" from $t = 0$ to $t = T$ using one of the difference schemes indicated above. In some cases the numerical solution values U_{ij} grow steadily or begin to oscillate with growing amplitude as j increases. This is referred to as *instability*. The approximation is said to be *stable* if the solution U_{ij} corresponding to the case in which p, q and S are all zero but f is not identically zero satisfies $|U_{ij}| < M$ for some constant M which is independent of i and j. If h and k must be related in some way for this condition to hold then we say the method is *conditionally stable*. There are two common means of determining stability of a difference equation approximation to an initial value problem.

VON NEUMANN STABILITY CRITERION

Let $D[U_{jk}] = S_{jk}$ denote a difference equation approximation to the differential equation in (12.7). Then we assume the equation $D[U_{jk}] = 0$ has a solution of the form $U_{jk} = z^k e^{ijb}$ where $i = \sqrt{-1}$. Here b denotes an arbitrary real number and the equation $D[U_{jk}] = 0$ induces a relationship between z and b. If it can be shown that $|z| \leq 1$ for all real b, then the method is said to be *stable in the Von Neumann sense*. Note that the Von Neumann criterion makes no reference to the boundary conditions and, thus, does not reflect their influence on stability. For two level difference equations like the forward and backward difference methods, the Von Neumann stability criterion is necessary and sufficient for stability.

MATRIX STABILITY CRITERION

For each j, if we let $U_j = [U_{ij}, U_{2j}, \ldots, U_{nj}]^{\mathsf{T}}$, then the forward and backward difference methods can be expressed in matrix notation $U_{j+1} = BU_j$ for an appropriate $n \times n$ matrix B (see Problem 1.3). The boundary conditions will affect the form of the matrix B slightly. Let $\lambda_1, \ldots, \lambda_n$ denote the eigenvalues of B. Then the method is *stable in the matrix stability sense* if $|\lambda_j| \leq 1$ for all j. Matrix stability is a necessary condition for stability of a two level difference equation. If the matrix B is symmetric, then it is also sufficient.

Theorem 12.3

Theorem 12.3. The forward and backward difference methods described above are both consistent with the differential equation (12.7). Both are convergent with $|U_{ij} - i_{ij}| \leq C(k + h^2)$ for a positive constant C. The backward difference method is stable and the forward difference method is conditionally stable if $dr \leq 1$, where d denotes the maximum value of $D(x,t)$ for $0 \leq x \leq 1, 0 \leq t \leq T$.

Hyperbolic Problems

Let $u = u(x, t)$ denote a solution of the first order initial value problem

$$L_2[u(x,t)] = \partial_t u(x,t) + a(x,t)\partial_x u(x,t) = S(x,t) \quad \text{all } x \text{ and } t > 0 \qquad (12.10)$$

$$u(x, 0) = f(x) \quad \text{all } x$$

where a, S, and f are given functions. Then $L_2[u]$ is an example of a first order hyperbolic partial differential operator.

APPROXIMATE SOLUTIONS

Let h and k denote fixed positive numbers and let $x_i = ih$, $t_j = jk$ for integer values of i and j. Then we can approximate the differential equation in (12.10) by any of the following difference equations. Here we let $r = k/h$.

Forward in Time, Forward in Space (FTFS)

$$\Delta_t U_{ij} + a_{j+1/2,j}\Delta_x U_{ij} = S_{ij}, \quad U_{i0} = f_i \quad \text{all } i$$

i.e.,

$$U_{i,j+1} = (1 + ra_{j+1/2,j})\, U_{ij} - ra_{j+1/2,j} U_{i+1,j} + kS_{ij}$$

Forward in Time, Backward in Space (FTBS)

$$\Delta_t U_{ij} + a_{i+1/2,j}\Delta_x U_{i-1,j} = S_{ij}, \quad U_{i0} = f_i \quad \text{all } i$$

i.e.,

$$U_{i,j+1} = (1 - ra_{i-1/2,j})\, U_{ij} - ra_{i-1/2,j} U_{i-1,j} + kS_{ij}$$

Forward in Time, Centered in Space (FTCS)

$$\Delta_t U_{ij} + a_{i,j}(2h)^{-1}(U_{i+1,j} - U_{i-1,j}) = S_{ij}, \quad U_{i0} = f_i \quad \text{all } i$$

i.e.,

$$U_{i,j+1} = U_{ij} + ra_{i,j}(U_{i+1,j} - U_{i-1,j})/2 + kS_{ij}$$

Lax-Friedrichs Method (LF)

$$U_{i,j+1} = (U_{i+1,j} - U_{i-1,j})/2 - ra_{ij}(U_{i+1,j} - U_{i-1,j})/2 + kS_{ij},\ U_{i0} = f_i \text{ all } i$$

Here $a_{i+1/2,j}$ can be chosen to be either the arithmetic or geometric mean of the quantities $a_{i+1,j}$ and a_{ij}.

Theorem 12.4

Theorem 12.3. Let U_{ij} denote an approximate solution for (12.10) generated by one of the methods above. Then

1. If U_{ij} is generated by FTFS, then the solution is stable if and only if $a_{ij} < 0$ and $|ra_{ij}| < 1$. In that case, $|U_{ij} - u_{ij}| \le C(k+h)$ provided $u(x, t)$ is smooth.
2. If U_{ij} is generated by FTBS, then the solution is stable if and only if $a_{ij} > 0$ and $|ra_{ij}| < 1$. In that case, $|U_{ij} - u_{ij}| \le C(k+h)$ provided $u(x, t)$ is smooth.
3. If U_{ij} is generated by FTCS, then the solution is unstable.
4. If U_{ij} is generated by LF, then the solution is stable if and only if $|ra_{ij}| < 1$. In that case, $|U_{ij} - u_{ij}| \le C(k+h)$ provided $u(x, t)$ is smooth.

CONSERVATION LAW EQUATIONS

Consider the initial value problem for the conservation law equation

$$\partial_t u(x, t) + \partial_x F(u) = 0, \quad u(x, 0) = f(x) \quad \text{all } x \qquad (12.11)$$

where F and f denote given functions. Then we have the following algorithms for generating approximate solutions for (12.11):

Lax-Friedrichs Method

$$U_{i,j+1} = \frac{1}{2}(U_{i+1,j} + U_{i-1,j}) - \frac{r}{2}(F(U_{i+1,j}) - F(U_{i-1,j}))$$

$$r = k/h$$

Lax-Wendroff Method

$$U_{i,j+1} = U_{ij} - \frac{r}{2}(F(U_{i+1,j}) - F(U_{i-1,j}))$$
$$+ \frac{r^2}{2}(a_{i+1/2,j}(F(U_{i+1,j}) - F(U_{i-1,j})) - a_{i-1/2,j}(F(U_{i,j}) - F(U_{i-1,j})))$$

where

$$a(u) = F'(u) \quad \text{and} \quad a_{i\pm 1/2,j} = a\left(\frac{U_{i\pm 1,j} + U_{i,j}}{2}\right)$$

MONOTONE METHODS

We saw in chapter 10 that solutions to conservation law equations may be discontinuous with sharp fronts that propagate along shock curves. In such cases, the numerical solution often exhibits oscillation in the neighborhood of the shock even if the exact solution is a monotone function. A difference equation of the form $U_{i,j+1} = G(U_{i-1,j}, U_{ij}, U_{i+1,j})$ is said to be *monotone* if

$$\frac{\partial G}{\partial U_{pj}} \geq 0 \quad \text{for} \quad p = i-1, i, i+1$$

The numerical solution U_{ij} is monotone increasing if $p > q$ implies $U_{pj} \geq U_{qj}$ and it is said to be monotone decreasing if $p > q$ implies $U_{pj} \leq U_{qj}$.

Theorem 12.5

Theorem 12.5. Suppose the numerical solution U_{ij} is generated by a monotone difference equation $U_{i,j+1} = G(U_{i-1,j}, U_{ij}, U_{i+1,j})$ approximating the conservation law equation (12.11). Then

1. If U_{ij} is monotone, it follows that $U_{i,j+1}$ is monotone of the same type.
2. $|U_{ij} - u_{ij}| \leq C(h+k)$ for positive constant C.
3. If U_{ij} tends to the limit $u(x_i, t_j)$ as h and k tend to zero with $r = h/k$ fixed, then $u(x,t)$ is a weak solution of (12.11) that satisfies the entropy condition across all discontinuities of u.

Theorem 12.6

Theorem 12.6. The Lax-Friedrichs approximation method for the conservation law equation (12.11) is monotone. The method is stable if $|ra(U_{ij})| \leq 1$ for all i, j.

Elliptic Equations

Let $u = u(x, y)$ denote a solution for the boundary value problem

$$\begin{aligned} &-\partial_x(a(x,y)\partial_x u) - \partial_y(b(x,y)\partial_y u) + c(x,y)u = S(x,y) \quad 0 < x, y < 1 \\ &u(x,0) = f(x) \quad \text{and} \quad u(x,1) = g(x), \quad 0 < x < 1 \\ &u(0,y) = p(y) \quad \text{and} \quad u(1,y) = q(y), \quad 0 < y < 1. \end{aligned} \tag{12.12}$$

where a, b, c, S, f, g, p, and q are all given functions. If $a > 0$ and $b > 0$, then this equation is an example of an elliptic partial differential equation. When $c = 0$ and $a = b =$ constant, then this equation reduces to the Poisson equation.

APPROXIMATE SOLUTIONS

Let (x_i, y_j) denote a square mesh on $(0,1) \times (0,1)$ with $x_i = ih$ and $y_j = jh$ for i and h running from zero to $n+1$ and $h = 1/(n+1)$. Approximating the partial derivatives in equation (12.12) by difference quotients leads to a difference equation that must be satisfied at each interior node of the mesh. The value of U_{ij} is given at each boundary node but is unknown at each interior node. Thus, we have $N = n^2$ unknowns and an equal number of equations. The system of difference equations approximating (12.12) has the form $AU = B$, where A is an $n^2 \times n^2$

matrix, $\boldsymbol{U}$ is a vector with n^2 entries, and $\boldsymbol{B}$ is a data vector containing the input from the boundary and the forcing term. The unknowns U_{ij} can be ordered in various ways within the vector $\boldsymbol{U}$. One possible ordering is

$$U = [U_{11}, U_{21}, \ldots, U_{n1}, U_{12}, U_{22}, \ldots, U_{n2}, \ldots, U_{nn}]^T \tag{12.13}$$

Problems 1.1 and 1.2 show two examples of problems of this form derived directly from the conservation principles instead of deriving the differential equation from conservation principles and then discretizing the derivatives to obtain the system of algebraic equations. We will show in the solved problems how to place the grid relative to the domain of the solution and how to incorporate the boundary conditions into the system of equations.

SOLUTION OF DISCRETE ELLIPTIC SYSTEMS

In chapter 1, we discussed the solution of linear systems of the form $AU = B$ by the Gaussian elimination algorithm. This algorithm applies elementary row operations to reduce the matrix A to upper triangular form. If, at some stage in the reduction process, a zero appears on the diagonal, the algorithm stops. When a zero appears on the diagonal, then either the matrix A is singular, in which case the system fails to have a unique solution, or the equations can be reordered to bring a nonzero entry onto the diagonal so that the reduction algorithm can proceed. This reordering by interchanging rows is referred to as *pivoting*. The algorithm of chapter 1 does not include a pivoting strategy. A system of equations that arises in connection with discretizing a well posed elliptic boundary value problem generally will not require pivoting.

An $n \times n$ matrix A is said to be *irreducible* if a change in any component of the vector $\boldsymbol{B}$ produces changes in every component of $\boldsymbol{U}$, where $A\boldsymbol{U} = \boldsymbol{B}$. Discretizing a well posed elliptic boundary value problem generally produces a system of equations with an irreducible coefficient matrix. Compare this fact with the so-called *organic behavior* discussed in Problem 9.1.

Theorem 12.7

Theorem 12.7. Consider the system $A\boldsymbol{U} = \boldsymbol{B}$ where the $n \times n$ matrix A is irreducible and satisfies for $i = 1$ to n:

(a) $a_{ii} < 0$ and $a_{ij} \geq 0$ for i not equal to j;

(b) $|a_{ii}| \geq \sum_j a_{ij}$, where the sum on j runs from 1 to n, skipping $j = i$ and strict inequality holds for at least one i

Then $A\boldsymbol{U} = \boldsymbol{B}$ has a unique solution which can be found by Gaussian elimination with no pivoting. The algorithm is stable with respect to the growth of roundoff errors.

LARGE SPARSE SYSTEMS

For problems in which the matrix A is 100×100 or smaller, Gaussian elimination is a perfectly adequate solution method. For larger problems, it may be advisable to take advantage of special aspects of the problem to improve the efficiency of the solution procedure. In Problems 1.1 and 1.2, we set up the problem $AU = B$ by generating the matrix A and the data vector B for two examples. In both of these examples, the $n^2 \times n^2$ matrices A are composed of three diagonal bands of smaller submatrices; the matrices are said to be *block tradiagonal*. For large values of n, this means that most of the entries of the coefficient matrices are zeroes. Such matrices are said to be *sparse matrices*. There are many special algorithms for solving sparse systems. We will describe three such methods that apply particularly to problems obtained from discretizing elliptic partial differential equations.

JACOBI ITERATION

For simplicity, consider (12.12) in the special case that $a = b = 1$ and $c = 0$. Then using the difference formula (c) in Theorem 12.2 to approximate the derivatives in the Poisson equation (12.12) leads to the following difference equation for the U_{ij}:

$$U_{ij} = \frac{1}{4}(U_{i-1,j} + U_{i+1,j} + U_{i,j-1} + U_{i,j+1} + h^2 S_{ij}) \tag{12.14}$$

This result (known as the *discrete mean value property* if $S = 0$) asserts that the value U_{ij} of the solution of the discrete Poisson equation is equal to the average of the values for U at the four adjacent nodes plus a contribution from the source term S. The form (12.14) of the discrete Poisson equation motivates the following iteration procedure known as *Jacobi iteration*: for each i, j, $1 \le i, j \le n$ and $k = 0, 1, \ldots,$

$$U_{ij}^{\ 0} = \text{initial guess (based on boundary information)}$$

$$U_{ij}^{k+1} = \frac{1}{4}(U_{i-1,j}^{k} + U_{i+1,j}^{k} + U_{i,j-1}^{k} + U_{i,j+1}^{k} + h^2 S_{ij}) \tag{12.15}$$

The iteration may be terminated when k exceeds some preselected iteration limit or when the difference $\left| U_{ij}^{k+1} - U_{ij}^{k} \right|$ between successive iterates is less than a previously chosen tolerance for all i and j. The method does not compute the solution for the system $AU = B$, but only generates a sequence of iterates that tend toward the solution.

GAUSS-SEIDEL ITERATION

Note that if the order in which the U_{ij} are calculated from (12.15) is the same as the ordering of the entries U_{ij} in the vector U is as indicated in

(12.13) then when U_{ij}^{k+1} is computed in (12.15) the values $U_{i-1,j}^{k+1}$ and $U_{i,j-1}^{k+1}$ are already known. Thus, we may define the so-called Gauss-Seidel iteration method:

$$\begin{aligned} U_{ij}^{0} &= \text{initial guess (based on boundary information)} \\ U_{ij}^{k+1} &= \frac{1}{4}\left(U_{i-1,j}^{k+1} + U_{i+1,j}^{k} + U_{i,j-1}^{k} + U_{i,j+1}^{k} + h^2 S_{ij}\right) \end{aligned} \tag{12.16}$$

The algorithm (12.16) assumes that the nodes (x_i, y_j) in the mesh are scanned according to the convention that the index i varies most rapidly. If a different scanning order is used then (12.16) must be modified accordingly. Both (12.15) and (12.16) are written for the case of Dirichlet type boundary conditions on all boundaries (i.e., the value of u is specified on all boundaries). The methods may be modified to accommodate other boundary conditions.

SUCCESSIVE OVERRELAXATION (SOR)

If both the Jacobi and Gauss-Seidel iterations converge toward the solution U_{ij}, then it is expected that the Gauss-Seidel will converge more rapidly since it incorporates updated information sooner. A method to accelerate convergence even more is called the method of *successive overrelaxation.*

Suppose values U_{ij}^{k} are used in (12.16) to calculate the next iterate which we denote by V_{ij}^{k+1}. Then the direction of convergence can be estimated from U_{ij}^{k} and V_{ij}^{k+1} and an improved update U_{ij}^{k+1} computed from

$$\begin{aligned} U_{ij}^{0} &= \text{initial guess (based on boundary information)} \\ V_{ij}^{k+1} &= \frac{1}{4}\left(U_{i-1,j}^{k+1} + U_{i+1,j}^{k} + U_{i,j-1}^{k+1} + U_{i,j+1}^{k} + h^2 S_{ij}\right) \\ U_{ij}^{k+1} &= sV_{ij}^{k+1} + (1-s)\,U_{ij}^{k} \qquad \text{for } 0 < s < 2 \end{aligned} \tag{12.17}$$

The constant s in (12.17) is called the relaxation parameter. When $s = 1$, the method reduces to Gauss-Seidel. The method is referred to as overrelaxation or underrelaxation according to whether $s > 1$ or $s < 1$, respectively.

SOLVED PROBLEMS

PROBLEM 12.1

Modify the Runge-Kutta algorithm or order 4 so that it applies to systems of the form

$$x'(t) = P(x, y, t), \quad x(0) = A \tag{1}$$
$$y'(t) = Q(x, y, t), \quad y(0) = B$$

Write a computer program to carry out the computations of the algorithm applied to the predator-prey system

$$x'(t) = x(t)\,(1 - y(t)), \quad x(0) = A \tag{2}$$
$$y'(t) = y(t)\,(x(t) - 1), \quad y(0) = B$$

SOLUTION 12.1

The 1-dimensional RK-4 algorithm is given by (12.5). For systems of the form (1), it takes the form

$$x_0 = A, \quad y_0 = B$$

$$XK_1 = hP(x_n, y_n, t_n)$$
$$YK_1 = hQ(x_n, y_n, t_n)$$

$$XK_2 = hP(x_n + XK_1/2, y_n + YK_1/2, t_n + h/2)$$
$$YK_2 = hQ(x_n + XK_1/2, y_n + YK_1/2, t_n + h/2)$$

$$XK_3 = hP(x_n + XK_2/2, y_n + YK_2/2, t_n + h/2)$$
$$YK_3 = hQ(x_n + XK_2/2, y_n + YK_2/2, t_n + h/2)$$

$$XK_4 = hP(x_n + XK_3/2, y_n + YK_3, t_n + h)$$
$$YK_4 = hQ(x_n + XK_3/2, y_n + YK_3, t_n + h)$$

$$x_{n+1} = x_n + (XK_1 + 2XK_2 + 2XK_3 + XK_4)/6$$

for $n = 0, 1, \ldots$

$$y_{n+1} = y_n + (YK_1 + 2YK_2 + 2YK_3 + YK_4)/6$$

The following is a BASIC computer program to carry out the steps of this algorithm applied to the predator-prey system (2):

```
REM2-DIMENSIONAL RK-4 METHOD
REM
DEF P (X, Y) = X(1 – Y)
DEF Q (X, Y) = Y(X – 1)
REM
INPUT "INITIAL VALUES FOR T, X, Y"; T, X, Y
INPUT "STEP SIZE h "; H
INPUT "NUMBER OF STEPS"; NSTEPS
REM
FOR N = 1 TO NSTEPS
    XK1 = H * P(X, Y)
    YK1 = H * Q(X, Y)
    XK2 = H * P(X + XK1/2, Y + YK1/2)
```

```
        YK2 = H * Q(X + XK1/2, Y + YK1/2)
        XK3 = H * P(X + XK2/2, Y + YK2/2)
        YK3 = H * Q(X + XK2/2, Y + YK2/2)
        XK4 = H * P(X + XK3, Y + YK3)
        YK4 = H * Q(X + XK3, Y + YK3)
        X = X + (XK1 + 2*XK2 + 2*XK3 + XK4)/6
        Y = Y + (YK1 + 2*YK2 + 2*YK3 + YK4)/6
        T = T + H
        PRINT " T = "; T;" X = "; X;" Y = "; Y
    NEXT N
    REM
    END
```

This program was executed with $h = 0.1$ for several choices of initial point (A, B). The pairs (A, B) were selected to produce distinct orbits, which have been plotted rather than printed and are displayed in Figure 5.7. These results are comparable to orbits generated by a 2-dimensional Euler's method with $h = 0.001$. Thus, the extra computational effort of the RK-4 algorithm allows the use of a step size that is up to 100 times as large as the one used to obtain similar results with the simpler Euler's algorithm.

PROBLEM 12.2

Suppose $u = u(x, y)$ is C^4 in a region D in the x plane. Then use Taylor's theorem in 2 variables to show that for any point (x,y) in D and for $h > 0$ sufficiently small that $(x \pm h, y \pm h)$ is in D, we have

$$\left|\frac{\partial^2 u}{\partial x^2} - \frac{u(x+h, y) - 2u(x, y) + u(x-h, y)}{h^2}\right| \le C_1 h^2 \text{ for some } C_1 > 0 \qquad (1)$$

and

$$\left|\frac{\partial^2 u}{\partial y^2} - \frac{u(x, y+h) - 2u(x, y) + u(x, y-h)}{h^2}\right| \le C_2 h^2 \text{ for some } C_2 > 0 \qquad (2)$$

hence for $C_3 = C_1 + C_2$

$$\left|u_{xx} + u_{yy} - h^{-2}\left(u(x, y+h) + u(x+h, y) - 4u(x, y) + u(x, y-h) + u(x-h, y)\right)\right| \le C_3 h^2 \qquad (3)$$

SOLUTION 12.2

According to Taylor's theorem, for u in C^4

$$u(x+h, y) = u(x, y) + u_x(x, y)\,h + u_{xx}(x, y)\frac{h^2}{2} + u_{xxx}(x, y)\frac{h^3}{6} + u_{xxxx}(x+ph, y)\frac{h^4}{24}$$

$$u(x-h, y) = u(x, y) - u_x(x, y)\,h + u_{xx}(x, y)\frac{h^2}{2} - u_{xxx}(x, y)\frac{h^3}{6} + u_{xxxx}(x-qh, y)\frac{h^4}{24}$$

Adding these two equations and solving for $u_{xx}(x, y)$ leads to

$$u_{xx}(x, y) - \frac{u(x+h, y) - 2u(x, y) + u(x-h, y)}{h^2}$$

$$- (u_{xxxx}(x+ph, y) + u_{xxxx}(x-qh, y)) \frac{h^2}{12}$$

Then for u in C^4, (1) follows for C_1 chosen such that

$$|u_{xxxx}(x+ph, y) + u_{xxxx}(x-qh, y)| \le 12C_1$$

A similar argument leads to

$$u_{yy}(x, y) - \frac{u(x, y+h) - 2u(x, y) + u(x, y-h)}{h^2}$$

$$- (u_{yyyy}(x, y+rh) + u_{yyyy}(x, y-sh)) \frac{h^2}{12}$$

and this yields (2). Then (1) and (2) together imply (3). The result (3) says that the *local truncation error* for the centered difference approximation for the Laplacian on a square grid is of order h^2.

PROBLEM 12.3

Apply the forward difference method to the problem (12.7) and write out the resulting equations for determining U_{ij}. Write a computer program to solve the equations and use it to solve the problem (12.7) when D is constant, S, f, and q are zero, and $p(t) = \sin \Omega t$.

SOLUTION 12.3

First define a mesh (x_i, t_j) with $x_i = ih$ and $t_j = jk$, where $h = 1/(n + 1)$ and k equal to a positive time step whose value will be chosen later. We can visualize the interval (0, 1) as divided into $n + 1$ "cells" by this mesh, with the nodes x_i located at the center of each cell (Figure 1.3). If we think of the problem in the context of diffusion, then the concentration U_{0j} in cell 0 at time t_j is given and equals $p_j = p(t_j)$ for each j. The diffusivity $D = D(x,t)$ varies with x and t and we define $D_{i+1/2,j}$ to be the diffusivity at time t_j on the face separating cells i and $i + 1$. We let

$$D_{i+1/2,j} = \frac{D_{i+1,j} + D_{ij}}{2} \quad \text{or} \quad \frac{2D_{i+1,j}D_{ij}}{D_{i+1,j} + D_{ij}} \tag{1}$$

The first choice in (1) is just the arithmetic mean of the diffusivities in the two consecutive cells, whereas the second choice is the geometric mean. In cases where one of the two D values is zero, using the arithmetic mean can cause conservation of mass to be violated. Using the geometric mean ensures that there is no flow into a cell in which $D = 0$. If $D = 0$ does not occur, then the arithmetic mean is a simpler choice.

In cells 2 through $n-1$, the discrete diffusion equation takes the form

$$\frac{U_{i,j+1}-U_{ij}}{k} = \frac{1}{h}\left(D_{i+1/2,j}\frac{U_{i+1,j}-U_{ij}}{h} - D_{i-1/2,j}\frac{U_{i,j}-U_{i-1,j}}{h}\right) + S_{ij}$$

That is, for $i = 2$ to n and $j = 0, 1, \ldots,$

$$U_{ij+1} = rD_{i+1/2,j}U_{i+1,j} + (1 + r(D_{i+1/2,j} + D_{i-1/2,j}))\,U_{ij} + rD_{i-1/2,j}U_{i-1,j} + S_{ij}k$$

For $i = 1$,

$$U_{1,j+1} = rD_{3/2,j}U_{2,j} + (1 + r(D_{3/2,j} + D_{1/2,j}))\,U_{1,j} + rD_{1/2,j}p_j + kS_{1,j}$$

For $i = n$, we have

$$D_{n+1/2,j}\frac{U_{n+1,j}-U_{nj}}{h} = q_j$$

hence,

$$U_{n,j+1} = rq_j + (1 - rD_{n-1/2,j})\,U_{n,j} + rD_{n-1/2,j}U_{n-1,j} + kS_{nj}$$

The following is a BASIC computer program to carry out the steps of this algorithm to solve the problem (12.7) with S, f, and q equal to zero, with D equal to the constant D_0 and $p(t) = A_0 \sin(\Omega t)$. The program is easily modified to deal with other choices.

```
REM  FORWARD DIFFERENCE ALGORITHM FOR A PARABOLIC
     PROBLEM
REM                         DIMENSION ARRAYS
DIM U(400), V(400)
REM                         DEFINE DATA FUNCTIONS
DEF D(X, T) = DO
DEF S(X, T) = 0
DEF F(X) = 0
DEF P(T) = A0*SIN(W*T)
DEF Q(T) = 0
REM                         INPUT DATA VALUES
REM  k = time step, TMAX = number of time steps, n = number of nodes
INPUT "ENTER VALUES FOR k, TMAX AND n"; K, TMAX, N
INPUT "ENTER VALUES FOR A0, D0"; A0, D0
H = 1/(N+1)
R = K/H/H
PRINT " R = "; R
REM                         INITIALIZE SOLUTION VECTOR
FOR I = 1 TO N
```

```
X = I*H
V(I) = F(X)
NEXT I
REM                      BEGIN TIME STEPPING
FOR J = 1 TO TMAX
T = J * K
V(0) = P(T)
X = H
REM                      COMPUTE PROFILE AT TIME T
FOR I = 1 TO N-1
X = I * H
D1 = (D(X-H, T) + D(X, T) ) / 2
D2 = (D(X, T) + D(X + H, T) ) / 2
U(I) = R*D1*V(I-1)+(1+R*(D1+D2))*V(I)+R*D2*V(I+1)+K*S(X,T)
NEXT I
U(N) = R*D2*V(N - 1) + (1 - R*D2)*V(N) + R*Q(T) + K*S(X + H, T)
U(0) = V(0)
REM                OUTPUT SOLUTION VECTOR U AT TIME T
FOR I = 1 TO N
V(I) = U(I)
NEXT I
NEXT J
END
```

For a fixed j the array $\{ U_{ij}\!: \ 1 \le i < n\}$ is called a *solution profile*. Plotting the profiles against i for successive values of j provides a visual picture of the evolution of the process that is being modeled. The backward difference implicit method is applied in Problem 1.7 to a system of equations derived directly from conservation laws in Problem 1.3.

PROBLEM 12.4

Use the Von Neumann stability criterion to show that the forward difference method of the previous problem is stable if and only if $2D_{ij}r \le 1$ for all i, j.

SOLUTION 12.4

We apply the method to the homogeneous equation ($S = 0$) and for simplicity, let us suppose $D(x, t)$ is a constant, $D > 0$. Then the difference equation generated by the forward difference method reduces to

$$U_{j,\,k+1} = rDU_{j+1,\,k} + (1 - 2rD)\,U_{jk} + rDU_{j-1,\,k} \qquad (1)$$

We assume a solution of the form $U_{jk} = z^k e^{ijb}$ for z, b real and i equal to the imaginary unit, $\sqrt{-1}$. Then substituting this into (1) gives

$$z = rDe^{ib} + (1 - 2rD) + rDe^{-ib}$$

$$= 1 + rD(e^{ib} - 2 + e^{-ib}) = 1 - 2rD(1 - \cos b)$$

$$= 1 - 4rD\sin^2(b/2)$$

Then it follows that $-1 \le z \le 1$ for all real b (in particular, for $b = \pi$) if and only if $4rD \le 2$; i.e., if $2kD \le h^2$.

PROBLEM 12.5

Use the Von Neumann method to show that the implicit backward difference method for parabolic problems is stable for all r, D.

SOLUTION 12.5

Under the assumptions of the previous problem, the difference equation that is generated by the backward difference method reduces to

$$-rDU_{j-1,k} + (1 + 2rD)\,U_{jk} - rDU_{j+1,k} = U_{jk} \tag{1}$$

Then substituting $U_{jk} = z^k e^{ijb}$ into (1) leads to

$$z(-rDe^{-ib} + (1 + 2rD) - rDe^{ib}) = 1$$

Then

$$z = (1 + 4rD\sin^2(b/2))^{-1} \le 1 \quad \text{for all } b$$

It follows that $|z| \le 1$ independent of r and D.

PROBLEM 12.6

Use the matrix stability analysis to show that the forward difference method applied to the problem

$$\partial_t u(x,t) = D\partial_{xx} u(x,t),\ u(x,0) = f(x),\ u(0,t) = u(1,t) = 0 \tag{1}$$

is stable if $2rD \le 1$.

SOLUTION 12.6

If we take the boundary conditions into account then the forward difference method leads to the following system of difference equations for U_{ij}

$$U_{1,j+1} = 0 + (1 - 2rD)\,U_{1,j} + rDU_{2,j}$$

$$U_{1,k+1} = rDU_{j-1,k} + (1 - 2rD)\,U_{jk} + rDU_{j+1,k} \qquad 1 < j < n$$

$$U_{n,j+1} = rDU_{n-1,j} + (1 - 2rD)\,U_{n,j} + 0$$

In matrix notation, this becomes

$$U_{j+1} = (I - rDA)\,U_j$$

where $U_j = [U_{1,j}, \ldots, U_{n,j}]^{\mathsf{T}}$, I denotes the $n \times n$ identity matrix and A is the $n \times n$ matrix from Problem 2.5. The matrices I and A are symmetric and, thus, the matrix $C = I - rDA$ is also symmetric. Then the eigenvalues $c_1, \ldots, c_n$ of C are real and the normalized eigenvectors of C form an orthonormal basis for R^n. If we denote these eigenvectors by $X_1, \ldots, X_n$ then $U_0 = [f(x_1), \ldots, f(x_n)]^T$ can be written as

$$U_0 = \sum_{k=1}^{n} U_0 \bullet X_k X_k$$

Then

$$U_{j+1} = CU_j = C^{j+1} U_0$$

$$= \sum_{k=1}^{n} U_0 \bullet X_k C^{j+1} X_k$$

$$= \sum_{k=1}^{n} c_k^{\,j+1} U_0 \bullet X_k X_k$$

and it is obvious from this last expression that $\|U_j\|$ does not grow without bound as j increases if and only if $|c_k| \le 1$ for every k. It is easy to show that if the eigenvalues of the matrix A are denoted by λ_m, then $c_m = 1 - rD\lambda_m$. In Problem 2.5, we showed that

$$\lambda_m = 2\left(1 - \cos\frac{m\pi}{n+1}\right) = 4\sin^2\frac{m\pi}{n+1}; \quad m = 1\ldots, n$$

Then

$$c_m = 1 - 4rD\,\sin^2\frac{m\pi}{n+1}; \quad m = 1, \ldots, n$$

and $-1 \le 1 - 4rD \le c_m$ for every m. This implies stability for $2rD \le 1$.

PROBLEM 12.7

Show that the difference equation

$$\frac{U_{i,j+1} - U_{i,j-1}}{2k} = \frac{U_{i+1,j} - (U_{i,j+1} + U_{i,j-1}) + U_{i-1,j}}{h^2} \tag{1}$$

is not consistent with the partial differential equation $\partial_t u(x, t) = \partial_{xx} u(x, t)$.

Show that the difference equation

$$\frac{U_{i,j+1} - U_{i,j}}{k} = \frac{U_{i+1,j} - 2U_{i,j} + U_{i-1,j}}{h^2} \tag{2}$$

is consistent with the heat equation.

SOLUTION 12.7

Using Taylor's theorem in two variables, we can show

$$U_{i,j+1} - U_{i,j-1} = 2k\partial_t u_{ij} + \frac{2}{6}\partial_{ttt} u_{ij} k^3 + \cdots$$

$$U_{i,j+1} + U_{i,j-1} = 2u_{ij} + \partial_{tt} u_{ij} k^2 + \cdots$$

$$U_{i+1,j} + U_{i,j-1} = 2u_{ij} + \frac{2}{6}\partial_{xx} u_{ij} h^2 + \cdots$$

Then

$$(U_{i,j+1} + U_{i,j-1})/2k - (U_{i+1,j} - (U_{i,j+1} + U_{i,j-1}) + U_{i-1,j})/h^2$$

$$= \partial_t u_{ij} + \frac{1}{6}\partial_{ttt} u_{ij} k^2 - \partial_{xx} u_{ij} + \partial_{tt} u_{ij} (k/h)^2 + \cdots$$

and it is clear that on a rectangular mesh with $k = bh$, as the mesh size tends to zero the difference equation (1) tends to the hyperbolic partial differential equation

$$\partial_t u(x,t) - \partial_{xx} u(x,t) + b^2 \partial_{tt} u(x,t) = 0 \tag{3}$$

On the other hand, Taylor's theorem also implies

$$U_{i,j+1} - U_{ij} = \frac{1}{2}\partial_{tt} u_{ij} k^2 + \ldots$$

$$U_{i+1,j} - 2U_{i,j} + U_{i-1,j} = \frac{1}{12}\partial_{xxxx} u_{ij} h^4 + \cdots$$

and, thus,

$$(U_{i,j+1} - U_{i,j})/k - (U_{i+1,j} - 2U_{i,j} + U_{i-1,j})/h^2$$

$$= \partial_t u_{ij} + \frac{k}{2}\partial_{tt} u_{ij} - \partial_{xx} u_{ij} - \frac{h^2}{12}\partial_{xxxx} u_{ij} + \cdots$$

$$= \partial_t u_{ij} - \partial_{xx} u_{ij} + \frac{1}{12} h^2 (r - \frac{1}{6})\partial_{tt} u_{ij} + \cdots$$

Then the difference equation (2) tends to the heat equation as the mesh size tends to zero regardless of how h and k may be related. It is important to know about consistency not only to be certain that the difference equation approximates the correct partial differential equation but also because of the so-called Lax equivalence theorem:

A consistent difference scheme for a well posed problem is convergent if and only if it is stable.

PROBLEM 12.8

Apply the Von Neumann stability test to detemine the conditions for stability of the following approximations to $\partial_t u(x,t) + a(x,t)\partial_x u(x,t) = 0$:

$$\text{(a)}\ U_{j,k+1} = \frac{1}{2}(U_{j+1,k}+U_{j-1,k}) - \frac{r}{2}a_{jk}(1(U_{j+1,k}-U_{j-1,k})),\quad r=\frac{k}{h}$$

$$\text{(b)}\ U_{j,k+1} = (1+ra_{j+1/2,k})U_{jk} - r_{j+1/2,k}U_{j+1,k}$$

$$\text{(c)}\ U_{j,k+1} = U_{jk} + \frac{r}{2}a_{jk}(U_{j+1,k}-U_{j-1,k})$$

SOLUTION 12.8

The Von Neumann stability criterion applies to constant coefficient equations but is used to analyze the stability of variable coefficient equations by "freezing" the coefficients to obtain an equation with constant coefficients. Then the results are viewed as giving an indication of stability or instability although they are not rigorous. Assuming a solution of the form $U_{jk} = z^k e^{ijb}$ for equation (a), the Lax-Friedrichs difference method leads to

$$\begin{aligned} z &= (e^{ib}+e^{-ib})/2 - ra(e^{ib}+e^{-ib})/2 \\ &= \cos b - ira\sin b \end{aligned}$$

Then

$$|z|^2 = \cos^2 b + (ra)^2 \sin^2 b \le 1 \quad \text{if} \quad (ra)^2 \le 1$$

Note that we have assumed $a_{jk} = a = \text{constant}$. The stability condition $(ra)^2 \le 1$ implies that the characteristic line passing through the point (x_j, t_{k+1}) extends backward to cut the line at $t = t_k$ in a point that lies between (x_{j-1}, t_k) and (x_{j+1}, t_k). Since the value of $U_{j,k+1}$ is based on U values on the interval (x_{j-1}, t_k) to (x_{j+1}, t_k), this means that the *analytic domain of dependence* is contained in the *numerical domain of dependence*.

Assuming $a_{jk} = a$ and $U_{jk} = z^k e^{ijb}$ in (b) leads to

$$z = (1+ra) - rae^{ib}$$

Then $|z| \le 1$ if $|1+ra-ra\cos\ b - ira\sin\ b| \le 1$ or, equivalently, if

$$(1+ra-ra\cos\ b)^2 + (ra\sin\ b)^2 \le 1;$$

i.e.,

$$1+2ra(1+ra)(1-\cos b) = 1+4ra\sin^2(b/2) \le 1$$

From this last inequality, it can be seen that $|z| \le 1$ if $a < 0$ and $|ra| \le 1$;

that is, the method is stable if $-1 \le ak/h < 0$.

Finally, for $a_{jk} = a$ and $U_{jk} = z^k e^{ijb}$, the difference equation (c) reduces to

$$z = 1 + ra(e^{ib} - e^{-ib})/2 = 1 + ira\sin b$$

Then $|z|^2 = 1 + (ra)^2 \sin^2 b > 1$ for all b different from zero. Then a solution of the form $U_{jk} = z^k e^{ijb}$ will grow without bound with increasing k. Fourier analysis can be used to write the solution corresponding to an arbitrary initial condition as a linear combination of solutions of this form; hence, the method is not stable.

PROBLEM 12.9

Show how to implement the *numerical method of characteristics* to solve

$$\partial_t u(x,t) + a(x,t)\partial_x u(x,t) = S(x,t), \qquad u(x,0) = f(x) \tag{1}$$

SOLUTION 12.9

The characteristic equations associated with (1) are

$$\frac{dx}{dt} = a(x,t) \quad \text{and} \quad \frac{du}{dt} = S(x,t) \tag{2}$$

We can define a partition of the x axis by $x_i = ih$ for all integers i. If $x_i(t)$ denotes the characteristic curve that originates at $(x_i, 0)$, then $x_i(0) = x_i$. Now applying Euler's method (12.3) to the ordinary differential equations in (2) leads to:

$$\text{for } j = 0, 1, \ldots : x_i(t_{j+1}) = x_i(t_j) + a_{ij}k, \quad x_i(0) = x_i$$

$$U_{i,j+1} = U_{ij} + S_{ij}k, \quad U_{i,0} = f(x_i)$$

where $t_j = jk$, $a_{ij} = a(x_i, t_j)$, and $S_{ij} = S(x_i, t_j)$.

A more accurate solution is obtained using the fourth order Runge-Kutta method (12.5). This produces

for

$$\begin{aligned} K_1 &= ka(x_i, t_j) \\ K_2 &= ka(x_i + K_1/2, t_j + k/2) \\ K_3 &= ka(x_i + K_2/2, t_j + k/2) \\ K_4 &= ka(x_i + K_3, t_j + k) \end{aligned}$$

then

$$\begin{aligned} x_i(t_{j+1}) &= x_{i,j} + (K_1 + 2K_2 + 2K_3 + K_4)/6 \\ x_{i,0} &= x_i \end{aligned}$$

and if

$$\begin{aligned} R_1 &= kS(x_i, t_j) \\ R_2 &= kS(x_i + K_1/2, t_j + k/2) \\ R_3 &= kS(x_i + K_2/2, t_j + k/2) \\ R_4 &= kS(x_i + K_3, t_j + k) \end{aligned}$$

then

$$U_{i,j+1} = U_{i,j} + (R_1 + 2R_2 + 2R_3 + R_4)/6$$
$$U_{i,0} = f_i$$

Note that even though the partition $x_i = ih$ of the x axis is uniform, this does not imply that the partition $x_j(t_j)$ of the line $t = t_j$ will be uniform unless $a(x, t)$ is constant. However, if $a(x, t)$ does not depend on u, then distinct characteristic curves cannot cross so $\{x_i(t_j)\}$ still forms a partition of $t = t_j$.

PROBLEM 12.10

Show that the Lax-Friedrichs method for conservation law equations is monotone if $|ra(U_{ij})| \le 1$ for all i, j, where $a(u) = F'(u)$.

SOLUTION 12.10

The Lax-Friedrichs method for conservation law equations has the form

$$U_{i,j+1} = (U_{i+1,j} + U_{i-1,j})/2 - \frac{r}{2}(F(U_{i+1,j}) - F(U_{i-1,j}))$$
$$= G(U_{i-1,j}, U_{ij}, U_{i+1,j})$$

Then

$$\frac{\partial G}{\partial U_{i-1,j}} = \frac{1}{2} + \frac{r}{2}F'(U_{i-1,j}) = \frac{1 + ra(U_{i-1,j})}{2}$$

$$\frac{\partial G}{\partial U_{i,j}} = 0$$

$$\frac{\partial G}{\partial U_{i+1,j}} = \frac{1}{2} - \frac{r}{2}F'(U_{i+1,j}) = \frac{1 - ra(U_{i+1,j})}{2}$$

If $|ra(U_{ij})| \le 1$ for all i, j, then $\partial G/\partial U_{ij}$ is non-negative and the method is monotone. A numerical method that is monotone preserves monotonicity; thus, if the solution profile contains no oscillation at t_j, then an application of a monotone method produces a nonoscillating profile at t_{j+1}.

PROBLEM 12.11

Let Ω denote the L-shaped region shown in Figure 12.1 and consider the following mixed boundary value problem

$$\partial_{xx}u(x, y) + \partial_{yy}u(x, y) = 0 \quad \text{in } \Omega \tag{1}$$

$$\partial_x u(1, y) = 0, \quad 1 < y < 2, \quad \partial_y u(x, 1) = 0, \quad 1 < x < 2 \tag{2}$$

$$u(0, y) = f_0(y), \quad 0 < y < 2, \quad u(2, y) = f_1(y), \quad 0 < y < 1 \tag{3}$$

$$u(x,0) = g_0(x),\quad 0<x<2, \qquad u(x,2) = g_1(x),\quad 0<x<1. \qquad (4)$$

Set up the mesh and the difference equations to solve the boundary value problem.

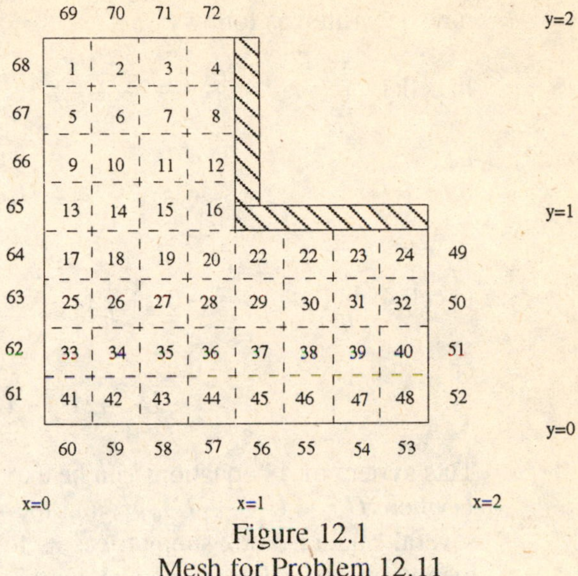

Figure 12.1
Mesh for Problem 12.11

SOLUTION 12.11

If we view this problem in the context of steady-state heat conduction, then the boundary conditions can be interpreted to mean that the L-shaped reentrant part of the boundary of Ω is insulated and the temperature is specified over the remainder of the boundary.

We define a mesh on Ω by letting $h = k = 1/4$ and

$$x_i = \frac{2i-1}{8}, \quad i = 1, \ldots, 8 \qquad \text{and} \qquad y_j = \frac{17-2j}{8}, \quad j = 1, \ldots, 8$$

Then there are 48 node points inside Ω. We visualize Ω as divided into 48 equal square cells, with a node at the center of each cell. The temperature U_{ij} at each node inside Ω is unknown and is to be determined from equation (1) and the boundary conditions. We postulate 24 exterior cells with edges along the four edges where the temperature is specified. The temperatures U_{ij} in these cells is given by the Dirichlet boundary conditions.

The difference equations are more conveniently expressed using a single index labelling for the nodes. To each double index labelled temperature $u_{i,j}$, we associated a U_p labelled with a single index. The association is defined as follows:

$$u_{1,1} = U_1, u_{2,1} = U_2, u_{3,1} = U_3, u_{4,1} = U_4, u_{1,2} = U_5, \ldots, u_{4,4} = U_{16}$$

$$u_{1,5} = U_{17}, \ldots, u_{8,5} = U_{24}, u_{1,6} = U_{25}, \ldots, u_{8,8} = U_{48}$$

The exterior cells where the temperature is specified by the boundary conditions are numbered U_{49} to U_{72} as shown in Figure 12.1.

The difference equations approximating (1) in each of the 48 cells can now be written as follows:

in cell 1: $$\frac{1}{h}\left(\frac{U_{68} - U_1}{h} - \frac{U_1 - U_2}{h}\right) + \frac{1}{h}\left(\frac{U_{69} - U_1}{h} - \frac{U_1 - U_5}{h}\right) = 0$$

i.e.,

$$-4U_1 + U_2 + U_5 = -U_{68} - U_{69}$$

in cell 4: $$\frac{1}{h}\left(\frac{U_3 - U_4}{h} - 0\right) + \frac{1}{h}\left(\frac{U_{72} - U_4}{h} - \frac{U_4 - U_8}{h}\right) = 0$$

or

$$U_3 - 3U_4 + U_8 = -U_{72}$$

This system of 48 equations can be expressed in matrix notation as $\boldsymbol{AU} = \boldsymbol{F}$ where $\boldsymbol{U} = [U_1, \ldots, U_{48}]^T$ and the 48×48 matrix A is composed of several smaller block submatrices as follows (here I_n denotes the $n \times n$ identity and 0 indicates a square matrix of all zeroes):

$$A = \begin{bmatrix} B_1 & I_4 & 0 & 0 & & & & & 0 \\ I_4 & B_1 & I_4 & 0 & & & & & \\ 0 & I_4 & B_1 & I_4 & & & & & \\ 0 & 0 & I_4 & B_1 & C & & & & \\ & & & C^T & B_2 & I_8 & 0 & & 0 \\ & & & & I_8 & B_3 & I_8 & & \\ & & & & & I_8 & B_3 & I_8 & 0 \\ & & & & & & & & I_8 \\ 0 & & & & & & & I_8 & B_3 \end{bmatrix}$$

where

$$B_1 = \begin{bmatrix} -4 & 1 & 0 & 0 \\ 1 & -4 & 1 & 0 \\ 0 & 1 & -4 & 1 \\ 0 & 0 & 1 & -3 \end{bmatrix}, \quad C = [I_4 \mid 0] \quad (C \text{ is } 4 \times 8)$$

$$B_2 = \begin{bmatrix} -4 & 1 & 0 & 0 & & & & 0 \\ 1 & -4 & 1 & 0 & & & & \\ 0 & 1 & -4 & 1 & & & & \\ 0 & 0 & 1 & -4 & 1 & 0 & 0 & 0 \\ & & & 1 & -3 & 1 & 0 & 0 \\ & & & & 1 & -3 & 1 & 0 \\ & & & & 0 & 1 & -3 & 1 \\ 0 & & & & 0 & 0 & 1 & -3 \end{bmatrix}$$

$$B_3 = \begin{bmatrix} -4 & 1 & 0 & 0 & & & & 0 \\ 1 & -4 & 1 & 0 & & & & \\ 0 & 1 & -4 & 1 & & & & \\ 0 & 0 & 1 & -4 & 1 & 0 & 0 & 0 \\ & & & 1 & -4 & 1 & 0 & 0 \\ & & & & 1 & -4 & 1 & 0 \\ & & & & 0 & 1 & -4 & 1 \\ 0 & & & & 0 & 0 & 1 & -4 \end{bmatrix}$$

The data vector
$F = [-U_{68}-U_{69}, -U_{70}, -U_{71}, -U_{72}, -U_{67}, 0, 0, 0, \ldots, -U_{52}-U_{53}]^T$
contains the temperature values specified by the boundary conditions. The 48×48 system $AU = F$ can be solved by Gaussian elimination without pivoting. Note that using a finer mesh, say $h = k = 1/8$, the number of unknowns increases to 192 and the blocks B_1 and B_2, B_3 increase in size to 8×8 and 16×16, respectively, but the structure of the matrix A remains the same.

PROBLEM 12.12

Set up the mesh and the difference equations to solve the Neumann problem

$$\partial_{xx} u(x, y) + \partial_{yy} u(x, y) = S(x, y) \quad \text{for } 0 < x, y < 1 \tag{1}$$

$$\nabla u \bullet N = 0 \quad \text{on the boundary} \tag{2}$$

SOLUTION 12.12

Note that by the result of Problem 9.15, this problem has no solution unless the integral of S over the square domain equals zero. Let us assume that this compatibility condition is satisfied. We can define a mesh on the square domain by choosing $h = k = 1/4$ and letting

$$x_i = \frac{2i-1}{8}, \; i = 1, \ldots, 4 \quad \text{and} \quad y_j = \frac{9-2j}{8}, \; j = 1, \ldots, 4$$

Then these 16 nodes can be viewed as the centers of 16 square cells covering the unit square. We introduce a one index numbering of the cells and the unknowns by defining cell 1 to be the cell with center (x_1, y_1) and proceeding in lexicographic order to cell 16 with center at (x_4, y_4). Then the difference equation approximations to the Poisson's equation (1) assume the form

$$\text{cell 1:} \quad \frac{1}{h}(0 - \frac{U_1 - U_2}{h}) + (0 - \frac{U_1 - U_5}{h}) = S_1$$

$$\text{i.e.,} \quad -2U_1 + U_2 + U_5 = h^2 S_1$$

Proceeding in this way we obtain a 16×16 system of equations $AU = S$, where A has the form

$$\begin{bmatrix} B_1 & I_4 & 0 & 0 \\ I_4 & B_2 & I_4 & 0 \\ 0 & I_4 & B_2 & I_4 \\ 0 & 0 & I_4 & B_1 \end{bmatrix}$$

where

$$B_1 = \begin{bmatrix} -2 & 1 & 0 & 0 \\ 1 & -3 & 1 & 0 \\ 0 & 1 & -3 & 1 \\ 0 & 0 & 1 & -2 \end{bmatrix} \quad \text{and} \quad B_2 = B_1 - I_4$$

and $S = h^2 [S_1, \ldots, S_{16}]^T$. Then the matrix A is singular since the sum of the elements in each row equals zero. The vector J whose entries are all 1's can be shown to be a basis for the null space of A, and since we have assumed that the integral of $S(x, y)$ over the unit square is zero, it follows that $S \bullet J = 0$; i.e., $S \bullet J$ is an approximation to the integral of S over the square. Then, by Theorem 1.12, the system has a nonunique solution U to which we can add any multiple of J and the result is still a solution. Physically, this means that the problem (1), (2) determines relative temperatures but not absolute temperatures inside the square. We can determine absolute temperatures if we specify the temperature in one of the cells. This is equivalent to adding to the system a single equation, say $U_1 = u$, where u is some constant. The new coefficient matrix A' has 17 rows and 16 columns; the last row has a 1 in the first entry and 0's elsewhere. Then the null space of A' has dimension zero since rank $A' = 16$ and, therefore, the system $A'U = S'$ has at most one solution. Here S' denotes the vector obtained by joining a 17th entry (equal to u) to the vector $\boldsymbol{S}$. The null space of A'^{T} has dimension 1 since the vector $\boldsymbol{J}'$ whose first 16 entries are 1's and whose last entry is a zero spans the null space of A'^{T}. But $\boldsymbol{J}' \cdot \boldsymbol{S}' = 0$ and, thus, $A'\boldsymbol{U} = \boldsymbol{S}'$ has a unique solution.

Discrete models for physical systems can be derived directly from discrete versions of conservation principles as described in chapter 1, or they may be obtained by discretizing continuous models using Taylor's theorem as illustrated in this chapter.

The simplest possible discretization of an initial value problem for an ordinary differential equation is known as Euler's method. The Runge-Kutta methods are improvements but still compute the solution value at the new time step using information from the previous time step only. Such single step methods apply to initial value problems for single equations or for systems of equations. We have not considered more sophisticated multistep methods nor have we considered numerical methods for boundary value problems.

Numerical methods for parabolic partial differential equations are either explicit (solution values at the new time level are given explicitly in terms of solution values at previous times) or they are implicit (new solution values are obtained by solving a system of equations at each step). Explicit methods apply equally well to initial value problems on unbounded domains or to initial-boundary value problems on bounded sets. The stability of these methods depends on the size of the time step. Implicit methods apply only to initial-boundary value problems on bounded domains. These methods are generally unconditionally stable; hence, they permit larger time steps.

Numerical solution of hyperbolic problems is generally more difficult than solving parabolic problems. Stability of the methods depends not only on the size of the time steps but on whether the method is compatible with the exact method of characteristics.

Stability is not an issue in solving elliptic boundary value problems numerically since the discrete problem assumes the form of a system of linear algebraic equations. If the boundary value problem is well posed then the system can generally be solved by Gaussian elimination without pivoting. If the system is very large, say more than 100 equations, then an iterative solution procedure may be more efficient than a direct method.

13

Introduction to the Finite Element Method

In the previous chapter we saw how replacing derivatives in differential equations by difference quotients transforms continuous initial value and boundary value problems into discrete problems. Then concepts from linear algebra, in conjunction with the computational power of the digital computer, can be used to construct numerical solutions for these problems.

In this chapter, we consider alternative means for transforming continuous boundary value problems to discrete problems (systems of simultaneous linear algebraic equations). This approach is reminiscent of the eigenfunction expansion methods of chapter 9 in that the solution is approximated by a linear combination of so-called trial functions. Then the continuous boundary value problem is replaced by the discrete problem of finding the unknown coefficients in the linear combination. There are various ways in which this replacement can be accomplished; here we describe two of the most common methods known as the Rayleigh-Ritz and Galerkin methods.

The success of these methods is heavily dependent on the choice of the trial functions used to approximate the solution. The advent of the computer has made it possible to make use of piecewise polynomial families of functions known as finite element families. These families of functions can be adjusted to conform to the geometry of the problem under consideration; hence, for certain problems with complicated geometry this finite element approach is preferred over the finite difference methods of the last chapter.

APPROXIMATION METHODS

Finite difference methods are not the only means of approximating the solution to a differential equation. We consider two methods that lend themselves to applications involving variational problems and weak formulations.

The Rayleigh-Ritz Method

Consider the problem of minimizing a functional J over a domain D that is a translate of subspace M that is dense in $H^0(\Omega)$. Such problems arise directly and it is also possible to reformulate certain boundary value problems as a functional optimization problem of this type.

Let p_0 be arbitrarily chosen in D and let $p_1, \ldots, p_n$ denote n linearly independent functions in M. Then for all constants $C_1, \ldots, C_n$,

$$u_n(x) = p_0(x) + \sum_{j=1}^{n} C_j p_j(x) \qquad (13.1)$$

belongs to D. If we denote by M_n the subspace spanned by the functions p_1 to p_n, then the set D_n of all functions of the form (13.1) is a translate of the subspace M_n.

Let $H(C_1, \ldots, C_n) = J[u_n]$ and note that H is a real valued function of n real variables defined on all of R^n. Thus, if H is smooth and if $(C_1^*, \ldots, C_n^*)$ minimizes H over R^n, then

$$\frac{\partial H}{\partial C_m}(C_1^*, \ldots, C_n^*) = 0 \quad \text{for } m = 1, \ldots, n \qquad (13.2)$$

Equations (13.2) form a set of n equations for the n unknowns C_1^* to C_n^*. If u^* in D minimizes J over D, then the function u_n^* obtained by substituting C_1^* to C_n^* in (13.1) satisfies

$$\| u^* - u_n^* \|_0 \le \| u^* - v \|_0 \quad \text{for all } v \text{ in } D_n \qquad (13.3)$$

The function u_n^* is referred to as the *Rayleigh-Ritz* approximation to u^* and the procedure of replacing the problem of minimizing J over D by the problem of solving the system of equations (13.2) is called the *Rayleigh-Ritz method* for approximating the minimum of J.

The Galerkin Method

Not every boundary value problem has a variational formulation; hence, the Rayleigh-Ritz cannot be applied in every case. Consider then a boundary value problem having a weak formulation of the form

Find u in D such that $B[u, v] = F(v)$ for all v in M (13.4)

Here D and M are as described above. Even boundary value problems not having any variational formulation have such weak formulations. Now let the functions $p_0, p_1, \ldots, p_n$ also be as described above and let u_n be given by (13.1). We will refer to the functions p_0 to p_n as *trial functions*. Let $q_1, q_2, \ldots, q_n$ denote n linearly independent functions in M (the q's may or may not be the same functions as the p's). We refer to the q's as *weight functions*. Then U_n is called the *Galerkin approximation* from D_n to the solution of (13.4) if U_n is given by (13.1) and satisfies

$$B[U_n, q_m] = F(q_m) \quad \text{for } m = 1, \ldots, n \tag{13.5}$$

While the weight functions need not be the same as the trial functions, in most applications we choose them to be the same. The method is known as the *Petrov-Galerkin* method when the p's differ from the q's, and when they are the same, it is called the *Bubnov-Galerkin* method.

Theorem 13.1

Theorem 13.1. Consider the boundary value problem (11.13) with weak formulation given by (11.14). Suppose the associated bilinear form $B[u, v]$ is symmetric and positive definite. Then the associated quadratic functional $Q[u]$ given by (11.16) has a unique minimum on D such that for each positive integer n and each choice of linearly independent trial functions $p_0, p_1, \ldots, p_n$, the system (13.2) has a unique solution; the corresponding approximation u_n^* satisfies (13.3). Similarly, if we choose weight functions q_m such that $q_m = p_m$ for $m = 1, \ldots, n$, then (13.5) also has a unique solution U_n and $U_n = u_n^*$.

In the cases where the bilinear form $B[u, v]$ in the weak formulation of the boundary value problem is not symmetric, then the Rayleigh-Ritz method does not apply. However, the Galerkin method still applies although existence theorems like Theorem 13.1 are more difficult to state and are much more difficult to prove.

THE FINITE ELEMENT METHOD

The success of approximation methods like Rayleigh-Ritz and Galerkin depends very much on making a good choice of trial functions. If the trial functions are chosen from families of piecewise polynomials called *finite element spaces*, then:

1. It is possible to deal systematically with irregular regions Ω (even those having curved boundaries).
2. The accuracy of the approximate solution can be estimated in a sys-

tematic way in terms of the adjustable parameters characterizing the finite element family.

3. The ingredients of the approximate problem including the coefficient matrix and data vector in the system of algebraic equations, and even the mesh, can be efficiently generated by computer.

Finite Element Spaces in 1 Dimension

We will consider families of functions defined on a closed bounded interval which we may as well take to be [0, 1]. Let $\{x_0, x_1, \ldots, x_n\}$ denote a partition of [0, 1], and, for convenience, let us take the partition to be uniform. Then $x_j = jh$ for $j = 0$ to n, where $h = 1/n$. Then it is customary to use the notation $S^h[k, r]$ to denote the set of all functions $\varphi(x)$ defined on [0, 1] such that:

1. $\varphi(x)$ is a polynomial of degree less than or equal to k on $[x_j, x_{j+1}]$.
2. $\varphi(x)$ is r times continuously differentiable on [0, 1].

Theorem 13.2

Theorem 13.2. For each admissible choice of h, k, and r, the family $S^h[k, r]$ is a finite dimensional vector space of functions of dimension $n(k-r)+r+1$; hence, $S^h[k, r]$ has a basis consisting of this number of functions.

EXAMPLE 1

(a) The space $S^h[1, 0]$ is a space of dimension $n = 1/h$. A simple basis is provided by the piecewise constant functions

$$\varphi_j(x) = \begin{cases} 1 & \text{if } x_{j-1} < x < x_j \\ 0 & \text{otherwise} \end{cases} \qquad j = 1, \ldots, n$$

(b) The space $S^h[0, -1]$ is the space of piecewise linear functions that are continuous but not continuously differentiable. The dimension of this space is $n + 1$ and a basis is provided by the so-called *hat functions*

$$\varphi_j(x) = \begin{cases} (x - x_{j-1})/h & \text{if } x_{j-1} < x < x_j \\ (x_{j+1} - x)/h & \text{if } x_j < x < x_{j+1} \\ 0 & \text{otherwise} \end{cases} \quad \text{for } j = 0, \ldots, n$$

The piecewise linear functions φ_j are uniquely determined by the conditions $\varphi_j(x_m) = \delta_{jm}$.

(c) The space $S^h[3, 1]$, referred to as the *cubic splines*, consists of piecewise cubics which are continuous and continuously differentiable with piecewise constant second derivatives; thus, $S^h[3, 1]$ is contained in $H^2[0, 1]$. A basis for $S^h[3, 1]$ is provided by the $2n + 2$ functions

$$\varphi_j(x) = \begin{cases} (|x - x_j| - h)^2 (2|x - x_j| + h)/h^3 & x_{j-1} < x < x_j \\ 0 & \text{otherwise} \end{cases}$$

$$\psi_j(x) = \begin{cases} (|x - x_j| - h)^2 (x - x_j)/h^2 & x_{j-1} < x < x_{j+1} \\ 0 & \text{otherwise} \end{cases}$$

for $j = 0$ to n. Note that the functions φ_j and ψ_j are uniquely determined by the conditions

$$\varphi_j(x_m) = \delta_{jm} \quad \text{and} \quad \varphi_j{}'(x_m) = 0$$

$$\psi_j(x_m) = 0 \quad \text{and} \quad \psi_j{}'(x_m) = \delta_{jm}$$

NEARLY ORTHOGONAL BASES

Only the piecewise constant basis is an orthogonal basis. The piecewise linear and piecewise cubic bases are "nearly orthogonal." We note for future reference that the piecewise linear basis satisfies

$$\int_0^1 \varphi_j(x)\,\varphi_i(x)\,dx = \begin{cases} 2h/3 & \text{if } i = j \\ h/6 & \text{if } |i - j| = 1 \end{cases} = (\varphi_i, \varphi_j)_0 \qquad (13.6)$$
$$\qquad\qquad 0 \quad \text{if } |i - j| > 1$$

and

$$(\varphi_i{}', \varphi_j{}')_0 = \begin{cases} 2/h & \text{if } i = j \\ -1/h & \text{if } |i - j| = 1 \\ 0 & \text{if } |i - j| > 1 \end{cases} \qquad (13.7)$$

$$(\varphi_i{}', \varphi_j{}')_0 = \begin{cases} -1/2 & \text{if } i = j - 1 \\ 1/2 & \text{if } i = j + 1 \\ 0 & \text{otherwise} \end{cases} \qquad (13.8)$$

Finite Element Spaces in the Plane

Let Ω denote a bounded domain in the plane and suppose Ω is subdivided into polygonal subdomains Ω_1 to Ω_n. Let h_j denote the length of the longest side of the polygonal subdomain Ω_j and let h equal the largest of the n numbers $h_1, \ldots, h_n$. Then h is the mesh size of the partition $\{\Omega_1, \ldots, \Omega_n\}$. Let Ω^h denote the union of the polygonal subdomains Ω_j and note that if the boundary of Ω is curved, then Ω^h may not coincide with Ω. Then $S^h[k, r]$ will denote the space of all functions $\varphi = \varphi(x, y)$ defined on Ω^h such that:

1. $\varphi(x)$ is a polynomial of degree less than or equal to k on each Ω_j.
2. $\varphi(x)$ is r times continuously differentiable with respect to x and y on Ω^h.

TRIANGULAR DECOMPOSITION

If Ω is decomposed into triangular subregions Ω_1 to Ω_n where no triangle has a vertex on a side of another triangle, then we say that the mesh $\{\Omega_1, \ldots, \Omega_n\}$ is a *proper triangulation* of Ω. Euler's formula for polyhedra implies that any proper triangulation of Ω into n triangles will contain m verices where

$$(n+5)/2 \leq m \leq n+2 \tag{13.9}$$

Since $S^h[k, r]$ is spanned by a basis of functions whose definition is based on the nodes (vertices) of the triangulation, (13.9) gives an estimate for the dimension of $S^h[k, r]$.

EXAMPLE 2

(a) Let $\{\Omega_1, \ldots, \Omega_n\}$ be a proper triangulation of Ω with nodes $\{z_1, \ldots, z_m\}$. Then a basis for the piecewise linear continuous functions $S^h[1, 0]$ is provided by the family $\{\varphi_1, \ldots, \varphi_m\}$ characterized by

(a) $\varphi_j(z_i) = \delta_{ij}$ for $i, j = 1, \ldots, m$,

(b) for $j = 1, \ldots, m$: $\varphi_j(x, y) = a_{jk} + b_{jk}x + c_{jk}y$ on Ω_k, $k = 1, \ldots, n$

The functions in $S^h[1, 0]$ are continuous with piecewise continuous (and hence square integrable) first derivatives.

(b) A basis for $S^h[3, 1]$ may be generated in more than one way. One basis is characterized by the conditions:

(a) for $j = 1, \ldots, m$: $\varphi_j, \partial_x\varphi_j, \partial_y\varphi_j$ are all continuous at each z_k $k = 1, \ldots, m$.

(b) for $j = 1, \ldots, m$: $\varphi_j(x, y) = a_{jk} + b_{jk}x + c_{jk}y + d_{jk}x^2 + e_{jk}xy$

$$+ f_{jk}y^2 + g_{jk}x^3 + p_{jk}(x^2y + xy^2) + q_{jk}y^3 \quad \text{on } \Omega_k,\ k = 1, \ldots, n$$

Since there are three nodes in each triangle, (a) provides nine conditions on each Ω_k, thus uniquely determining the nine constants $a_{jk}, b_{jk}, \ldots, q_{jk}$ which define $\varphi_j(x, y)$ on Ω_k. The functions so determined form a basis for $S^h[3, 1]$.

RECTANGULAR DECOMPOSITION

The bounded domain Ω can also be decomposed into rectangular subdomains Ω_k. The decomposition is proper if no corner of any rectangle lies on any side of another rectangular subdomain. Then bases for the finite element spaces $S^h[k, r]$ can be obtained by forming products $\varphi_j(x)\varphi_k(y)$, where φ_j is a basis function from a 1-dimensional finite element space; i.e., the 2-dimensional basis is a *tensor product* of two 1-dimensional bases.

The Finite Element Method

The finite element method for approximating the solution to a boundary value problem consists of applying a Rayleigh-Ritz or Galerkin method in which the families of trial and weight functions are chosen to be finite element spaces $S^h[k, r]$.

1-DIMENSIONAL PROBLEMS

For problems in 1 dimension, the finite element method represents a solution method for boundary value problems, whereas the finite difference techniques presented in chapter 12 have been applied only to initial value problems. We can adapt the finite difference methods to apply to boundary value problems to produce what are called *shooting methods*, but space does not permit the inclusion of these methods here. For 1-dimensional problems, the finite element method and finite differences are very comparable in terms of accuracy for effort required in constructing the solution. Although the two methods are based on quite different concepts it is not unusual for the two methods to lead to identical discrete problems when applied to the same 1-dimensional problem for a differential equation.

PROBLEMS IN MORE THAN ONE DIMENSION

For certain boundary value problems in two or three independent variables, the finite element method may provide distinct advantages over finite difference methods. The finite element method is particularly effective for problems on irregular domains with complicated geometry. Computer programs for automatic mesh generation and for computing a basis for the trial functions are readily available. Without such aids, the amount of programming required to implement the finite element method may be overwhelming. The extensive problem preparation required by the finite element method is one of its most serious disadvantages. In general, the finite difference methods can be applied more quickly and easily for simple problems, but for problems with complicated geometry, the finite element method is preferred.

SOLVED PROBLEMS

Approximation Methods

PROBLEM 13.1

Let Ω denote a bounded region in $\Re^m$ having smooth boundary S consisting of complementary parts S_1 and S_2 and consider the boundary value problem

$$-\text{div}\ (a(x)\nabla u(x)) + q(x)u(x) = f(x) \quad \text{for } x \text{ in } \Omega \tag{1}$$

$$u(x) = g(x) \quad \text{for } x \text{ on } S_1, \quad \nabla u(x) \bullet N = 0 \ \text{for } x \text{ on } S_2 \tag{2}$$

where a, q, and f are defined on Ω with $a(x) > 0$ and g is defined on S_1. Then describe the construction of the Rayleigh-Ritz and the Galerkin approximations for the solution of this boundary value problem.

SOLUTION 13.1

Let $G(x)$ denote a function in $H^1(\Omega)$ such that the restriction of G to S_1 is equal to g. If S is smooth, such a G can always be found. We will show in subsequent examples how G can be constructed using finite elements. Let M denote the set of all functions in $H^1(\Omega)$ that vanish on S_1 and let D denote the set of all the functions u in $H^1(\Omega)$ that are of the form $u = G + v$ for v in M. Then M is dense in $H^0(\Omega)$ and D is a translate of M.

The boundary value problem (1), (2) is equivalent to minimizing the functional

$$J[u] = \int_\Omega (a(x)\nabla u \bullet \nabla u + q(x)u(x)^2 - 2f(x)u(x))\,dx \tag{3}$$

over D. Note that J is quadratic in u, and since $a > 0$, J is positive definite as well. Then u in D minimizes J if and only if

$$\delta J[u, v] = 2(B[u, v] - (f, v)_0) = 0 \quad \text{for all } v \text{ in } M \tag{4}$$

where $$B[u, v] = \int_\Omega a(x)\nabla u \bullet \nabla v + q(x)uv\ dx \tag{5}$$

Note that if u in D solves the boundary value problem, then

$$B[u, v] = (f, v)_0 \quad \text{for all } v \text{ in } M \tag{6}$$

and u is a weak solution of the boundary value problem if u solves (6).

Now for positive integer n, let M_n denote the subspace spanned by p_1 to p_n where the p's are linearly independent functions chosen from M. Let

$$u_n(x) = G(x) + \sum_{j=1}^{n} c_j p_j(x) \tag{7}$$

Then u_n belongs to $D_n = G + M_n$ for all choices of the constants c_1 to c_n. Then $J[u]$ is minimized over D_n if and only if the quadratic function of n real variables $Q(c_1, ..., c_n) = J[u_n]$ is minimized over $\Re^n$; i.e., if and only if

$$\frac{\partial Q}{\partial c_m} = 2(B[u_m, p_m] - (f, p_m)_0) = 0 \quad \text{for } m = 1, \ldots, n$$

i.e.,

$$\sum_{j=1}^{n} c_j B[p_j, p_m] = (f, p_m)_0 \quad \text{for } m = 1, \ldots, n \tag{8}$$

This is a system of n linear algebraic equations in the n unknowns c_1 to c_n. If the c's satisfy (8), then u_n given by (7) is the Rayleigh-Ritz approximation to the solution of the boundary value problem.

The Galerkin approximation

$$U_n(x) = G(x) + \sum_{j=1}^{n} C_j p_j(x) \tag{9}$$

for trial functions $p_1(x)$ to $p_n(x)$ and weight functions $r_1(x)$ to $r_n(x)$ is obtained by solving

$$B[U_n, r_m] = (f, r_m)_0 \quad \text{for } m = 1, \ldots, n$$

i.e.,

$$\sum_{j=1}^{n} C_j B[p_j, r_m] = (f, r_m)_0 \quad \text{for } m = 1, \ldots, n \tag{10}$$

This system (10) is also a set of n linear equations in n unknowns. If the weight functions are different from the trial functions, then solving the system (10) will lead to an approximation that is, in general, different from the approximation u_n. However, if we choose the weight functions to be the same as the trial functions, then it is clear that $U_n = u_n$.

Note that this discussion applies to the case $m = 1$ where Ω is just an interval $(0, L)$ and (1), (2) reduces to

$$-(a(x)u'(x))' + q(x)u(x) = f(x), \quad u(0) = g \quad \text{and} \quad u'(L) = 0$$

PROBLEM 13.2

Let Ω denote a bounded region in $\Re^m$ having smooth boundary S consisting of complementary parts S_1 and S_2 and consider the boundary value problem

$$-\operatorname{div}\,(a(x)\nabla u(x)) + q(x)u(x) = \lambda r(x)u(x) \quad \text{for } x \text{ in } \Omega \tag{1}$$

$$u(x) = 0 \quad \text{for } x \text{ on } S_1, \qquad \nabla u(x) \bullet N = 0 \quad \text{for } x \text{ on } S_2 \tag{2}$$

where a, q, and r are defined on Ω with $a(x)$, $r(x) > 0$, and $q(x) \geq 0$. Then show how this eigenvalue problem can be formulated as a variational problem.

SOLUTION 13.2

This is a Sturm-Liouville problem with separated boundary conditions when $m = 1$; thus, the eigenvalues form an increasing sequence of real numbers $\lambda_1 < \lambda_2 < \cdots$, and the corresponding eigenfunctions are a family of functions orthogonal with respect to the weight function $r(x)$. These statements are true even when m is greater than one.

Let M denote the subspace of the previous problem and define functionals

$$J[u] = \int_{\Omega} (a(x)\nabla u \bullet \nabla u + q(x)u(x)^2)\,dx \quad \text{and} \quad K(u) = \int_{\Omega} r(x)u(x)^2 dx$$

Then for the quotient functional $Q[u] = J[u]/K[u]$, we can show easily

$$\delta Q[u, v] = \frac{\delta J[u, v]K[u] - J[u]\delta K[u, v]}{K[u]^2} \qquad u, v \text{ in } M$$

If u_1 in M minimizes Q over M, then $\delta Q[u_1, v] = 0$ for all v in M. This implies

$$\delta J[u_1, v] - Q[u_1]\delta K[u_1, v] = 0 \quad \text{for all } v \text{ in } M \tag{3}$$

But

$$\delta J[u, v] = 2\int_{\Omega} (a(x)\nabla u \bullet \nabla v + q(x)u(x)v(x))\,dx$$

$$\delta K[u, v] = 2\int_{\Omega} ruv\ dx$$

hence, (3) implies u_1 satisfies

$$\int_{\Omega} (a(x)\nabla u \bullet \nabla v + q(x)u(x)v(x) - \lambda_1 ruv)\,dx \quad \text{for all } v \text{ in } M \tag{4}$$

where $\lambda_1 = Q(u_1) = \min\,\{Q[u] : u \text{ in } M\}$. Then (4) is the statement that u_1 is a weak solution of the eigenvalue problem. Integration by parts, using the boundary conditions together with the fact that M is dense in $H^0(\Omega)$, shows

$$\nabla Q[u_1] = -\text{ div } (a(x)\nabla u_1(x)) + q(x)u_1(x) - \lambda_1 r(x)u_1(x)$$

provided u_1 belongs to $D = \{u \text{ in } H^2(\Omega) : u = 0 \text{ on } S_1\}$; D is the domain of ∇Q. Then the eigenfunction u_1 minimizes Q over M, and, conversely, if u_1 minimizes Q over M, then u_1 is an eigenfunction of the weak problem. If u_1 minimizes Q over M and belongs to D, then u_1 is an eigenfunction for (1), (2).

Note that eigenvalue/eigenfunction pairs $\{\lambda_n, u_n\}$ for larger eigenvalues can be obtained by minimizing Q over M subject to the constraints

$$K_j[u] = \int_\Omega ruu_j\, dx = 0 \quad \text{for } j = 1, \ldots, n-1 \tag{5}$$

PROBLEM 13.3

Describe how the eigenvalues and eigenfunctions of the problem

$$-\text{ div } (a(x)\nabla u(x)) + q(x)u(x) = \lambda r(x)u(x) \quad \text{for } x \text{ in } \Omega \tag{1}$$

$$u(x) = 0 \quad \text{for } x \text{ on } S_1, \quad \nabla u(x) \bullet N = 0 \text{ for } x \text{ on } S_2 \tag{2}$$

can be approximated by the Rayleigh-Ritz method.

SOLUTION 13.3

For positive integer n, let M_n denote the subspace spanned by p_1 to p_n, where the p's are linearly independent functions chosen from M. Let

$$u_n(x) = \sum_{j=1}^{n} c_j p_j(x) \tag{3}$$

Then (4) of the previous problem implies that

$$Q[u_n] = \frac{J[c_1, \ldots, c_n]}{K[c_1, \ldots, c_n]}$$

is minimized over M_n if and only if the numbers $c_1, \ldots, c_n$ satisfy

$$\sum_{i=1}^{n} c_i B[p_i, p_j] = \Lambda_n \sum_{i=1}^{n} c_i H[p_i, p_j] \quad \text{for } j = 1, \ldots, n \tag{4}$$

where

$$B[u, v] = \int_\Omega (a(x)\nabla u \bullet \nabla v + q(x)u(x)v(x))\, dx$$

$$H[u, v] = \int_\Omega ruv\, dx$$

$$\Lambda_n = Q[c_1, \ldots, c_n]$$

In matrix notation (4) becomes

$$BC = \Lambda_n HC, \quad C = [c_1, \ldots, c_n]^T$$

where $B = (B[p_i, p_j])$ and $H = (H[p_i, p_j])$ are symmetric $n \times n$ matrices. The smallest eigenvalue of this matrix eigenvalue problem is an approximation to the smallest eigenvalue of (1), (2) and the corresponding eigenvector $\boldsymbol{C}$ when substituted into (3) provides an approximation for the corresponding eigenfunction.

The Finite Element Method in 1-dimension

PROBLEM 13.4

Compute the Rayleigh-Ritz finite element approximation to the solution of the boundary value problem

$$y''(x) + 4y(x) = \sqrt{x}, \quad y(0) = 2, \; y(1) = -3 \tag{1}$$

SOLUTION 13.4

We define a uniform partition on (0, 1) by letting $x_i = i/4$ for $i = 0, \ldots, 4$ and we let $\varphi_0(x)$ to $\varphi_4(x)$ denote the piecewise linear continuous functions of Example 13.1(b). Then $G(x) = 2\varphi_0(x) - 3\varphi_4(x)$ is a function in $H^1(0, 1)$ whose restriction to the boundary (the two points $x = 0$ and $x = 1$) satisfies the boundary conditions and

$$y_4(x) = 2\varphi_0(x) + \sum_{j=1}^{3} c_j \varphi_j(x) - 3\varphi_4(x)$$

belongs to $D_4 = G + M_3$ for all choices of the constants c_1 to c_3. Here M_3 is spanned by φ_1, φ_2, and φ_3. The c_j's are required to satisfy

$$\sum_{j=1}^{3} c_j B[\varphi_j, \varphi_m] = (f, \varphi_m)_0 \quad \text{for } m = 1, 2, 3$$

where

$$B[u, v] = \int_0^1 (u'(x)v'(x) + 4u(x)v(x))\,dx \quad \text{and} \quad f(x) = \sqrt{x}$$

Then (13.6) and (13.7) imply

$$B[\varphi_j, \varphi_m] = \begin{cases} 2/h + 8h/3 = 49/6 & \text{if } m = j \\ -1/h + 2h/3 = -23/6 & \text{if } |m - j| = 1 \\ 0 & \text{if } |m - j| > 1 \end{cases}$$

and we can compute

$$(f, \varphi_1)_0 = \int_0^{1/4} \sqrt{x}\, \varphi_1(x)\, dx + \int_{1/4}^{1/2} \sqrt{x}\, \varphi_1(x)\, dx = 1/8$$

$$(f, \varphi_2)_0 = 0.177, \quad (f, \varphi_3)_0 = 0.217$$

Then

$$\frac{1}{6}\begin{bmatrix} 49 & -23 & 0 \\ -3 & 49 & -3 \\ 0 & -23 & 49 \end{bmatrix} \begin{bmatrix} c_1 \\ c_2 \\ c_3 \end{bmatrix} = \begin{bmatrix} 0.125 \\ 0.177 \\ 0.217 \end{bmatrix}$$

and $c_1 = 0.05$, $c_2 = 0.074$, $c_3 = 0.061$; i.e.,

$$y_4(x) = 2\varphi_0(x) + 0.05\varphi_1(x) + 0.074\varphi_2(x) + 0.061\varphi_3(x) - 3\varphi_4(x)$$

PROBLEM 13.5

Use the procedure of Problem 13.3 to approximate the smallest eigenvalue of the Sturm-Liouville problem

$$-(xy'(x))' = \lambda xy(x), \quad y(0) = \text{finite}, \quad y(1) = 0 \tag{1}$$

for which the exact eigenfunctions are $y_n(x) = J_0(\lambda_n x)$, where J_0 is the Bessel function of the first kind of order zero and λ_n denotes the nth zero of $J_0(x)$.

SOLUTION 13.5

If we let M_2 denote the subspace spanned by the trial functions $p_1(x) = \cos \pi x/2$ and $p_2(x) = \cos 3\pi x/2$, then $y_2(x) = c_1 p_1(x) + c_2 p_2(x)$, where $C = [c_1, c_2]^T$ satisfies $(B - \Lambda_2 H)C = \mathbf{0}$ for B and H two 2×2 symmetric matrices with entries

$$B_{ij} = B[p_i, p_j] = \int_0^1 xp_i'(x) p_j'(x)\, dx, \quad i, j = 1, 2$$

$$H_{ij} = H[p_i, p_j] = \int_0^1 xp_i(x) p_j(x)\, dx, \quad i, j = 1, 2$$

We compute

$$B = \begin{bmatrix} \dfrac{\pi^2 + 4}{16} & -3/4 \\ -3/4 & \dfrac{9\pi^2 + 4}{16} \end{bmatrix}, \quad H = \begin{bmatrix} \dfrac{\pi^2 - 4}{4\pi^2} & \dfrac{-1}{\pi^2} \\ \dfrac{-1}{\pi^2} & \dfrac{9\pi^2 - 4}{36\pi^2} \end{bmatrix}$$

and then the quadratic equation $\det(B - \Lambda H) = 0$ has roots $\Lambda = 5.790$, 30.578. Compare these values with the exact values $\lambda_1 = 5.784$ and $\lambda_2 = 30.470$ which are the first two zeroes of $J_0(x)$.

PROBLEM 13.6

Compute the Rayleigh-Ritz finite element approximation to the solution of the boundary value problem

$$EIy^4(x) = f(x), \quad y(0) = y'(0) = 0, \quad y''(L) = y^{(3)}(L) = 0 \tag{1}$$

for the transverse deflection $y = y(x)$ of an elastic beam of length L under the loading $f(x)$. The boundary conditions reflect that the end $x = 0$ is clamped and the end $x = L$ is free.

SOLUTION 13.6

Solving the boundary value problem (1) is equivalent to minimizing the potential energy functional

$$J[y] = \int_0^L \left(\frac{1}{2}EIy''(x)^2 - f(x)y(x)\right)dx \tag{2}$$

over the domain $M = \{y = y(x) \text{ in } H^2(0, L) : y(0) = y'(0) = 0\}$, see Problem 11.13. The boundary conditions at $x = L$ are natural boundary conditions and will be satisfied automatically by the solution, but they are not incorporated into the definition of M.

We begin the approximation procedure by defining a partition $\{x_0, \ldots, x_n\}$ on $(0, L)$ with $x_m = Lm/n$ for $m = 0, \ldots, n$. Since the functional J involves the second derivative of y, the piecewise linear finite element basis is not suitable for approximating the solution of this problem. Instead let $\{\varphi_0, \varphi_1, \ldots, \varphi_n; \psi_0, \ldots, \psi_n\}$ denote the cubic spline basis from Example 13.1(c). Then

$$y_n(x) = a_0\varphi_0(x) + \sum_{j=1}^{n} a_j\varphi_j(x) + b_0\psi_0(x) + \sum_{j=1}^{n} b_j\psi_j(x) \tag{3}$$

belongs to M for all a_j, b_j $(j = 1, \ldots, n)$ if $a_0 = b_0 = 0$. We let M_n denote the subspace of M that is spanned by the function $\varphi_j, \psi_j, j = 1,\ldots,n$.

In order to minimize J over M_n we require the constants a_j, b_j to be such that $\delta J[y_n, \varphi_j] = \delta J[y_n, \psi_j] = 0$ for $j = 1, \ldots, n$. But

$$\delta J[y_n, \varphi_j] = \int_0^L \left(\frac{1}{2}EIy''(x)\varphi_j''(x) - f(x)\varphi_j(x)\right)dx$$

$$= \sum_{i=1}^{n} a_i\left(EI\int_0^L \varphi_i''(x)\varphi_j''(x)\,dx\right) + \sum_{i=1}^{n} b_i\left(EI\int_0^L \psi_i''(x)\varphi_j''(x)\,dx\right) - (f, \varphi_j)_0$$

$$= \sum_{i=1}^{n} a_iA_{ij} + \sum_{i=1}^{n} b_iB_{ij} - f_j \quad \text{for } j = 1, \ldots, n.$$

Similarly,

$$\delta J\,[y_n, \psi_j] = \sum_{i=1}^{n} a_i B_{ij} + \sum_{i=1}^{n} b_i C_{ij} - g_j \quad \text{for } j = 1, \ldots, n.$$

where

$$C_{ij} = EI \int_0^L \psi_i''(x)\, \psi_j''(x)\, dx \quad \text{and} \quad gj = (f, \psi_j)_0$$

Then the $2n$ constants a_i, b_i are required to satisfy the $2n$ equations

$$\sum_{i=1}^{n} a_i A_{ij} + \sum_{i=1}^{n} b_i B_{ij} = f_j \quad \text{for } j = 1, \ldots, n,$$

$$\sum_{i=1}^{n} a_i B_{ij} + \sum_{i=1}^{n} b_i C_{ij} = g_j \quad \text{for } j = 1, \ldots, n.$$

The matrices A, B, and C are all symmetric and at least A and C are positive definite. Then the system of equations has a unique solution which, when substituted into (3), leads to the approximate solution of the boundary value problem. Since this is an approximate solution, the natural boundary conditions may not be satisfied exactly but $y_n''(L)$ and $y^{(3)}(L)$ should be small.

Finite Element Method in the Plane

PROBLEM 13.7

Let a square of sidelength 2 be triangulated as shown in Figure 13.1. Then compute a basis for $S^h[1, 0]$, the piecewise linear continuous functions on Ω^h.

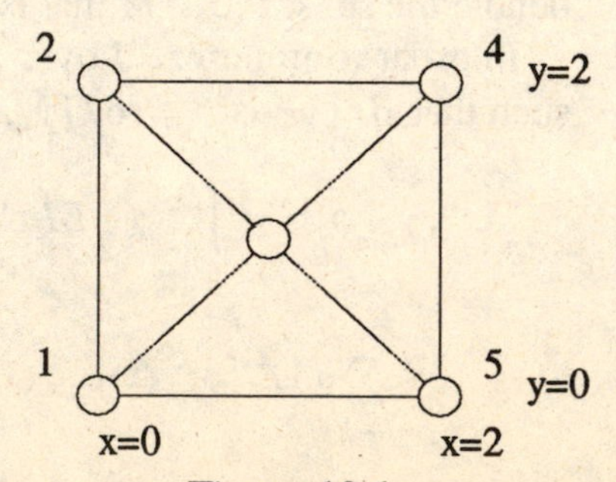

Figure 13.1
Sketch for Problem 13.7

SOLUTION 13.7

Note that the triangulation consists of five nodes z_1 to z_5 and four triangles Ω_1 to Ω_4. The basis functions $\varphi_j(x, y)$ satisfy the conditions

$$\varphi_j(z_i) = \delta_{ij} \quad \text{for } i, j = 1, \ldots, 5. \tag{1}$$

and

$$\varphi_j(x, y) = a_{jk} + b_{jk}x + c_{jk}y \quad \text{on } \Omega_k \quad j = 1 \text{ to } 5, \quad k = 1, \ldots, 4 \tag{2}$$

That is, each function is equal to 1 at one node and is 0 at all the other nodes and is piecewise linear and continuous on all triangles. It follows that the number of basis functions equals the number of nodes and that each basis function is identically zero on all triangles that do not contain the node at which the basis function equals 1. Thus, each basis function in this example is different from zero on just two triangles and it is given by (different) linear expressions of the form (2) of these triangles.
The coordinates (x_j, y_j) for node z_j are given by

j	1	2	3	4	5
x_j	0	0	1	2	2
y_j	0	2	1	2	0

Then on Ω_1 (which contains the node z_1), the function $\varphi_1(x, y)$ satisfies

$$\begin{aligned} \varphi_1(z_1) &= a_{11} + b_{11}(0) + c_{11}(0) = 1 \\ \varphi_1(z_2) &= a_{11} + b_{11}(0) + c_{11}(2) = 0 \\ \varphi_1(z_3) &= a_{11} + b_{11}(1) + c_{11}(1) = 0 \end{aligned} \quad \text{or} \quad \begin{bmatrix} 1 & x_1 & y_1 \\ 1 & x_2 & y_2 \\ 1 & x_3 & y_3 \end{bmatrix} \begin{bmatrix} a_{11} \\ b_{11} \\ c_{11} \end{bmatrix} = \begin{bmatrix} 1 \\ 0 \\ 0 \end{bmatrix}$$

Then $\quad \varphi_1(x, y) = 1 - x/2 - y/2 \quad$ for (x, y) in Ω_1

Similarly, $\varphi_1(x, y) = 1 - x/2 \quad$ for (x, y) in Ω_4

This last result is obtained by solving

$$\begin{bmatrix} 1 & x_1 & y_1 \\ 1 & x_3 & y_3 \\ 1 & x_5 & y_5 \end{bmatrix} \begin{bmatrix} a_{11} \\ b_{11} \\ c_{11} \end{bmatrix} = \begin{bmatrix} 1 \\ 0 \\ 0 \end{bmatrix}$$

since Ω_4 is the second triangle on which φ_1 is different from zero and Ω_4 contains the vertices z_1, z_3, and z_5. More generally, we can organize the procedure for computing the φ_j's into the following algorithm:

Enter Data: N = number or triangles in the decomposition of Ω
NV = number of vertices

Table 1. Coordinates of vertex z_j for $j = 1, \ldots, NV$

j	1	...	N
x_j	x_1	...	
y_j	y_1	...	

Table 2. List of vertices contained in triangle Ω_k for $k = 1, \ldots, N$

k	1	...	N
$z_1(k)$	$z_1(1)$	...	$z_1(N)$
$z_2(k)$	$z_2(1)$	...	$z_1(N)$
$z_3(k)$	$z_3(1)$	...	$z_1(N)$

each column lists (in increasing order) the nodes contained in triangle k

Compute:

For $k = 1$ to N

For $i = 1$ to 3

find $j = j(i, k)$ such that $z_j = z_i(k)$

load $(1, x_j, y_j)$ into ith row of the 3×3 coefficient matrix M

For $i = 1$ to 3

Solve: $MC = E_i$ where $C = [a_{ik}, b_{ik}, c_{ik}]^T$
E_i = unit 3 vector with 1 in ith place

Save: a_{ik}, b_{ik}, c_{ik}

Output:

Three $3 \times N$ matrices $A = [a_{ik}]$, $B = [b_{ik}]$, $C = [c_{ik}]$ containing the coefficients needed to form

$$\varphi_{ik}(x, y) = a_{ik} + b_{ik}x + c_{ik}y \quad \text{for } i = 1 \text{ to } 3 \quad \text{and} \quad k = 1 \text{ to } N$$

which is the expression on Ω_k for φ_j such that $j = j(i, k)$.

For the present example, $N = 4$ and $NV = 5$. Table 1 has been given and

Table 2 is the following:

k	1	2	3	4
$z_1(k)$	1	2	3	1
$z_2(k)$	2	3	4	3
$z_3(k)$	3	4	5	5

Then the algorithm gives the following piecewise definitions for the functions φ_j:

on Ω_1: $\varphi_1(x,y) = 1-x/2-y/2$, $\varphi_2(x,y) = -x/2+y/2$, $\varphi_3(x,y) = x$

on Ω_2: $\varphi_2(x,y) = -x/2+y/2$, $\varphi_3(x,y) = 2-y$, $\varphi_4(x,y) = -1+x/2+y/2$

on Ω_3: $\varphi_3(x,y) = 2-x$, $\varphi_4(x,y) = -1+x/2+y/2$, $\varphi_5(x,y) = x/2-y/2$

on Ω_4: $\varphi_1(x,y) = 1-x/2$, $\varphi_4(x,y) = y/2$, $\varphi_5(x,y) = x/2-y/2$

For a function $u = u(x, y)$ defined on the square with values u_1 to u_5 at the nodes, the function

$$U(x,y) = \sum_{j=1}^{5} u_j \varphi_j(x,y)$$

is the piecewise linear continuous approximation to $u(x, y)$ on the square and $u = U$ at the nodes.

PROBLEM 13.8

Let Ω denote the irregular plane region shown in Figure 13.2. For the indicated triangulation compute:

(a) the definitions for all φ_j in $S^h[1, 0]$ that are different from zero on triangles 1 to 4,

(b) the complete definition φ_1 and φ_{13}.

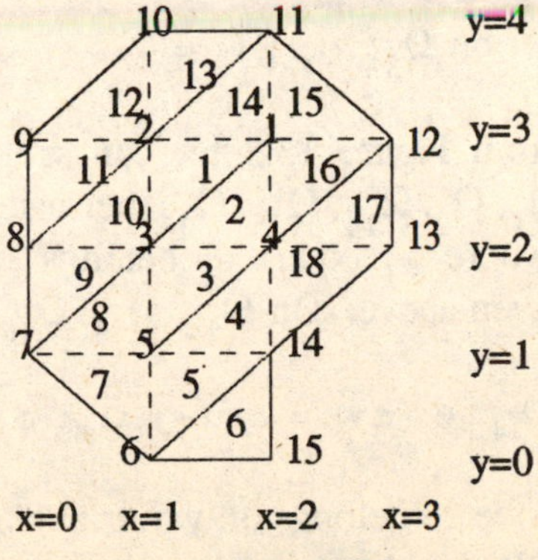

Figure 13.2
Sketch for Problem 13.8

SOLUTION 13.8

For the triangulation shown in Figure 13.2 we have $N = 18$, $NV = 15$ and Tables 1 and 2 read as follows:

Table 1:

j	1	2	3	4	5	6	7	8	9	10	11	12	13	14	15
x_j	2	1	1	2	1	1	0	0	0	1	2	3	3	2	2
y_j	3	3	2	2	1	0	1	2	3	4	4	3	2	1	0

Table 2:

k	1	2	3	4	5	6	7	8	9	10	11	12	13	14	15	16	17	18
$z_1(k)$	1	1	3	4	5	6	5	3	3	2	2	2	2	1	1	1	4	4
$z_2(k)$	2	3	4	5	6	14	6	5	7	3	8	9	10	2	11	4	12	13
$z_3(k)$	3	4	5	14	14	15	7	7	8	8	9	10	11	11	12	12	13	14

Submitting this data to the algorithm of the previous problem leads to the generation of piecewise formulas for the piecewise linear continuous functions $\varphi_j(x, y)$ for $j = 1, \ldots, 15$. From Table 2 we can see that on triangles 1, 2, and 3 the functions that are different from zero are:

$$\text{on } \Omega_1: \ \varphi_1, \varphi_2, \varphi_3; \quad \text{on } \Omega_2: \ \varphi_1, \varphi_3, \varphi_4; \quad \text{on } \Omega_3: \ \varphi_3, \varphi_4, \varphi_5$$

From the algorithm we find:

$$\text{on } \Omega_1: \ \varphi_1(x, y) = -1 + x, \ \varphi_2(x, y) = -1 - x + y, \ \varphi_3(x, y) = 3 - y$$

$$\text{on } \Omega_2: \ \varphi_1(x, y) = -2 + y, \ \varphi_3(x, y) = 2 - x, \ \varphi_4(x, y) = 1 + x - y$$

$$\text{on } \Omega_3: \ \varphi_3(x, y) = -x + y, \ \varphi_4(x, y) = -1 + x, \ \varphi_5(x, y) = 2 - y$$

From Figure 13.2 we can see that φ_1 will be different from zero on $\Omega_1, \Omega_2, \Omega_{14}, \Omega_{15}, \Omega_{16}$ because these are the triangles to which node 1 (where φ_1 equals 1) belongs. The definition of φ_1 on Ω_1 and Ω_2 is given above. On $\Omega_{14}, \Omega_{15}, \Omega_{16}$, we have

$$\Omega_{14}: \ \varphi_1(x, y) = 2 + x - y; \Omega_{15}: \ \varphi_1(x, y) = 6 - x - y; \ \Omega_{16}: \ \varphi_1(x, y) = -x + y$$

Node 13 belongs only to triangles 17 and 18. Then φ_{13} differs from zero only on these triangles where we have

$$\Omega_{17}: \ \varphi_{13}(x, y) = x - y \quad \text{and} \quad \Omega_{18}: \ \varphi_{13}(x, y) = -2 + x$$

PROBLEM 13.9

Show how to construct the Galerkin approximation to the weak solution of the boundary value problem

$$-\operatorname{div}\,(a(x,y)\nabla u(x,y)) + b(x,y)\,\partial_x u + u(x,y) = f(x,y) \quad \text{in } \Omega \tag{1}$$

$$u = g(x,y) \text{ on } S_1 \quad \text{and} \quad a\nabla u \cdot N = q(x,y) \text{ on } S_2 \tag{2}$$

where Ω denotes the region shown in Figure 13.2, S_2 is the part of the boundary containing the nodes 6 through 10 and S_1 is the remainder of the boundary of Ω. Here f is defined on Ω and g and q are defined on S_1 and S_2 respectively. Also the coefficients $a = a(x, y)$ and $b = b(x, y)$ are given functions defined on Ω with $a(x, y) > 0$.

SOLUTION 13.9

If we let M denote the subspace of $H^1(\Omega)$ containing those functions that are zero on S_1 then M is dense in $H^0(\Omega)$. If we multiply both sides of (1) by an arbitrary v in M and integrate over Ω then (1) reduces to

$$B[u, v] = F(v) \quad \text{for all } v \text{ in } M \tag{3}$$

where

$$B[u, v] = \int_\Omega a\nabla u \bullet \nabla v + vb\partial_x u + uv \tag{4}$$

and

$$F(v) = (f, v)_0 + \int_{S_2} qv \tag{5}$$

We used the divergence theorem here to conclude that

$$-\int_\Omega \operatorname{div}(a\nabla u)\,v = \int_\Omega a\nabla u \bullet \nabla v - \int_S va\nabla u \bullet N$$

$$= \int_\Omega a\nabla u \bullet \nabla v - \int_{S_2} va\nabla u \bullet N \quad \text{for all } v \text{ in } M$$

$$= \int_\Omega a\nabla u \bullet \nabla v - \int_{S_2} vq \quad \text{if } u \text{ satisfies (2)}$$

Note that the bilinear form $B[u, v]$ is not symmetric and so there is no variational formulation for this boundary value problem. However, the problem does have a weak formulation. The weak solution of the boundary value problem is a function u in $D = \{u \text{ in } H^1(\Omega) : u = g \text{ on } S_1\}$ satisfying (3). The Galerkin approximation U to the weak solution u, based on the triangulation of the previous problem, is defined to be

$$U(x,y) = \sum_{j=1}^{10} C_j \varphi_j(x,y) + \sum_{j=11}^{15} g(z_j)\,\varphi_j(x,y) = U_h(x,y) + G_h(x,y) \tag{6}$$

Here $G_h(x, y)$ is piecewise linear and continuous on Ω [hence G_h is in $H^1(\Omega)$] and $G_h = g$ at nodes 11 through 15. Thus, G_h is an approximation, based on the present triangularization, to a function G in $H^1(\Omega)$ whose restriction to S_1 equals g. If we let M_h denote the subspace of $H^1(\Omega)$ that is spanned by φ_1 to φ_{10} then M_h is a subspace of M and for all choices of the constants C_j, $j = 1, \ldots, 10$, U belongs to $D_h = G + M_h$, the set of all functions of the form $G + v$ for v in M_h.

Now we define approximations for f and q in terms of the functions φ_j

$$f_h(x, y) = \sum_{j=1}^{15} f(z_j)\,\varphi_j(x, y)$$

$$q(x, y) = \sum_{j=6}^{10} q(z_j)\,\varphi_j(x, y)$$

Substituting these expressions for U, f, and q into $B[u, v]$ and $F(v)$ leads to

$$B[U, \varphi_m] = B[U_h + G_h, \varphi_m] = B[U_h, \varphi_m] + B[G_h, \phi_m]$$

$$= \sum_{j=1}^{10} C_j B[\varphi_j, \varphi_m] + \sum_{j=11}^{15} g(z_j) B[\varphi_j, \varphi_m]$$

and

$$F(\varphi_m) = \sum_{j=1}^{15} f(z_j)\,(\varphi_j, \varphi_m)_0 + \sum_{j=6}^{10} q(z_j) \int_{S_2} \varphi_j \varphi_m \, ds$$

Using these in (3) produces the system of equations which determine the constants C_1 to C_{10}. For $m = 1, \ldots, 10$

$$\sum_{j=1}^{10} C_j B[\varphi_j, \varphi_m] = \sum_{j=1}^{15} f(z_j)\,(\varphi_j, \varphi_m)_0 + \sum_{j=6}^{10} q(z_j) \int_{S_2} \varphi_j \varphi_m \, ds$$

$$- \sum_{j=11}^{15} g(z_j) B[\varphi_j, \varphi_m]$$

Continuous boundary value problems on bounded domains can be transformed to discrete problems by means of so-called variational approximation methods. If the boundary value problem is equivalent to the problem of minimizing some functional over a suitable domain, then the Rayleigh-Ritz method can be used to replace the boundary value problem by the problem of solving a system of simultaneous algebraic equations in which the unknowns are the coefficients in a linear combination of trial functions intended to approximate the solution of the two equivalent problems. If the boundary value problem is not equivalent to any variational problem (i.e., if the associated bilinear form is not symmetric) then the Galerkin method can be used to replace the weak form of the boundary value problem by a system of linear algebraic equations.

The variational approximation methods as described in this chapter apply only to boundary value problems and not to initial value problems. In some cases, it is possible to combine the variational and finite difference approaches in constructing a numerical solution to an initial boundary value problem by discretizing in the time variable and using variational methods on the space variables. Such hybrid methods can be considered in a more extensive treatment of approximation methods for problems in partial differential equations.

The Rayleigh-Ritz and Galerkin methods are not the only ways of passing from the continuous problem to the discrete. Other popular methods include collocation and least squares approaches. These have not been discussed here due to space limitations.

The finite element families, $S^h[k, r]$, *of piecewise polynomials tailored to a given mesh have many more properties than have been described here. In particular, we have not mentioned the many estimates of solution error that can be derived in the context of these families. Numerous exhaustive texts on the finite element method are available in which this and other theoretical and practical information can be found.*

Index

F

M

N

O

P

Q

R

S

T

U

V

W

Y